"十三五"普通高等教育本科部委级规划教材

服饰配件设计与应用

| 中英双语教材 |

谢 琴 主 编
陈于依澜 邵翃恩 朱 文 副主编

中国纺织出版社有限公司
国家一级出版社
全国百佳图书出版单位

内 容 提 要

根据设计专业本科教学基础的要求,对服饰配件的设计与应用知识的"点"和"面",进行较为全面的阐述。在对运用纤维、金属、珠宝、玻璃、陶瓷等材料进行设计创作的同时,结合制作的工艺来实现作品的创意与价值。

本书中英文对照,具有较高的学术价值,适用于服装院校的师生以及广大服装爱好者学习与参考。

Abstract

This textbook is based on the requirements of the undergraduate teaching foundation of the design major, and provides a comprehensive interpretation of the breadth and depth of the knowledge "point"and "face" of the design and application of apparel accessories. while design and creation could be made by the application of fiber, metal, jewelry, glass, ceramics and other materials, creativity and value of work can be achieved via professional craftsmanship and production.

This textbook has both Chinese and English versions with high academic values, applying to teachers and students in garment colleges and universities, as well as fashion enthusiasts.

图书在版编目(CIP)数据

服饰配件设计与应用/谢琴主编. -- 北京:中国纺织出版社有限公司,2019.11(2022.2 重印)
"十三五"普通高等教育本科部委级规划教材
ISBN 978-7-5180-6545-5

Ⅰ.①服… Ⅱ.①谢… Ⅲ.①服饰—配件—设计—高等学校—教材 Ⅳ.① TS941.3

中国版本图书馆 CIP 数据核字(2019)第 179479 号

策划编辑:孙成成　　　责任编辑:杨 勇
责任校对:王花妮　　　责任印制:王艳丽

中国纺织出版社有限公司出版发行
地址:北京市朝阳区百子湾东里 A407 号楼　邮政编码:100124
销售电话:010 — 67004422　传真:010 — 87155801
http://www.c-textilep.com
中国纺织出版社天猫旗舰店
官方微博 http://weibo.com/2119887771
唐山玺诚印务有限公司印刷　各地新华书店经销
2019 年 11 月第 1 版　2022 年 2 月第 3 次印刷
开本:787×1092　1/16　印张:16.25
字数:417 千字　定价:58.00 元

凡购本书,如有缺页、倒页、脱页,由本社图书营销中心调换

前言 PREFACE

服饰配件与服装的起源是紧密相连的,它不仅具备了人类生理方面需求的物质属性特征,还涵盖来自心理方面需求的精神属性特征。服饰的起源则更多地出于人类追求美的向往,或是崇尚某种事物的精神需求。随着科技和社会的不断发展,服饰成了人类生活不可分割的一部分,因为它们在人类的生产、生活、社交等各类社会活动中起着十分重要的作用,它的内涵随着时代的变迁,在不断地变化和加强。因此,服饰配件是人们生活着装搭配中不可或缺的一部分。服饰配件不仅装点着人们的穿衣风格,还能从整体上提升人们的形象和气质。因此,服饰配件与服装搭配,现已成为着装形态的重要一环。日复一日,感知、图像、音律不停地催生了灵感,它对于声色的感知与领悟滋养了设计师们的想象力。它们赋予服饰配件以提升人体的外形美,在塑造外形的同时,也在强化或弱化着服装的风格,从而达到让配饰与服装互补的作用,这样,能让设计师有更大可发挥的空间和自由度。

Clothing accessories are closely related to the origin of clothing. It not only possesses the material properties of human physiological needs, but also the spiritual properties from psychological requirements. The origin of the clothing and accessories is more out of the desire for human pursuit of beauty, or the mighty spiritual needs of advocating something. With the continuous development of science and technology and society, clothing has become an inseparable part of human life, because they play an important role in various social activities such as human production, life, and social activities. Its connotation changes with the times, constantly changing and strengthening. Therefore, clothing accessories are the indispensable part of people's life. Dress and accessories not only decorate people's dressing style, but also enhance people's image and temperament as a whole. Thus, the matching of clothing accessories and clothing has become an important part of the dress pattern. Day after day, perception, image, and temperament continue to inspire, and the perception and understanding of sensuality nourishes the imagination of designers. They enhance the beauty of the human body via the clothings and accessories. While shaping the outline, it also strengthens or weakens the style of the clothing, to complement the accessories and clothing, so that the designer can have more space to exert their talents with less limits.

服饰配件的设计与制作对于整个服装与配饰领域来说,是一件至关重要的事。因为它与经济、社会、文化、风俗、习惯"丝丝入扣",在不同的历史时期,有不同艺术设计表现。因此我们根据服饰配件发展的趋势,以设计教学的知识点和学生须掌握的工艺、制作的要求编撰这本书,并以中英文"双语"呈现。

The design and production of apparel accessories is not a trivial matter for the entire apparel and accessories sector. Because it is "striving" with economy, society, culture, customs, and habits, it has different artistic design performances in different historical periods. Therefore, according to the development trend of apparel accessories, we have compiled this textbook with the knowledge of design teaching and the craftsmanship and production requirements that students must master, and present it in both Chinese and English.

第一章绪论对服饰配件进行综述，分别解说服饰配件的定义、特征、作用，并对服饰配件所用的材质，按地区、按民族、按风格、按佩戴部位进行分类；第二章从服饰配件的内容、文化内涵、艺术要素及服饰配件设计的底蕴，进行全面的阐述；第三章对服饰配件的设计思考、设计原理等基础理论分别进行介绍，同时对制作的材料，以及纺织纤维、金属、玻璃、陶瓷及其他材料的分类和工艺、应用特点，进行分门别类地论述；第四章重点对纺织纤维材料、金属材料、玻璃陶瓷材料及各类材料的设计案例进行解析；第五章针对服饰配件——首饰的设计进行了常用首饰的分类、首饰设计元素、首饰创意与表达及首饰材料的搭配运用。另外重点介绍首饰工艺与制作；第六章讲述服饰配件搭配，常用服饰配件及其特点与服装的搭配必须注意的问题，同时附有案例解析。第七章对世界不同地域：亚洲、欧洲、大洋洲、美洲、非洲服饰配件的设计特点进行描述，并介绍世界著名服饰配件的品牌。

The first chapter of the textbook is an introduction to the accessories, to explain the definition, characteristics and functions of the accessories, and to classify the materials used in the accessories, by region, ethnicity, style and wearing parts. The second chapter is a comprehensive exposition from the content, cultural connotation, artistic elements and the design of clothing accessories. The third chapter introduces the basic theories such as the design considerations, design principles and design principles of the apparel accessories, and classifies the materials, textile fibers, metals, glass, ceramics and other materials, as well as the process and application characteristics discussions. The fourth chapter focuses on the design case analysis of various materials like textile fiber materials, metal materials, glass ceramic materials. The fifth chapter is the classification of common jewelry, jewelry design elements, jewelry creativity and expression, and the use of jewelry materials for the design of apparel accessories.Also highlights the jewelry craftsmanship and production.

The sixth chapter discusses about the matching of clothing accessories, the common clothing accessories and their characteristics and the issues in matching of clothing, as well as the case studies.

The seventh chapter introduces the characteristics of the design of clothing accessories in different regions of the world: Asia, Europe, Oceania, America, and Africa, together with the introduction of the well-known brands in fashion accessories.

●本书适应本专业的本科教学、其他专业可用作服饰设计师设计时的参考用书。

●本书共分六章：第一章、第二章由东华大学陈于依澜撰写；第三章、第六章由上海视觉艺术学院邵翃恩撰写。第四章由上海视觉艺术学院秦岭、上海工程技术大学王书利、上海视觉艺术学院谢琴撰写；第五章由上海视觉艺术学院王渊撰写。

●本书由上海工程技术大学朱文翻译。由上海工程技术大学郭昳菁校审。

●本书由谢琴修改、调整、完成。

●本书在撰写过程中得到东华大学刘晓刚教授、中国美术学院凌雅丽副教授、夏晓洁老师、上海视觉艺术学院王烨老师的支持与帮助，在此表示深切的感谢！

This textbook is suitable for undergraduate teaching, and also serves as reference book for the related majors, and fashion designers.

The textbook is divided into six chapters: The first and second chapters are written by Chen Yuyilan, Donghua University. The third and sixth chapters are written by Qinling, Shanghai Institute of Visual Arts; Wang Shuli, Shanghai University of Engineering and Science; Xie Qin, Shanghai Institute of Visual Arts. The fourth are written by Shao Hongen, Shanghai Institute of Visual Arts. The fifth chapter is written by

Wang Yuan, Shanghai Institute of Visual Arts. This textbook was translated by Zhu Wen from Shanghai University of Engineering and Science. The English language was amended by Guo Yijing from Shanghai University of Engineering and Science. This textbook was revised, adjusted and completed by Xie Qin.

In the process of editing, we received the support of Prof. Liu Xiaogang from Donghua University, Associate Professor Ling Yali from the China Academy of Art, Ms. Xia Xiaojie and Ms. Wang ye from the Shanghai Institute of Visual Arts. We would like to extend the deep gratitude to them!

目录
CONTENTS

第一章 绪论
Chapter 1　Introduction

第一节　服饰配件概述 / 002
Section 1　Clothing accessories overview

　　一、服饰配件的概述 / 003
　　　A. The concept of apparel accessories
　　二、服饰配件的特征 / 003
　　　B. Features of clothing accessories
　　三、服饰配件的作用 / 007
　　　C. The Functions of clothing accessories

第二节　服饰配件的分类 / 011
Section 2　Classification of clothing accessories

第二章 服饰配件的设计基础
Chapter 2　Design basics of clothing accessories

第一节　服饰配件的内容 / 015
Section 1　Clothing accessories content

　　一、服饰配件的物质内容 / 015
　　　A. Material content of clothing accessories
　　二、服饰配件的精神内容 / 024
　　　B. Spiritual content of clothing accessories

第二节　服饰配件文化内涵 / 027
Section 2　Cultural connotation of costume accessories

　　一、服饰配件中的艺术形态 / 027
　　　A. Artistic form of clothing accessories
　　二、服饰配件中的设计风格 / 031
　　　B. Design styles in apparel accessories
　　三、服饰配件中的技术美感 / 043
　　　C. Technical beauty in apparel accessories

四、服饰配件设计的底蕴 / 045
D. The backgrounds of apparel accessories design

五、服饰配件的发展趋势 / 052
E. Development trends of fashion accessories

第三章　常用服饰配件设计
Chapter 3　The design for Daily fashion and fewelries

第一节　西方首饰发展简史 / 060
Section 1　The brief history of western jewelries

一、石器时代 / 060
A. The stone age

二、苏美尔时期 / 061
B. The Sumer period

三、古埃及时期 / 062
C. The Ancient Egypt era

四、古希腊时期 / 064
D. The Ancient Greek age

五、古罗马时期 / 065
E. The ancient Roman age

六、拜占庭时期（8世纪）、14～15世纪时期 / 066
F. The Byzantine period (8th century)、14th~15th century

七、文艺复兴时期 / 068
G. The Renaissance period

八、17世纪首饰 / 068
H. The 17th Century jewelry.

九、18～19世纪的首饰 / 069
I. The jewelries from 18th century to 19th century

十、英国"手工艺运动"对首饰的影响 / 070
J. The impact of english "Hand making activity"

十一、新艺术主义与装饰主义时期 / 070
K. The Art Nouveau and Art Deco Time

十二、现当代首饰（1910年至今） / 070
L. The contemporary jewelry (1910 till now)

第二节　中国珠宝首饰发展简史 / 071
Section 2　The brief overview of chinese jewelry

一、商周时期：小巧简约 / 071
A. Shang and Zhou time: simple and refined

二、魏晋南北朝时期：清新活泼 / 072

B. The Wei Jin, Northern and Southern Dynasties: Fresh and lively

三、隋唐时期：绚丽多彩 / 072

C. The Sui Tang dynasty: vividly colorful

四、宋元时期：清丽典雅 / 073

D. The Song Yuan dynasty elegant

五、明清时期：华丽繁缛 / 073

E. The Ming Qing dynasty: luxuriously complicated

第三节　中国古代首饰分类 / 074

Section 3　The classification of Chinese ancient jewelries

一、笄【jī】/ 074

A. Ji

二、簪【zān】/ 075

B. Zan (hairpin)

三、钗【chāi】/ 075

C. Chai (Hair Fork)

四、步摇【buyao】/ 076

D. Buyao (step swings)

五、扁方 / 076

E. Bian Fang

六、梳篦【shū bì】/ 076

F. Shu Bi(Comb)

七、华胜 / 077

G. Hua Sheng

八、抹额 / 078

H. Mo E (Frehead Bnd)

九、花钿【diàn】/ 078

I. Dian

十、珥珰【ěr dāng】/ 078

J. Er Dang(Earring)

十一、玉玦【yù jué】/ 079

K. Yu Jue

十二、项圈 / 079

L. Necklace

十三、璎珞【yīng luò】/ 079

M. Ying Luo

十四、胸饰 / 080

N. Brooch

十五、腰饰 / 080

O. The waist jewelries

十六、禁步 / 081

P. Jin Bu (step cautious)

十七、臂钏【bì chuàn】/ 081

Q. Bi Chuan (Armlet)

十八、手镯 / 081

R. Bracelet

十九、戒指 / 082

S. Ring

二十、指甲套 / 082

T. Nail shield

第四节　首饰设计 / 082

Section 4　Jewelry Design

一、常用首饰的分类 / 082

A. Classification of Commaonly used jewelry

二、首饰设计元素 / 083

B. Jewelry design elements

三、首饰材料 / 087

C. Jewelry material

四、首饰设计常用材料、工具与设备 / 094

D. Materials、tools and equipment

五、首饰设计与制作的基本工艺 / 096

E. The basic process of jewelry design and production

六、首饰设计与制作的基本步骤 / 098

F. The basic steps of jewelry design and production

第五节　潮流与趋势——3D打印对首饰设计行业的影响 / 099

Section 5　Trends and directions—the impact of 3D printing on the jewelry design industry

一、3D打印技术与珠宝首饰行业 / 099

A. 3D printing technology and jewelry industry

二、珠宝首饰设计相关软件简介 / 101

B. The introduction of jewelry design related softwares

第四章　非金属材料在服饰配件领域的制作与工艺

Chapter 4　Production and Process of Non-metallic Materials in the Field of Apparel Accessories

第一节　纺织纤维材料在服饰配件上的制作与工艺 / 106

Section 1　Fabrication and process of textile fiber materials on clothing accessories

一、常用服饰用纺织纤维的分类及性能 / 106

　　A. Classification and performance of textile fibers for common apparel

　　二、纺织类纤维服饰配件的设计 / 111

　　B. Design of textile fiber clothing accessories

　　三、纺织纤维类服饰配件的设计原则 / 115

　　C. Design principles for textile fiber apparel accessories

第二节　运用纺织纤维类材料制作服饰配件的案例解析 / 119

Sectio 2　Casestudy of making clothing accessories with textile fiber materials

　　一、首饰 / 119

　　A. Jewelry

　　二、包袋 / 122

　　B. Bag

　　三、帽饰 / 129

　　C. Hat decoration

第三节　玻璃材料和陶瓷材料 / 135

Section 3　Glass and ceramic materials

　　一、玻璃材料的分类及应用特点 / 137

　　A. Classification and application characteristics of glass materials

　　二、陶瓷材料的分类及应用特点 / 139

　　B. Classification and application characteristics of ceramic materials

　　三、玻璃材料制作服饰配件工艺 / 142

　　C. Glass material manufacturing and apparel accessories techniques

　　四、陶瓷材料制作服饰配件工艺 / 145

　　D. Ceramic materials to make clothing accessories process

　　五、陶艺服饰配件的装饰与烧成工艺 / 153

　　E. Decoration and burning process of ceramic art accessories

第四节　颜色釉 / 155

Section 4　Color Glaze

　　一、上釉工艺 / 156

　　A. Glazing Process

　　二、烧成工艺 / 156

　　B. Burning Process

第五节　运用陶瓷材料制作服饰配件案例解析 / 158

Section 5　Case study of using ceramic materials to make clothing accessories

　　一、绘制图纸 / 159

　　A. Drawing

　　二、制作泥片 / 159

　　B. Making mud pieces

三、泥片烧制 / 159

C. Mud film burning

四、陶瓷彩绘 / 159

D. Ceramic painting

五、彩绘烧制 / 160

E. Painted firing

六、最终完成 / 160

F. Final completion

第五章　服饰配件搭配
Chapter 5　The collocation of accessory and attire

第一节　配件及其特点 / 161

Section 1　Accessories and their features

　　一、帽 / 164

　　A. Hat

　　二、包 / 165

　　B. Bag

　　三、鞋、袜 / 168

　　C. Shoes and hosieries

　　四、围巾 / 172

　　D. Scarf

　　五、领带、领结 / 174

　　E. Tie, bow tie

　　六、腰饰 / 177

　　F. Belt

　　七、花饰品 / 182

　　G. Floral accessory

　　八、伞 / 182

　　H. Umbrella

　　九、手表 / 183

　　I. Watch

　　十、眼镜 / 184

　　J. Glasses

　　十一、首饰 / 185

　　K. Jewelries (Discussed with Pictures)

第二节　服饰品与服装搭配的关系及协调 / 194

Section 2　Relationship and coordination between clothing items and clothing collocation

一、风格上相呼应 / 195

　　A. Style echoes

　　二、体积上的对比 / 196

　　B. The contrast of volume

　　三、肌理上相对比 / 196

　　C. Texture contrast

　　四、色彩的配合 / 197

　　D. The combination of color

第六章　世界不同地域的历史、文化背景与其服饰配件的设计

Chapter 6　The history and cultural background of different regions of the world and the design of their accessories

第一节　亚洲及其四国的服饰配件特色 / 200

Section 1　Clothing accessories in Asia and its four countries

　　一、中国的文化 / 201

　　A. Chinese culture and its accessories' features

　　二、服饰形制与色彩分析 / 206

　　B. The analysis of shape system and color of clothing and accessories

　　三、服饰配件特征 / 206

　　C. Clothing accessories features

第二节　日本的文化及其服饰配件特色 / 217

Section 2　Japanese culture and its clothing and accessories features

　　一、文化背景 / 217

　　A. Cultural background

　　二、服饰形制与色彩分析 / 217

　　B. The analysis of clothing shape system and color

　　三、和服的种类 / 218

　　C. The categories of Kimon

　　四、和服的配饰 / 220

　　D. The accessoria for Kimono

第三节　韩国的文化及其服饰配件特色 / 222

Section 3　Korean culture and its clothing and accessories' features

　　一、文化背景 / 222

　　A. Cultural background

　　二、服饰配件形制与色彩分析 / 222

　　B. The analysis of the shape system and color of the clothing an accessories

　　三、服饰配件的特征 / 223

　　C. The features of accessories

第四节　印度的文化及其服饰配件特色 / 226

Section 4　Indian culture and its clothing and accessories' features

　　一、文化背景 / 226

　　A. Cultural background

　　二、服饰配件形制与色彩分析 / 226

　　B. The analysis of the shape system and color of clothing and accessories

第五节　欧洲服饰配件特色 / 228

Section 5　The features of European clothing and accessories

　　一、文化背景 / 228

　　A. Cultural background

　　二、服饰配件形制与色彩分析 / 229

　　B. The analysis of shape system and color of the clothings and accessories

　　三、影响现代服饰潮流分析 / 233

　　C. Affecting the trend analysis of modern clothing

第六节　美洲服饰配件特色 / 235

Section 6　The features of american clothing and accessories

　　一、文化背景 / 235

　　A. Cultural background

　　二、服饰配件形制与色彩分析 / 236

　　B. The analysis of shape system and color of clothing and accessories / 236

第七节　非洲服饰配件特色 / 240

Section 7　The features of African clothing and accessories

　　一、文化背景 / 240

　　A. Cultural backgroud

　　二、服饰配件的形制与色彩分析 / 240

　　B. The analysis of shape system and color of clothing and accessories

参考文献 / 245

第一章 绪论
Chapter 1 Introduction

课程名称：绪论

Course Name: Introduction

学习目的：通过本章的学习，使学生明确服饰配件设计的概念、范畴；了解服饰配件设计的从属性、凸显性、服用性、个性化等特征；清晰服饰配件设计的作用；深刻理解服饰配件的分类；确定对服饰配件设计的基础认知。在之后的学习中能够客观地设计和研究服饰配件设计。

本章重点：服饰配件的基本概念；服饰配件的特征、作用；服饰配件的分类方法。

课时参考：4课时。

Learning Objectives: Through the study of this chapter, students can define the concept and scope of clothing accessories design; understand the characteristics of the accessories, highlighting, taking, and personalization of clothing accessories design; clear the role of clothing accessories design; profound understanding of clothing accessories Classification; establish a basic understanding of the design of apparel accessories. In the subsequent study, the design of apparel accessories can be objectively designed and studied.

The focus of this chapter: the basic concept of clothing accessories; the characteristics and functions of clothing accessories; the classification method of clothing accessories.

Class time reference: 4 class hours.

服饰配件是人们着装搭配中不可缺少的部分，有着独特的时尚魅力和广阔的市场潜力。从着装效果看，服饰配件装点着人的穿衣风格，它能提升人的整体形象、气质。从市场角度分析，服饰配件占整体服饰市场的比重越来越大，而且在价格上趋于"攀升"状况。如个别奢侈品品牌的服饰配件，已经在价格上超过了服装。时尚产业发达的国家，将服饰配件列为专业教学已经多年，为此培养了一大批世界知名的服饰配件设计师以及相应的品牌。我国在这方面的教学，尚未形成独立、完整的服饰配件专业的教学体系，仅是以一门专业课程的形式，依附于服装设计学科中。本课程尊重教学规律，以教学服务于实践为理念，理论知识与实践知识并举，以符合服饰发展的客观规律为依托，以丰富翔实的教学内容为载体，提升学生的服饰配件审美与设计水平和实操能力。

Clothing accessories are an indispensable part of people's clothing collocation, which has unique fashion charm and broad market potential. From the perspective of dressing effect, clothing accessories decorate a person's dressing style, which can improve the overall image and temperament of people. From the perspective of the market, the proportion of apparel accessories in the overall clothing market is increasing, and the price tends to "climb". For example, clothing accessories of individual luxury brands have already exceeded clothing in price. In countries with developed fashion industry, clothing accessories have been listed as professional teaching for many years, which has cultivated a large number of world-renowned fashion accessories designers and corresponding brands. China's teaching in this aspect has not formed an independent and complete teaching system for the clothing accessories major, which is only in the form of a professional course, attached to the clothing design discipline. This course respects the law of teaching, takes teaching service in practice as the concept, combines theoretical knowledge with practical knowledge, relies on the objective law of costume development, and takes rich and detailed teaching content as the carrier, so as to improve students' aesthetic and design level and practical ability of costume accessories.

第一节　服饰配件概述
Section 1　Clothing accessories overview

　　服饰配件与服装搭配，已成为人们着装形态的重要部分。追溯服饰配件与服装的起源，二者是紧密相连不可分割的。服装的起源具备了人类生理需求的物质属性特征，还涵盖了来自心理方面需求的精神属性特征。因此服饰品既出于人类追求美的向往，也是崇尚某种事物的精神需求。正是这些需求的存在，使得人类对自身进行各种艺术装饰，或用所处环境中的天然材料来装饰自己，从而形成服饰配件的雏形。

Clothing accessories and clothing matching have become an important part of people's clothing patterns. Tracing the origin of clothing accessories and clothing, the two are closely linked and inseparable. The origin of clothing has the characteristics of material attributes of human physiological needs, and also includes the characteristics of spiritual attributes from psychological needs. Therefore, clothing is not only out of the human pursuit of beauty, but also the spiritual pursuit of something. It is these requirements that human beings decorate themselves with various kinds of art, or decorate themselves with the natural materials in their environment, thus forming the prototype of costume accessories.

　　随着科技和社会的不断发展，服饰常常会成为人类生产、生活、社交等各类社会活动中"亮点"，同时，它的内涵作用还在不断升华。在现代社会中，服饰配件与服装组合而成的人类着装状态，直接影响着一个人的风度、气质、文化与修养。因此，在生活中，人们除了注重着装的选择与搭配，还越来越重视与服装配套的服饰配件。由于服饰配件的种类及穿着佩戴方式多样，而有些是不直接接触人类皮肤的，所以在材质选择上会有更大的空间和自由度。

With the continuous development of science and technology and society, clothing often becomes the "highlight" in various social activities such as human production, life and social interaction, and its connotation function is constantly sublimated. In the sublimation of modern society, the human dressing state formed by the combination of costume accessories and costume directly affects one's demeanor, temperament, culture and cultivation. Therefore, in life, people not only pay attention to the choice and collocation of clothing, but also pay more and more attention to the matching accessories of clothing. Due to various types of clothing accessories and wearing styles, some of them do not directly touch human skin, so there will be more space and freedom in material selection.

　　同时服饰配件具有工艺品的特征，不少服饰配件本身，就是一件相对独立的工艺品或艺术品。尤其是利用贵重材料制作的服饰配件，会包含使用者的情怀故事，具有"传世"价值。因此，服饰配件的设计，通常都需要设计师的情感注入。在个人着装状态中，一件配饰的价值，高于整套服装的价值也是常见的。这是因为服装对于消费者来说大都作为消费品，使用周期短，更换频率高，而服饰配件有时候却会成为人们喜爱的艺术品，而长期使用，乃至珍藏。为此，当今服饰配件及设计受到了越来越多的关注和重视。

Meanwhile, costume accessories have the characteristics of handicrafts. A lot of dress fittings themselves are relatively independent handicrafts or artworks. In particular, the use of rare and precious materials to make clothing accessories, will have the feelings of the user story, which has the value of "passing on the world". Therefore, we can say that the design of clothing accessories usually requires the emotional input of designers. It is also common for an accessory to be worth more than the entire outfit in a personal dressing state. This is because clothes are mostly consumer goods for consumers, with a short service cycle and a high frequency of replacement. Sometimes, clothing accessories will become popular works of art, making them be used for a long time and even treasured. Therefore, nowadays, more and more attention has been paid to garment accessories and their design.

一、服饰配件的概述
A. The concept of apparel accessories

服饰配件又称服饰品、配饰物、服装配饰。从广义上说，是指服装以外的所有附属在人体上的装饰。服饰配件的起源与使用都早于服装，远古时代，在人们还不懂穿衣时，为了传递某种美或标志性信息，就已经会使用石头、贝壳、羽毛等天然材料披挂于身，还会在人体的各个部位文身、穿刺，这些成为服饰配件的初始形态。从狭义上讲，服饰配件是指服装以外的所有附属、装饰于人体的物品。因为"服饰"，它表示"服装、服饰、衣着、穿着"的"配件"，从而体现了一种从属性，即服饰配件从属于服装的装饰品，是穿戴于人体上，如首饰、鞋袜、箱包、雨伞、帽子、丝巾、胸针等具有装饰作用的物品。这些是设计学所研究的关于服饰配件的主体内容，本课程主要以此概念展开。

Clothing accessories are also called clothing products, accessories, clothing accessories. In a broad sense, it refers to all the decorations attached to human body other than clothing. The origin and use of clothing accessories are earlier than clothing. In ancient times, people would hang themselves with stones, shells, feathers and other natural materials in order to convey certain beauty or symbolic information before they knew how to wear clothes. In a narrow sense, apparel accessories refer to all accessories other than clothing, decoration in the human body of goods. Because it is "dress", it expresses "dress, dress, clothing, clothing". the "fittings" reflected a kind of subordinate attribute thereby, namely dress fittings belongs to the adornment of dress. They are to wear on the human bodies, including articles with decorating functions such as jewelry, shoe socks, box bag, umbrella, hat, silk scarf, brooch and so on to have adornment effect the article. These are the main contents of the study on clothing accessories of the institute of design. This course is mainly based on this concept.

二、服饰配件的特征
B. Features of clothing accessories

服饰配件作为一种具有艺术审美价值的商品，有其相应的特征。纵观历史进程，服饰配件的变化和发展，都记载着每一个地域的文化和历史。小小的服饰配件，却融汇了各个地区的生存环境、生活习俗、宗教信仰、审美认知和科技发展。因此，可以说服饰配件是生动的历史，记载了人类的发展和进步。法国作家、诺贝尔文学奖获得者阿纳托尔·法朗士说："如果死后能在未来的图书馆选一本书，我将不会选择文字作品和历史教材，而是一本时尚杂志，一个时代女性的衣着打扮，比一切哲学家、小说家、预言家都更能够告诉我未来世界的样子。"比较不同时代、不同民族的服饰配件，可以清楚地看到历史的轨迹、文化的特征，特别是这些特征，决定了服饰配件的艺术价值和服饰配件所具有的完整概念。服饰配件具体有以下四个特征：

As a kind of commodity with artistic aesthetic value, costume accessories have their corresponding characteristics. Throughout the course of history, the changes and development of clothing accessories have recorded the culture and history of each region. Small clothing accessories, have integrated the living environment, life customs, religious beliefs, aesthetic cognition, scientific and technological development of various regions. Therefore, we can say that clothing accessories are a vivid history, having recorded human development and progress. French writer and Nobel Prize winner, Anatole France, Anatole, said: "If after death, I can pick a book of the library in the future, I will not choose literature and history textbooks, but a fashion magazine—because an era of women's dress, more than all the philosopher, novelist, prophet can tell me the way for the future of the world." We can clearly see the historical track and cultural characteristics, especially these characteristics, when comparing the clothing accessories of different ages and nationalities, which determines the artistic value of the clothing accessories and the complete concept of the clothing accessories. There are four specific features:

1. 从属性
Attribute

服饰配件中的"配"字，有"配合""配套""般配"的意思，在某种程度上暗示着一种并非"主角"的意味，具有从属性的特性。人们的衣着装扮，展现着自身的风格、气质、修养和社会地位，这种外在的表现不仅取决于服装本身，还需要对应的配饰与其相得益彰的烘托、搭配、结合，才能够更加生动地打造整体效果。服饰配件既要符合原材料的特性又要充分发挥其设计特性，根据服装款式和着装对象的特点而定❶。无论是从体积大小还是从服用性能的角度来看，服饰配件都是依附于服装的从属物品，这是服饰配件的基本特征之一。

但从属不等于一致，它可以有不同的表现程度，对服饰配件的从属性的理解，随时尚流行风潮的变化而变化。比如，近年来盛行的"混搭风格"，可以把与风格上无关的、甚至对立的服装与配饰集合在一起，使得服饰配件比以往少了很多通常意义上的从属度。

The word "match" in costume accessories has the meanings of "match", "agree" and "suit", which implies to some extent that it is not the meaning of "leading role" and has the characteristics of subordination. When people dress up, they show their tastes in style, temperament, culture and social status. This kind of exterior expression not only depends on dress itself, but still needs corresponding deserve to act the role of its complement each other foil, tie-in, union, which can more vividly make integral effect. Clothing accessories should not only conform to the characteristics of raw materials but also give full play to their design characteristics, according to the characteristics of clothing styles and clothing objects. From the point of view of size and performance, clothing accessories are attached to the subordinate articles of clothing, which are accessories of clothing

Attributes do not mean "agreement". Clothing accessories from the understanding of attributes can show their different expressions and various interpretations, and may change in accordance with different fashions. For example, in the popular "mash-up style" in recent years, we can combine clothes and accessories that have nothing to do with style, or even with opposite styles, so that the accessories are much less dependent than before.

---

❶ 陈东升，王秀艳. 新编服装配饰学［M］. 北京：中国轻工业出版社，2004.11：3.

第一章 绪论

2. 凸显性
Highlighting

服饰配件本身可以单独的形式存在，配件本身的独立性决定了服饰配件常常在造型中起到"画龙点睛"的作用，从而凸显了整体的格调和气韵。从造型效果方面来看，服装与服饰配件之间，是很难判定究竟是哪一方起到了更大的作用，简洁的服装搭配、夸张的服饰配件可以成为一种独特的时尚造型方式。比如，香奈儿2013春／夏秀场上的巨型包、时尚潮人巨大的围脖和独特的草编蝴蝶结发饰，其凸显度丝毫不亚于服装本身（图1-1～图1-3）。

Clothing accessories themselves can exist in a separate form, and the independence of the accessories themselves determines that apparel accessories often play the role of "making the finishing point" in the modeling, thus highlighting the overall style and charm. From the perspective of modeling effect, it is difficult to determine which side has played a greater role between clothes and accessories. Concise clothing collocation and exaggerated clothing accessories can become a unique fashion modeling way. For example, in Chanel 2013 spring and summer show, "the giant bag" fashion hipster huge neck and unique straw braid bowknot hair ornaments are no less prominent than the clothing itself (figure 1-1~ figure 1-3).

另一方面，服饰配件在漫长的发展历程中，逐渐形成了属于自身的艺术感和审美价值。许多服饰配件做工精湛、选材精美，象征着极致的"奢华"和"无上"的权利，已经作为独立的艺术品，被展示、欣赏、收藏，这也更加体现了服饰配件的凸显性这一特征。比如，历史上的宫廷服饰配件，在历史的长河中洗尽铅华，依旧风采卓然（图1-4）。

On the other hand, during the long course of development, apparel accessories gradually formed their own artistic and aesthetic values. Many of them are exquisitely crafted and exquisitely selected, symbolizing the ultimate "luxury" and "above all" rights. As independent artworks, they have been displayed, appreciated, and collected. This is also a manifestation of the salient features of apparel accessories. For example, historical palace costume accessories still have outstanding style through the long historic journeys (Figure 1-4).

3. 服用性
Wearability

服饰配件的服用性是其适合穿着佩戴的性能，从功能上来说就是能够满足人类审美而且具有实用价值的特性。随着时代的变迁，服用性的含义也在变化，不同的历史阶段有不同的服饰文化，从最早

图1-1 香奈儿2013春／夏秀场上的巨型包
Figure 1-1 The giant handbag of Chanel's 2013 spring and summer show

图1-2 时尚潮人巨大的围脖
Figure 1-2 Fashionable People's Huge Collar

图1-3 草编蝴蝶结发饰
Figure 1-3 Straw Bow Hair Accessories

图1-4 作为头饰的大英帝国皇冠不失为一件艺术品
Figure 1-4 Crown of the British Empire as a headdress, an artwork

要求的穿戴舒适，逐步过渡到既要美观又要卫生、健康、安全。如今，服饰配件的服用性可概括为舒适性、美观性、安全性三个方面的"集合"。舒适性是指人通过感觉（触觉、听觉、嗅觉、视觉）和知觉，对所穿戴服饰配件的综合体验，包括生理上的舒适感，如人们佩戴帽子、墨镜保护头部和眼睛不受到太阳光或雨水的伤害。在人类的早期文明，我们的祖先就开始发明鞋子来避免脚在陆地上行走时可能遇到的损伤。心理上的愉悦感，如某些来自神话、宗教、人类想象等美好的寓意而给人带来的幸福感，或是作为战利品佩戴时给人带来的荣誉感、自豪感以及来自社会方面的自我实现、自我满足感。❶美观性是指服饰配件必须具有符合社会或个人审美要求的美感。安全性是指穿戴服饰配件时，其造型、材料和表面处理均不给穿戴者或他人带来伤害。

当今社会每个人都扮演着不同的社会角色，也有不同的需求，人们常常在辛勤忙碌的职场和日常生活的活动中不停地转换，有时是穿着硬底皮鞋的白领，有时是穿着登山鞋的旅行者，或是穿着运动鞋在健身房挥汗如雨的锻炼者，或是佩戴一身华丽的珠宝出席精致的晚宴的娱乐者，由此可见，服饰配件的服用性能使其自身成为适应不同环境的好帮手。

The wearability of clothing accessories is its suitability for wearing. From the perspective of function, it is a feature that can satisfy human aesthetics and has practical value. With the change of The Times, the meaning of wearability is also changing. Different historical stages have different costume cultures, from the earliest requirement of wearing comfort, to the gradual transition from being beautiful to being hygienic, healthy and safe. Nowadays, the wearability of clothing accessories can be summarized as a "collection" of comfort, beauty and safety. Comfort refers to a person's comprehensive experience of wearing accessories, including physical comfort, through feeling (senses of touching, hearing, smelling, seeing) and perception, such as wearing hats and sunglasses to protect the head and eyes from the sun or rain. For example, in the early days of human civilization, our ancestors began to invent shoes to avoid possible foot injuries while walking on land. Psychological sense of pleasure, such as the sense of happiness brought by some beautiful connotations from myths, religions and human imagination, or the sense of honor, pride and social self-fulfillment and self-satisfaction brought by wearing as trophies. Esthetic quality is to point to dress fittings must have the esthetic sense that accords with society or individual esthetic requirement. Safety refers to the wearing of clothing accessories, its shape, materials and surface treatment do not bring harm to the wearer or others.

In today's society everyone plays a different social role, and displays different needs. People are constantly transforming their styles between hard busy work dressing style and daily life dressing codes. For example, sometimes they are hard bottom shoes-wearing white-collars, sometimes they are hiking shoes-wearing travelers, or sports shoes-wearing gym sweating exercisers, or gorgeous jewelry-wearing evening entertainers. Thus, apparel accessories play a great role to make themselves become the good assistant of the adapted to different environment.

4. 个性化

Individuality

服装配饰是个性化色彩浓烈的商品。选择服装配饰时，需根据使用者的年龄、职业、爱好、经济状况等个人因素挑选服饰配件。同时不同性格的个体会使用不同风格的服饰搭配。服饰配件有时能凸显穿戴者的个性。服饰配件的多重组合搭配，使得有时只要仅仅改变服饰配件，就可使服装有不同的风格倾向，从而适应多种穿着场合。服饰配件能体

---

❶ 王明葵. 服装品牌、价值及服用性之间关系的探讨［J］. 中国纤检，2011（10）：66-67.

现一个人的品位，即使在相同的场合，不同的人也会选择不同的服饰配件来打造自己的个人形象，彰显自己的与众不同，或者是表达属于自身的艺术感受和时尚态度。

也可以这么说，服饰配件是辨识穿着者个性的"捷径"。我们可以用服饰品，装点出一个理想的形象，给人以深刻的印象。当然，服饰也会暴露深藏在心底的"奥秘"。因此，服饰社会心理学理论把服饰反映个性这一现象，区分为表现公开的自我和流露隐藏的自我两个不同的侧面。一个人的个性有不同的侧面和丰富的内涵，服饰形象有多种不同的表现形式，服饰配件的个性化特征在其中起到了"推波助澜"的作用。

Clothing accessories are personalized and strongly colored goods. When choosing clothing accessories, you must choose clothing accessories according to the age, occupation, hobbies, financial status and other personal factors of the user. At the same time, different personalities will use different styles of clothing matching. Clothing accessories sometimes accentuate the wearer's personality. The multiple combination and collocation of costume accessories make it possible to have different styles and tendencies of clothing as long as the clothing accessories are changed. Clothing accessories can reflect a person's taste. Even on the same occasion, different people will choose different clothing accessories to create their own personal image, showing their own uniqueness, or expressing their own artistic feelings and fashion attitudes.

It can also be said that clothing accessories are a "shortcut" to identify the wearer's personality. We can use the dress goods to decorate an ideal image and give a person with deep impression; Of course, the clothes will also expose the hidden "secret" inside. Therefore, the theory of costume social psychology divides the phenomenon of costume reflecting personality into two different aspects: expressing the open self and revealing the concealed self. A person's personality has different sides and rich connotations, and there are many different forms of expression in costume images. The personalized features of costume accessories play a role of "pushing forward".

三、服饰配件的作用
C. The Functions of clothing accessories

衣着服饰是人类非常重要的生活内容之一，在漫长的服饰发展历程中，服饰配件起到的作用是不容置疑的。而服饰配件具有典型的代表性，在服装设计中也发挥着重要的作用，人们通常会根据所处时代的流行趋势，进行符合时宜的服装搭配。合理的搭配不仅可以弥补某些服装的不足，而且还可以给人带来视觉冲击力，达到完善审美的效果，因此，服饰配件的作用包含四个方面：点缀、标识、补正、保护。

Clothing and apparel accessories are one of the most important aspects of life for human beings. In the long process of clothing development, the role of clothing accessories is evident. From ancient times to now, clothing accessories have typical meanings, which caused our attentions. Clothing accessories also play an important role in the fashion design. People usually carry out clothing matching according to the fashionable trend of the times. An appropriate combination can not only make up for the lack of certain clothing, but also provide the visual impacts, to achieve perfect aesthetic results. Therefore, the role of apparel accessories includes four aspects: embellishment, identification, correction, protection.

1. 点缀
Embellishment

服饰配件作为从属于服装的一部分，具有丰富服装内容的作用。在起着提升个体形象的作用时，既能强化个性风格，又可以完善人的整体效果。在着装的过程中，通过选择适当的服饰配件加以点缀，使得服饰整体效果避免了单调，让着装风格更加鲜明，格调得到进一步的升华。一方面，通过选择适当的服饰配件加以点缀，可以使得服装更有层次感。比如，女性的裙装上加一根腰带，冬季里使用的围脖和围巾，都会令服装展现出更好的状态。另一方面，服饰配件作为服饰的"点睛之笔"，具有提升着装者个体形象的效果。例如，职业服装常表现端庄、稳重，但略显拘谨。如果适当地在职业装上点缀一枚别致的胸针或戴一副精致的耳环，就能够增添其楚楚动人的一面；郊游的服装往往是轻松随意的，如在颈间或辫梢上束一方漂亮飘逸的丝巾，会尽显青春浪漫的风采。另外，点缀常常体现为一种色彩的渲染效果，人们在服装的选择上，往往会倾向于选择同一色系之间的搭配。而使用其他色系的服饰配件加以点缀，是丰富着装的绝佳选择，比如，使用与服装不同色彩的帽子、手套、围巾、鞋袜可作为整体造型的点缀，弥补色彩的单调（图1-5）。

图1-5　GUCCI 2017秋/冬秀场上的服装配饰在整体造型中的点缀作用
Figure 1-5　The embellishment of the clothing accessories on the GUCCI 2017 fall and winter show

As a subordinate part of clothing, apparel accessories have the function of enriching the content of clothing. They have the effect of enhancing the individual image, which can strengthen the personality style and improve the overall effect as well. In the process of dressing, embellishment can be achieved by choosing the right accessories to avoid monotony, so that the style of dress is more distinct and the style is further sublimated. On one side, apparel accessories, as a part of clothing, have the effect of enriching the content of clothing. In the process of dressing, by selecting the appropriate accessories to embellish, the clothing itself can be more layered. For example, a belt added to skirt or the collars and scarves used in winter, will make the garments look better. On the other hand, costume accessories, as the "pointing touch" of costumes, have the effect of enhancing the individual image of the wearers. For example, professional clothing often shows dignity and stability, but it is slightly restrained. If a chic brooch or an exquisite earring can be worn in the business occasion, it will add its vivid side; the clothes on the outings are often relaxed and casual, so a beautiful and graceful scarf on the neck or binding the hair, will present the elegance of youthful romantic sense. In addition, embellishment is often embodied as a color rendering effect. People often tend to choose the one color system for the selection of clothing. The use of other color accessories to embellishment, is a great choice for a rich dress, such as the use of different colors and clothing, hats, gloves, scarves, shoes and socks can be used as a whole modeling embellishment, to make up for the monotony of color (Figure 1-5).

2. 标识

Identification

服饰配件具有展现某种特定内容的标识作用。无论是东方还是西方，在服饰史中，都表明服饰无一例外地经历过被政治"干预"的阶段。统治阶级制订了详细的规范和礼法来限制谁着打扮，并且利用衣着打扮来划分尊卑等级，通过服饰配件的标识作用，使得一些特定内容的象征意义更加明显。普列汉诺夫在《论艺术》中提到："这些东西最初只是作为勇敢、灵巧和有力的标志而佩戴的，到了后来，也正是因为他们是勇敢、灵巧而有力的标志，所以开始形成审美的共鸣，收入装饰品的范围。"至此，人们开始利用装饰品来展示自己的能力、身份和权利。❶唐代阎立本绘制的《历代帝皇图》中晋

---

❶ 普列汉诺夫. 论艺术［M］. 上海：生活·读书·新知出版社，1964.

武帝司马炎冕服上的服饰配件多达13种。并且古时哪一级别佩戴哪些饰品都有着严格的规定，任何人不得僭越。❶比如，大臣朝服上的补子和帽子上的顶戴花翎，区分了文武官员以及官位等级，虎符、勋章、肩章分别象征了权利、荣誉、地位。这些具有特殊意义的服饰配件是为了其标识作用而产生的，而佩戴者也因此获得了相应的标识价值（图1-6、图1-7）。

Apparel accessories have the function of presenting specific content. Whether it is the East or the West, in the history of costume development without exception, it has been shown that costumes have experienced political "intervention". The ruling class has formulated detailed norms and ritual laws to restrict the dress and wearing, as to classify the social status. Through the marking of clothing accessories, the symbolic meaning of certain specific contents becomes more obvious. Plekhanov mentioned in "On Art": "These things were originally worn as a sign of bravery, dexterity, and strength. Later, it was precisely because they were brave, dexterous, and powerful signs that they started. Form the aesthetic resonance within the scope of decorations." So far, people began to use decorations to demonstrate their abilities, identities and rights. [Plekhanov. On Art [M]. Shanghai: Life, Reading, Xinzhi Publishing House, 1999, 1973]. In the Tang dynasty Yan Liben's drawing of the "Emperors' portraits", Jin Emperor Sima Yan's accessories are as many as 13 types in ancient times, what accessories should to be worn had strict regulations, no one could disobey [Edited by Sheng Yu. Apparel Design [M]. Zhengzhou: Henan Fine Arts Publishing House, 2010.6:6]. For example, the supplements on the minister's imperial dress and the top wear on the hats distinguish the civil and military officials and the rank of the official. The tiger symbol, the medal, and the insignia symbolize rights, honors and status respectively. These special clothing accessories are created for their logo function, and the wearer carry them for that value of Logo (Figure 1-6, Figure 1-7).

图1-6　大臣朝服上的补子
Figure 1-6　The supplements on the minister's court garment

图1-7　帽子上的顶戴花翎
Figure 1-7　Top feather accessories on a hat

人们开始会"以貌取人"，根据不同的衣着打扮判断一个人所具有的影响力。时至今日也不例外，人们依旧会根据一个人的"行头"来大致判断这个人的某些特征，如消费水平、品味、财富状况等。正因如此，许多人也不约而同地依靠一些奢侈的或个性的服饰配件来标识和展示自己的财富、地位及独特的品位，如限量发售的昂贵的珠宝首饰、名表、箱包、手机（图1-8）。

---

❶ 盛羽. 服饰品设计［M］. 郑州：河南美术出版社，2010.6：6.

People start to "judge from appearances"; they judge people's influence from their dressing and wearing. Today is no exception. People will still roughly judge a certain person's characteristics based on a person's wearing, such as the level of consumption, taste, wealth, For this reason, many people also rely on some luxury or personalized clothing accessories to display their wealth, status, unique taste, such as limited sale of expensive jewelry, watches, bags, mobile phones (Figure 1-8).

3. 补正
Correction

补正即指补充、修正。服饰配件的补正作用，主要是强调对整体造型的补充和修正，其作用一般可以从以下的两个方面来体现。第一，扬长避短。服饰配件能够弥补人体或服装的不足，掩盖人的缺陷。例如，一些剪裁简洁大气的服装单独穿着往往会显得单调乏味，如果加上适当的帽子和项链，就刚好弥补了服装过于简单的缺憾；腰带的正确使用，可以美化人体比例（图1-9）；爱美的女生使用过膝的高筒靴来掩盖腿粗的缺点，或是利用戴帽子的办法来遮挡额头泛滥的青春痘。第二，形式美学上的补正。为了达到视觉上的和谐统一，配饰往往能够起到调和的作用，比如，为了达到色彩平衡而使用的手提包就是利用补色原理来调和整体的造型效果（图1-10）；为了达到视觉平衡，在不对称剪裁的上衣设计中佩戴的臂环或丝巾，使得造型达到富有变化又轻重平衡的境界（图1-9~图1-11）。

Correction refers to supplements and amendments. The correcting function of apparel accessories is mainly to emphasize the complement and correction of the overall model, and its role can generally be reflected from the following two aspects. First, To avoid disadvantages. Clothing accessories can cover or make up for the visual disadvantages of human body. For example, some clothing that simply tailored is often boring. If appropriate hats and necklaces added, it will just make up for the shortcomings of overly simple clothing; if a waist belt used correctly, it cvan optimize the proportion of the human body (Figure 1-9). Another example is the high-heeled knee-high boots for girls to cover up the shortcomings of thick legs, or use hats to cover the acne on the forehead. Second, the form of aesthetic correction. In order to achieve visual harmony and unity, accessories can often play a role in reconciliation. For example, the handbag used to achieve color balance is to use the complementary color principle to reconcile the overall modeling effect (Figure 1-10); for example,

图1-8 昂贵的奢侈品象征财富与地位
Figure 1-8 Expensive luxury goods symbolize wealth and status

图1-9 腰带的补正作用 巴宝莉·珀松伦敦2016秋/冬秀场
Figure 1-9 Correction of the belt. Burberry Prorsum London 2016 Fall Winter Show

图1-10 箱包在整体造型中的色彩补正作用 巴宝莉·珀松伦敦2016秋/冬秀场
Figure 1-10 Color correction effect of the bag in the overall shape. Burberry Prorsum London 2016 Fall Winter Show

图1-11 臂环在形式构成上的补正作用 普罗恩萨·施罗官网
Figure 1-11 Correction of the form of the arm ring. Proenza schuler official website

in order to achieve visual balance, the armbands or scarves worn in the asymmetrically tailored tops design allow the shape to reach the realm of richness and balance (Figure 1-9~Figure 1-11).

4. 保护
Protection

保护作用是指服饰配件对人体生理上的一种防御作用，也是服饰配件服用性能的一种主要的体现。人们在漫长的时间里逐渐形成了用于护身、御寒、防晒、防虫等品类固定的服饰配件。服饰配件的保护作用往往包括了（1）御寒保暖，如靴子、围巾等。（2）防晒，如墨镜、帽子等。（3）避免身体受到伤害，如鞋子、护膝、护肘等。（4）方便携带和省力，如手提包、箱包、肩包、背包。（5）防虫、防臭，如佩戴一些香包、锦囊。进入电子信息时代的今天，更是有了不胜枚举的保护功能，比如，许多科技公司开发的电子可穿戴产品，可以监测身体状态、记录人体数据、操作移动设备、辅助人体行走等，亦成为新型服饰配件。谷歌实验室Google X研发的医用手环可以测脉搏、心跳与皮肤温度，还有外界的光照射情况与噪音水平（图1-12、图1-13）。

Protection refers to a defensive effect of apparel accessories on human bodies, also a major manifestation of clothing accessories taking performance function. The clothing accessories that for body protection, cold protection, sun protection, and insect protection have been gradually formed from long time ago. The protective functions of clothing accessories often include (1) Keep warm, such as boots, scarves and so on. (2) Sun protection, such as sunglasses, hats, etc. (3) Avoid physical injuries such as shoes, knee pads, elbow pads, etc. (4) Easy to carry, such as handbags, bags, shoulder bags, and backpacks. (5) Anti-insect, deodorant, such as wearing some sachets, tips. In today's era of electronic information, there are numerous protection functions included. For example, electronic wearable products developed by many technology companies can monitor the physical status, record human data, operate mobile devices, and assist the human body to walk. It also becomes a new type of clothing accessories. for example, the Google X Labs Google developed medical wristbands that measure pulse, heartbeat, and skin temperature, as well as ambient lighting and noise levels (Figure 1-12, Figure 1-13).

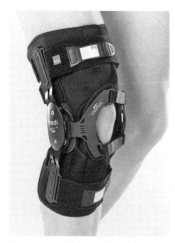

图1-12 强调保护功能的服饰配件，护膝可以保护膝关节减少受到的部分伤害
Figure 1-12 Accessories with emphasis on protection functions. Knee protectors can protect knees from partial injuries

图1-13 强调保护功能的服饰配件，监测身体数据的智能手环
Figure 1-13 Emphasis on protective function accessories, smart bracelet for monitoring physical data

第二节　服饰配件的分类
Section 2　Classification of clothing accessories

服饰配件的种类繁多，从不同的角度可以有不同的分类分法。随着人文和科技生活的进步发展，

越来越多的服饰配件应运而生，不同的分类方法有助于人们客观地认识、设计和研究服饰配件。

There are various types of apparel accessories, which can be divided into different categories from different perspectives. With the development of the humanities and science and technology, more and more clothing accessories have emerged. Different classification methods help people to objectively understand the accessories; design and research in different categories.

1. 按用途分类
Classifying by purpose

服饰品能够满足人类常用需求：

装饰美化：装扮是服装配饰重要的用途之一，它体现了人类自始至终对美的永恒向往，如胸针、耳环、丝巾等。

防御保护：御寒保暖、防风防晒是许多服装配件的重要功能，人类利用各式各样的配件来保护自身、尽可能地抵御来自外界的伤害。如护膝、手套、凉帽等。

收纳物品：用来携带零碎、分散的物品，尽可能减轻双手的负担以便适应生活中的各种变化，如钱包、拎袋、公文箱等。

固定造型：服饰配件有时会完全作为服装的从属物出现，帮助服装的造型完整或变化，如领带夹、腰带、绳结等。

Apparel products can be used to satisfy the general needs of human beings with various purposes.

Decoration: Decorating is one of the most important uses of clothing accessories, embodying the eternal longing for the beauty from beginning to end, such as brooch, earrings, scarves and so on.

Defensive protection: Warm-keeping, wind-breaking, and sun-protecting are an important functions of many clothing accessories. Humans use various accessories to protect themselves and resist the harm from the outside. Such as knee pads, gloves, cool hats and so on.

Storage items: Accessories can be used to carry pieces, scattered items, as much as possible to reduce the burden on the hands in order to adapt to changes in daily life, such as wallets, bags, briefcase.

Fixed styling: Apparel accessories sometimes appear completely as clothing affiliates, helping to complete or change the styling of the clothing, such as tie clips, belts, knots, etc.

2. 按材质分类
Classifying by material

服饰配件的材质从根本上影响着外观风格，体现人物的身价，决定制作工艺的难易。设计者通常会在综合考虑材质的审美表现力和制作加工难度后，再进行艺术设计。

The material of costume accessories fundamentally affects the appearance style, reflects the value of the wearers, and determines the difficulty of the production process. Designers usually consider the aesthetic expression of materials and the difficulty of production and processing before performing artistic design.

常用材料制作的服饰配件：

The commonly used materials made accessories:

纺织品类：布腰带、布帽、帆布鞋等。

Textile categories: cloth belts, cloth caps, canvas shoes ,etc.

毛皮类：皮包、皮鞋、皮带等。

Fur: leather bags, shoes, belts,etc.

金属类：金手镯、银头饰、铜手杖等。

Metals: gold bracelets, silver headdresses, copper walking sticks, etc.

植物类：竹编帽、木屐、檀香扇等。

Plants: bamboo hats, clogs, sandalwood fans, etc.

有机物类：贝壳镜架、珍珠项链、牛骨发髻等。

Organics: shell frames, pearl necklaces, bovine hair buns, etc.

矿石类：钻石戒指、蓝宝石挂坠、珐琅手表等。

Ore categories: diamond rings, sapphire pendants, enamel watches, etc.

其他类：树脂手串、玻璃项链等。

Other categories: resin bracelets, glass necklaces and so on.

3. 按性别分类
Classification by gender

服饰配件通常按使用者的性别进行设计。很多服饰配件的品类是男女相同的，只是在尺寸、样式等方面有所区分。作为商品，按性别分类的同时，还是会考虑使用者的年龄因素。

男性饰品：常用于男性的饰品，如手杖、领带夹、烟斗等。

女性饰品：常用于女性的饰品，如发带、花饰、高跟鞋等。

中性饰品：男女样式比较模糊或均可使用的饰品，几乎分布于所有非性别专属的品类中。

Clothing accessories are usually designed according to the user's gender. Many types of apparel accessories are the same for men and women, but they differ in size, style, and other aspects. As a commodity, the so-called gender-disaggregation here would take the user's age factors into consideration.

Male accessories: commonly used for men, such as walking sticks, tie clips, pipes and so on.

Women's Accessories: commonly used for women, such as hair bands, floral ornaments, high heels and so on.

Neutral accessories: vaguely styled accessories are found in almost all unisex exclusive categories.

4. 按地区分类
Classifying by region

不同地区的文化孕育了不同的服饰品风格，其服饰配件都会带有浓浓的当地特色。整体上看，由于国家、人种、气候、宗教、艺术等方面的差异，使得各地区在服饰配件的设计上均有不同，划分范畴可依据需求决定，小至一个具体的县城、地区；大至笼统的方位性划分，如东西方的服饰配件，就有着明显的地区差异。

西方服饰配件：设计题材比较广泛，创意大胆，形式多样，风格跨度较大，材质与工艺均能依靠先进制造技术的支撑、创新。

东方服饰配件：设计题材相对传统，常采用对称的设计，选择天然的本色，崇尚"天人合一"的哲学思想，制造技术比较传统。

非洲服饰配件：设计题材崇尚大自然，造型质朴，以天然材质为主，色彩比较鲜亮。

The culture of different regions breeds different clothing styles, and their accessories will have strong local characteristics. From the comprehensive aspect, there are differences in the country, race, climate, religion, and art, which will be reflected on the design, The Classfication can be done in line with the needs from a specific townor region to a general location. For example, there are obvious regional differences between East and West clothing accessories.

Western apparel accessories: The design theme is relatively wide, creative and bold; the forms are diverse and style span is large; materials and technology are based on the support of advanced manufacturing technology and dare to innovate.

Oriental costume accessories: Design theme is more traditional, often using symmetrical design, choosing the natural qualities, and advocating the "integration of heaven and man" philosophy, with more traditional manufacturing technology.

African dress accessories: Design theme advocates nature, simple shape, mainly natural materials, and bright colors.

5. 按民族分类
Classified by ethnic group

不同民族的风情、习俗文化、信仰，使其所属的服饰品具有不同的形制和特色。中国是一个多民族国家，每个民族都有着自己的文化底蕴、服饰特征、色彩偏好、图腾崇拜，例如，苗族劈线绣的荷包，水族花丝工艺的银压领，以及土家族织锦的背带等；世界其他民族也拥有着别具一格的服饰配件，例如，印第安部落的动物羽毛装饰和贝壳装饰、澳大利亚土著的帽饰与颈饰都反映了不同民族的特色。

The different customs, culture, and beliefs of different ethnic groups make the different forms and features of costume products. China is a multi-ethnic country, each nation has its own national culture connotation , costume features, color preferences, totem worship, such as the pockets of the Miao and Yi line embroidery, the silver pressure collar of the aquatic silk filigree, and the straps of the Tujia brocade. Other nations of the world also have unique clothing accessories. For example, animal feather decoration and shell decoration of Indian tribes, Australian indigenous hat ornaments and neck ornaments, all reflecting different ethnic characteristics.

6. 按风格分类
Classifying by style

人类在漫长的历史变迁中形成的特征鲜明的艺术表现被总结成为不同的风格，服饰配件作为一种带有艺术特质的装饰物，也展现着不同的艺术风格。比如，经典风格、民族风格、华丽风格、浪漫风格、前卫风格、田园风格、简约风格、古埃及风格、古希腊及古罗马风格、拜占庭风格、哥特风格、文艺复兴风格、巴洛克风格、洛可可风格等，都有与之匹配的服饰配件。

The distinctive artistic expression formed by mankind in the long historical changes has been summed up into different styles. As an accessory with artistic characteristics, it demonstrates different artistic styles, such as classic style, ethnic style, gorgeous style, romantic style, avant-garde style, rustic style, simple style, ancient Egyptian style, ancient Greek and Roman style, Byzantine style, Gothic style, Renaissance style, Baroque style, Rococo style and so on. All styles have their matching clothing accessories.

7. 按佩戴部位分类
Classifying by wearing parts

按照佩戴部位的分类方法，是服饰配件中比较普遍的常用分类方式。由于服饰配件是一种用于佩戴在人体上的物品，按照佩戴部位分类的方式也更加便于人们使用和搭配。

Classification by wearing parts is a common method for classifying fashion accessories. Since the clothing accessory is an item for wearing on the body, the classification according to the wearing part is also more convenient for people to use and match.

头饰：用于下巴以上部位的服饰配件，如帽子、头巾、眼镜、耳环、鼻环、发饰等。

Headwear: accessories for the upper part of the chin, such as hats, headscarves, eyeglasses, earrings, nose rings, hair accessories.

颈肩饰：用于脖子及肩部的服饰配件，如项链、项圈、领带、领结、围巾、披肩等。

Neck and Shoulder Accessories: accessories for the neck and shoulders, such as necklaces, collars, ties, bow ties, scarves, shawls, etc.

胸饰：用于胸部的服饰配件，如胸针、胸花、领带夹、徽章、手巾等。

Chest accessories: accessories for the chest, such as brooches, corsages, tie clips, badges, and towels.

腰饰：用于腰部前后的服饰配件，如腰封、腰带、腰链、香包、玉佩、腰牌等。

Waist accessories: apparel accessories for the waist before and after, such as girdle, belt, waist chain, sachet, jade, waist and so on.

手饰：用于前臂及手部的服饰配件，如手链、手表、手镯、手环、手套、臂章等。

Hand accessories: accessories for the forearms and hands, such as bracelets, watches, bracelets, bracelets, gloves, armbands, etc.

足饰：用于小腿及以下部位的服饰配件，如鞋子、脚环、脚链、袜子等。

Foot accessories: apparel accessories for lower legs and below, such as shoes, foot rings, anklets, socks.

第二章 服饰配件的设计基础
Chapter 2　Design basics of clothing accessories

课程名称：服饰配件的设计基础

学习目的：通过本章的学习，使学生明确服饰配件的设计所包含的精神内容和物质内容，初步了解服饰配件设计应具备的知识结构、工艺技能和审美素养，并能知晓综合运用设计元素与材料设计服饰配件的知识；提高自身的审美素养；结合服饰配件设计所需的工艺技能；了解服饰配件设计的未来发展方向,完成设计实践。

本章重点：服饰的配件的基本内容和文化内涵；服饰配件设计的艺术要素和设计风格；服饰配件设计的技能要求、审美要求；服饰配件的未来发展趋势。

课时参考：8课时。

Course Name: Design basis for apparel accessories

Learning Objectives: Through the study of this chapter, students will be able to clarify the spiritual content and material content of the design of apparel accessories, and initially understand the knowledge structure, craft skills and aesthetic qualities of apparel accessories design, and be aware of the comprehensive use of design elements and materials. Design the knowledge of clothing accessories; improve their own aesthetic quality; combine the craft skills required for the design of apparel accessories; understand the future development direction of apparel accessories design and complete the design practice.

The focus of this chapter: The basic content and cultural connotation of clothing accessories; the artistic elements and design style of apparel accessories design; the skill requirements and aesthetic requirements of apparel accessories design; the future development trend of apparel accessories.

Class time reference: 8 class hours.

第一节　服饰配件的内容
Section 1　Clothing accessories content

一、服饰配件的物质内容
A. Material content of clothing accessories

服饰配件的物质内容是其在物质方面的本质和表现。物质内容是设计师关注的内容和产品立足的支点，也是服饰配件表现精神内容的基础，只有掌握了服饰配件的物质内容，才能在此基础上去探寻和追求其丰富多彩的精神内涵。本课程主要从形制、材料、工艺这几个与设计关系最为密切的内容进行探讨。

The material content of apparel accessories is the presentation of its nature. Material content is the fulcrum of designers' attention on the content and products, also the basis for expressing the spiritual content of clothing accessories. Only mastering the material content of the clothing accessories enables us to explore and pursue its rich and colorful spiritual connotations on this basis. This course focuses on terms of shape system, materials, and processes that are most closely related to design.

1. 形制
The Shape

"形"指形式、形状、轮廓,"制"指制式、样式。服饰配件的形制主要是指服饰配件根据其本身不同的服用性能、佩戴方式、尺寸大小而形成的比较固定的、规律的样式。服饰配件在设计的过程中有一些相对可以把握的外形规律或构成要素,例如,珠宝、首饰、手表都是一些环状或者可缠绕悬挂的物品;又如箱包,即使是一些形状很特殊的箱包,也有一定的收纳空间,可以被拎提或背拷;帽子,需要能够形成一个贴合头颅的半球形等。服饰配件的形制往往是依据其功能性而逐渐形成的,而后又成为一种构成要素作用于物品本身。如今有许多大胆的艺术创作想要摒弃或者颠覆人们心中约定俗成的形制特征,例如,简·葛森·韦茨曼的图册《艺术与鞋底》中出现的150种千奇百怪的鞋子(图2-1),所涉及的材质有羽毛、白纸、陶瓷、金属、树脂、卡片,还有用鲜花、起皱的硬纸板、施华洛世奇的水晶,甚至是糖粉,每一件作品都极富创造力与想象力。这些设计摒弃了鞋子常规的材质,使其丧失了部分功能性,但保留了鞋子的形制,所以这些作品依旧被辨识为鞋子,这就是艺术设计的精妙之处。不难发现,在创新设计中要有所保留地"颠覆",才能够保持服饰配件本身的辨识度,这也就是此物之所以为此物的标准。

"Form" refers to the form, shape, outline, while "system" refers to the standard, style. The shape and system of the clothing accessories mainly refer to the relatively fixed and regular patterns formed by the clothing accessories according to their own different wearing performances, wearing manners, and sizes. In the process of designing, there are some relative controllable rules or elements for the accessories. For example jewelries, and watches are all items that are looped or can be hung and suspended; if they are bags, even some bags in special shapes, providing a certain storage space, which can be lifted or carried; Take hats as another example, a hemisphere form fitting the skull is needed. The form of apparel accessories is often formed on the basis of its functionality, and later becomes a constituent element acting on the article itself. Today, there are many bold artistic creations that attempt to abandon or subvert the conventional features of people's minds, such as the 150 kinds of strange shoes that appear in Jane Gershon Weitzman's album *Art & Soles* (Figure 2-1). The materials involved are feathers, white paper, ceramics, metals, resins, cards, and flowers, wrinkled cardboard, Swarovski crystals, and even sugar powder. Every single piece is full of creativity and imagination. These designs abandon the conventional materials of the shoes, causing them to lose some of their functionality, while retaining the shape of the shoes. Therefore, these works are still recognized as shoes. This is the subtlety of art design. It is not difficult to find that in innovative design, it is necessary to subvert it with reservations in order to be able to maintain the recognition of the apparel accessories itself. This is the reason why this object is the standard for this product.

图2-1 《艺术与鞋底》书中简·葛森·韦茨曼设计的部分鞋子
Figure 2-1 Parts of Shoes Designed by Jane Gershon Weitzman in 'Art & Sole'

2. 材料
Materials

材料是服饰配件最直观的物质内容。材料的不同选择，决定了服饰配件展现出来的质地和效果，而对于设计者来说，不同的材料还决定了制作时不同的加工工艺和最终的设计风格。当今的艺术观念在不断拓展，服饰配件的材料范围也不断扩大，越来越多的新材料出现在人们的视野中，也极大地丰富了服饰配件的艺术形态。❶服饰配件的材料使用所受到的制约要比服装相对较小，除了金属、玉石、皮草、纺织品等常规材料，经过处理的植物、纸张、塑料等具有可塑性的材料也出现在了服饰配件的应用当中。另外许多先锋设计师更乐于对材料进行再创造，也就是指对材料的二次加工，即通过解构、重组来改变材料原本的特征。对材料的二次设计，必须建立在对备选材质的性能有足够了解和驾驭能力的基础上，才能够准确地把握设计目标和材料变化，塑造出独特的、个性的、视觉效果强烈的外观效果。

Materials are the most intuitive material content of apparel accessories. Different choices of materials determine the texture and effect exhibited by the accessories. To the designers, different materials also determine the different processing techniques and the final design style. Today's concept of art is constantly expanding, and the range of materials for apparel accessories is also expanding. More and more new materials are appearing in people's eyes, and it also greatly enriches the art form of clothing accessories [Edited by Lu Xiaoyun. Decorative Art Design [M]. Beijing: Peking University Press, 2011.1: 90]. The use of materials for apparel accessories is relatively broader compared to clothing, apart from the conventional materials such as metal, jade, fur, and textiles, treated plants, paper, plastics and other plastic materials also appear in the application of apparel accessories. In addition, many pioneer designers are more willing to recreate the material, that is, the secondary processing of the material, that is, through the deconstruction and reorganization to change the original characteristics of the material. The secondary design of materials must be based on a sufficient understanding of the properties of the candidate materials and the ability to manage them. Only then can they accurately grasp the design goals and material changes, and create a unique, individualized, visually powerful appearance.

（1）服饰配件材料的分类。

Classification of apparel accessories.

材料的加工工艺难易程度，决定了批量化生产的成本问题。比方说对于外观同样是透明体的塑料、玻璃、水晶、钻石等材质，其材料成本和工艺难度有着天壤之别。由此可见，材料这个物质内容对于整个服饰配件产业链会产生巨大的影响。材料可分为天然材料、人工材料两大类。

天然材料是指在自然界天然形成的材料。其中包括有机材料，如竹、木、丝、橡胶、动物皮毛等；无机材料，如矿石、黏土、金银。天然材料往往由于地域的差异而呈现不同的性质和纯度。

人工材料是指由人工合成的材料。人工材料往往分为两类：一类是为了弥补天然材料的缺点而产生的，比如合金、人造水晶、有机玻璃、黏胶纤维；另一类则是在自然界中提炼出来的材料，比如，塑料、橡胶；随着时代的进步，特别是信息技术的发展，电子材料、光学材料、磁性材料、能源材料也成为服饰配件研发的新热点。

The difficulty of material processing determines the cost of mass production. For example, for plastic, glass, crystal, diamond and other materials that are also transparent in appearance, the material cost and process difficulty are quite different. It can be seen that the material content will have a great impact on the whole clothing accessories industry chain. Materials can be divided into natural materials and artificial

---

❶ 陆晓云. 装饰艺术设计 [M]. 北京：北京大学出版社，2011.1: 90.

materials.

Natural materials refer to materials formed naturally in nature. These include organic materials, such as bamboo, wood, silk, rubber, animal fur. Inorganic materials, such as ore, clay, gold and silver. Natural materials often have different properties and purity due to regional differences.

An artificial material is a material that is artificially synthesized. Artificial materials are usually divided into two categories. One is to make up for the shortcomings of natural materials, such as alloy, artificial crystal, organic glass and viscose fiber. The other is materials extracted from nature, such as plastics and rubber. With the progress of the times, especially the development of information technology, electronic materials, optical materials, magnetic materials, and energy materials have also become new hot spots in the research and development of apparel accessories.

（2）服饰配件材料的审美。

Apparel Accessories Material Aesthetics.

材料作为一种物质内容，可以凭借生理感知，唤起视觉、嗅觉、听觉和触觉的体验并产生不同的知觉，转化为精神上对美的体验。材料是工艺设计和制作过程中审美信息的转化和传递的载体，这些美感主要来源于其中的质感美和肌理美。

质感美是人对材质产生的审美感觉，这种感觉往往来源于材质表面的色彩、结构、光泽、纹理、质地等因素。质感本身包含了两个基本属性：一是生理属性，即物体表面作用于人的触觉和视觉系统的刺激性信息，比如软硬、粗细、冷暖、干湿、滑涩等；二是物理属性，即物体表面传递给人的知觉系统的信息，比如物体材质的类别、价值、功能等。材料的不同会产生不同的触觉、联想、心理感受和审美情趣，人们将这些带入到设计中去，使得作品或多或少地带有感情色彩和情感倾向。

肌理美实际上是质感美的一种代表性体现。肌理是材料的主要表面特征，可以在不改变材料性质的情况下，加工而得。肌理丰富了材料的形态，令材质更有层次。不同于图案和纹样，肌理往往能唤起触觉体验，而视觉上也相对更加立体，是空间范畴的审美体验，对风格、题材的表现也具有很强张力，能唤起更加具体、真实的记忆或者想象。因此，肌理的形态及加工方法成为服饰配件设计师必须掌握的专业表现内容之一（图2-2）。

As a stuff, material can use physical sensations to evoke visual, olfactory, auditory, and tactile experiences and produce different perceptions, turning them into spiritually beautiful experiences. Material is the carrier of transformation and transmission of aesthetic information in process design and production engineering. These aesthetics are mainly derived from the beauty of texture and its organization.

The beauty of texture is the aesthetic sensation that people have on the material. This sensation often comes from the color, structure, luster, texture,etc. The texture itself contains two basic properties: one is the physiological property, that is, the irritating information that the surface of the object acts on the human tactile and visual system, such as soft and hard, thickness, cool, warm, wet and dry, and slippery; the second is the physical property, that is, the information conveyed by the surface of the object to the human perception system, such as the type, value, function. of the material of the object. The different correspondence of materials will produce different touch, association, psychological feelings and aesthetic tastes. People will bring these into the design, making the works more or less emotional and sensational.

Beauty of texture organization is actually a representation of the beauty of texture. Texture organization is the main surface feature of the material and can be processed without changing the properties of the material. The texture organization enriches the material's shape and makes the material more layered. Unlike patterns and patterns, texture organization can often evoke haptic experiences, but it is also visually more dimensionally. It is an aesthetic experience of the spatial category. It also has a strong tension in the expression of style and theme, and can evoke more specific and real memories or imaginations. Therefore, the form and processing method of texture organization become one of the professional performance contents that designers of fashion accessories must master (Figure 2-2).

图2-2 纸质的包
Figure2-2　Paper Package

（3）服饰配件材料的语义。

The semantics of the clothing accessories materials.

材料作为服饰配件的物质载体，有一定的情感象征所指，就是这里所说的语义。语义对于传递创作者的观念、思想、服饰配件的内涵有着积极的意义。每一种材料都有着自身的基本语义，这种语义可能会随着环境的改变和时间的推移以及不同的文化背景的差异而不同。下面是服饰配件中比较常用的七种材料的语义分析。

金属材料坚固耐久、锃亮、耐磨、坚硬、耐腐蚀，给人刚毅、冷酷、持久、时代感等语义。其中黄金、白银、铂金等则象征了财富，具有高贵感。自2015年以来，玫瑰金又重新出现在人们的视野中，象征着时尚、娇媚、儒雅。

皮革材料柔韧、厚实、弹性，带有优雅、气派、保暖、野性的语义。稀有的天然皮革是高贵身份的标志，动物保护主义及人造革的大量出现和使用则逐渐减弱了这层语义。这正是随着时代的变化和新旧事物的更迭，人对事物语义感受也随之产生了变化。

纤维材料柔软、飘逸、温暖，有呵护、传统、随和、温馨等语义。比如棉、麻、丝、毛会让人联想到自然的生活体验和质朴的品质感受。冬季的毛可以给人温暖、柔和的感觉；夏季的丝又可以给人凉爽、轻盈的印象。

木材环保、韧性高、易加工，花纹自然、色泽朴素，给人温暖、朴素、大方、天然、不拘一格的语义。

纸张轻、薄、价格低廉、可塑性强、易损坏，给人脆弱、环保、平和、朴素、自然的语义。

塑料可塑性强、强度高、耐腐蚀、易加工，带有价格便宜、时尚、特立独行的语义。塑料能够很好地满足设计师的造型需求，在普及性服饰配件中常见。

玻璃坚硬、透明、纯粹，带有浪漫、梦幻、光怪陆离的语义。玻璃既可以隔离又可以产生视觉上的穿透感，能够表现若隐若现的朦胧美，具有丰富的创意表现力。

As a material carrier of costume accessories, material bears some emotional symbols—semantics. Semantics has a positive significance for the transmission of creators' ideas, thoughts, and the contents of the accessories. Each kind of material has its own basic semantics, which may have certain variances with the environment and time and different cultural backgrounds. The following is a semantic analysis of seven materials that are commonly used in apparel accessories.

Metal materials are strong, durable, bright, wear-resistant, hard and corrosion-resistant, presenting the impression of fortitude, coldness, longevity, and sense of the times. Among them, gold, silver, platinum, etc. symbolize wealth and have a sense of nobleness. Since 2015, rose gold has reappeared in people's sights, symbolizing fashion, coquetry and elegance.

Leather material is flexible, thick, elastic, with

elegant, stylish, warm, wild semantics. Rare natural leather is a hallmark of noble status, and the emergence and use of animal protectionism and artificial leather has gradually weakened this layer of semantics. This is precise because of the change of the times and the replacement of old and new things, people's semantic perception of things has also changed.

The fiber material is soft, elegant, warm, with care, tradition, easy-going, warm and other semantics. For example, cotton, linen, silk, and wool are reminiscent of natural life experiences and pristine quality feelings. Winter fur can convey a warm, soft feeling; summer silk can give a cool, light impression.

The wood is environmentally friendly, high in toughness, easy to process, with natural and simple patterns, giving people warm, simple, generous, natural and eclectic semantics.

Paper is light, thin, inexpensive, malleable, easily damaged, and gives people a sense of vulnerability, environmental protection, peace, simplicity, and naturalness.

Plastic plasticity, high strength, corrosion resistance, easy processing, with cheap, stylish, unique semantics. Plastics can well meet the styling needs of designers and are common in popular clothing accessories.

Glass is hard, transparent and pure, with romantic, dreamy, bizarre semantics. Glass can be isolated, produce a visual sense of penetration, express the looming beauty, and has a rich creative expression.

3. 工艺
Craftsmanship

工艺是利用各类生产工具对原材料、半成品进行加工制作，最终使之成为成品的技术手段和处理方法。服饰配件的创作离不开工艺，而工艺的发展又离不开技术的革新。《考工记》是中国目前所见年代最早的手工业技术文献，它在中国古代思想中，第一次率先提出了一个朴素的工艺观和艺术观，其中"天有时，地有气，材有美，工有巧，合此四者，然后可以为良"的哲学观，精辟分析了在设计艺术中如何依据自然规律、空间条件、材料质地和熟练技艺来完善设计作品。当材料的个性特征得到了恰如其分的发挥，材料的工艺美就自然而然地显露出来。所以，材料的审美是工艺加工的结果。❶古时的人们主要依靠人力来生产制作，劳动者花费大量的时间和精力完成服饰品。随着时代的进步和发展，机器的诞生彻底革新了人类的生产方式，随之而来的是更加丰富的工艺选择，在这样的时代背景下，制作的精良和经济的合理构成了工艺选择需要考虑的两大要素。将工艺涉及的内容分开来理解，其中包括：（1）加工工具的选择；（2）材料的处理手段；（3）处理进程中的外界因素控制。

Craftsmanship is to apply various types of production tools to process raw materials, semi-finished products, and ultimately make it a finished product of the technical means and processing methods. The creation of fashion accessories is inseparable from the craft, and the development of craftsmanship can not be separated from technological innovation. "Kao Gong Ji" is the earliest craftsmanship technique literature in China dating back to the present era. It was the first time in ancient Chinese thoughts to propose a simple craft and art view. The philosophical idea "God provides the timing, the earth provides the energy, the material provides the beauty, the workers contribute the craftsmanship, and together they can then be good." It gives an incisive analysis of how to improve the design work based on natural laws, space conditions, material textures, and skilled techniques in the design art. When the personality characteristics of the material are properly exerted, the beauty of the craftsmanship of the material naturally emerges. Therefore, the aesthetic of the material is the result

---

❶ 陆晓云. 装饰艺术设计［M］. 北京：北京大学出版社，2011.1: 90.

of the process [Lu Xiaoyun. Art Deco Design [M]. Beijing: Peking University Press, 2011.1 : 90]. In ancient times, people mainly depended on manpower to produce and produce with tremendous laborers, time and energy to complete furnishings. With the progress and development of the times, the birth of the machine has completely revolutionized the mode of man's production. With the richer technological choices, under the circumstances of this era, the production of sophisticated and economical rationality constitutes the need for process selection, which constitute into two major factors to consider. The contents involved in the process are understood separately, including: (1)Selection of processing tools; (2)Processing methods of materials; (3)Control of external factors in the process of processing.

（1）加工工具的选择。

Selection of processing tools.

工具加工是根据所加工的材料、加工的形状和加工的工序等因素而决定的。服饰配件所处的行业跨度很大，分散在箱包、鞋帽、首饰、眼镜、皮革等行业，这些行业面对的加工材料千姿百态，采用的对应工具也有天壤之别。不同的材料有相应的加工工具，比如，木质饰品应选用木工工具，金属饰品应选用金工工具。即便是同一种材料，也会因加工不同的形状、尺寸、表面处理而选择不同的工具，比如，打孔会使用台钻、手电钻等，研磨会使用砂轮机、抛光机等。工具的不同形成了不同的工种，比如，木工、车工、钳工、铣工、油漆工、注塑工、模具工等。因此，加工工具的种类繁多、性能迥异，一个人难以全部熟练操作。

加工工具看似并非服饰配件的内容，但作为实现工艺的先决条件，是学好本课程的基本内容之一。设计师应该熟悉和掌握各自领域内主要工具的使用方法，只有真正了解了工具的性能，才能让设计构思得以完美实现。除了经验之外，工具的优劣往往是加工效果的决定性因素，所谓"工欲善其事，必先利其器"。比如，要将一块金属表面手工打磨到镜面效果，手中没有5000目以上砂纸是会有难度的（图2-3）。

Tools are determined based on factors such as the material being processed, the shape of the machining, and the processing steps. The clothing accessories are in a wide range of industries, scattering the bags, shoes, hats, jewelry, glasses, leather and other industries. The processing materials faced by these industries are varied, and the corresponding tools used are also very different. Different materials have corresponding processing tools. For example, wooden jewelry should use woodworking tools, and metal jewelry should use metalworking tools. Even the same material, different tools will be selected for processing different shapes, sizes, and surface treatments. For example, drills and hand drills will be used for drilling, and grinders and polishers will be used for grinding. Different tools have formed different types of work, such as woodworking, lathe, fitter, milling, painter, injection molding, mold and so on. Therefore, there are many types of processing tools and the performance is very different. It is difficult for one person to fully operate them all.

Processing tools do not seem to be the content of apparel accessories, but as a prerequisite for the realization of the process, it is one of the basic contents of this course. Designers should be familiar with and master the use of the main tools in their areas. Only by truly understanding the performance of the tools can the design ideas be perfectly realized. In addition to experience, the pros and cons of tools are often decisive factors in the processing results. The so-called "workers must first sharpen their tools if they want to do good things". For example, to grind a piece of metal surface by hand to mirror effect, it is difficult to achieve it without the sandpaper more than 5,000 mesh (Figure 2-3).

（2）材料的处理手段。

Materials processing methods.

对材料的处理，是工艺的重要内容。服饰配件

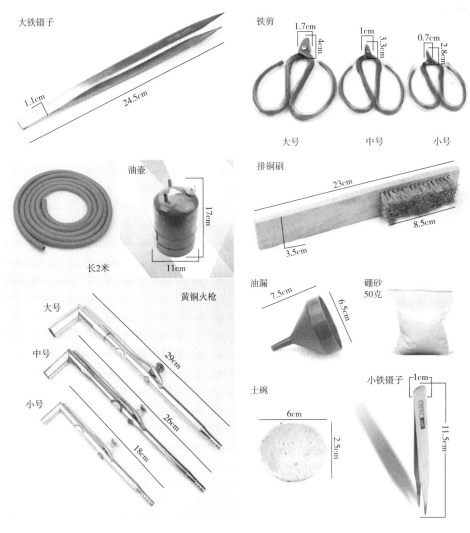

图2-3 部分手工小工具
Figure2-3 Some Manual Gadgets

在经过设计构思之后，可以绘制出风格鲜明的设计图。将设计构思实物化的重要步骤是选择合适的材料，配合相应的工具，经过一定的工艺程序之后，实现设计的预期效果。比如，要在缝纫、刺绣、模压、雕刻、编结、镶嵌、焊接、琢磨、镂空、热塑定性、烧制等可供选择的工艺中实现一个装饰图案，首先要确定承载这一图案的材料是否在工艺上可行，其次是运用某种工艺，完成这一图案的表现。整个过程看似制作一个图案，实质上是开展针对材料的处理。

即便是完全相同的材料，对其处理手段的不同，可以出现截然不同的效果。比如，上述图案在黄铜的表面上錾刻而成，或是腐蚀而成，两者效果自然大相径庭。除了设计意图以外，选择材料处理的因素往往取决于工具的限制或者对成本的考量。比如，刺绣工艺是比较常见常用的服饰配件制作工艺之一，为了达到刺绣工艺效果，可以选择手工绣制、机器绣制、刺绣成品的热转印，甚至用印花的方式模仿刺绣效果等不同的处理手段（图2-4、图2-5）。每一种处理手段都有各自的利弊因素，同时也影响着服饰配件的价值。

The processing of materials is an important part of the craftsmanship. After the design of the apparel accessories, you can draw a distinctive styled sketch. The important step of producing the design concept physically is to select the appropriate materials and

图2-4 机器绣制古驰官网
Figure2-4 Machine embroidered Gucci official website

图2-5 刺绣成品的热转印古驰官网
Figure2-5 Embroidery finished heat transfer Gucci official website

cooperate with the corresponding tools to achieve the intended effect of the design after a certain process procedure. For example, to realize a decorative pattern in the sewing, embroidering, embossing, engraving, braiding, inlaying, welding, honing, hollowing, thermoplastic qualification, firing, and other alternative processes, the first step is to determine whether the material carrying the pattern feasible in the process, and use a certain process to complete the performance of this pattern. The whole process seems to create a pattern, but essentially carrying out the processing of the material.

Even the exact same material can have a completely different effect on the processing methods. For example, if the above pattern is engraved on the surface of brass, or corroded, the effect of the two is naturally quite different. In addition to design intent, the choice of material handling factors often depends on tool limitations or cost considerations. For example, the embroidery process is one of the most commonly used clothing accessories manufacturing processes. In order to achieve the effect of the embroidery process, you can choose different processing means such as hand-embroidered, machine-embroidered, thermal transfer of embroidery products, or even imitation of embroidery effects with printing. Processing means. (Figure 2-4、Figure 2-5). Each treatment method has its own advantages and disadvantages, and it also affects the value of apparel accessories.

（3）处理进程中的外界因素控制。

External factors in process control.

采用同样的工具、同样的工艺对同样的材料进行处理时，外界因素的改变会影响结果的最终表现。最为典型的几种外部因素包括：时间因素、温度因素、技术因素等。在技术相同的前提下，同样的工艺因实施时间不同而结果各异，所谓"慢工出细活"。比如，对某种材料进行快速切割或缓慢切割，其截面精度往往不尽一致。油漆的成膜质量与环境温度有很大关系，将会影响漆膜的耐磨度和咬合度。而相同的工艺流程由技术等级不同的操作人员实施，其结果更是显而易见。因此，外界控制因素的微妙变化往往是影响最终品质优劣的关键因素。

When the same material is processed using the same tool and the same process, changes in external factors will affect the final performance of the results. The most typical external factors include: time, temperature, techniques and so on. Under the same premise of technology, the same process has different results due to different implementation times, as it is so called "refined works from slow working paces". For example, when a material is cut quickly or slowly, its cross-section accuracy is often not consistent. For another example, the filming quality of the paint has a great relationship with the ambient temperature, which will affect the wear resistance and bite degree of the paint film. The same process flow is implemented by operators with different technical levels, and the results are even more obvious. Therefore, the subtle changes in external control factors are often the key factors affecting the quality of the final product.

二、服饰配件的精神内容
B. Spiritual content of clothing accessories

服饰配件从诞生之初就被人类赋予了丰富的精神内涵，人类在精神上的需求是服饰配件产生的重要原因，本课程以人类本质的精神需求作为切入点，以科学辩证的角度，从审美、象征、寓意三个方面来探讨服饰配件所蕴含的精神内容。

Clothing accessories have been endowed with rich spiritual connotations from the their birth. The spiritual needs of human beings are an important reason for the production of apparel accessories. This course takes the spiritual needs of human nature as the starting point, and try to explore the spiritual content contained in apparel accessories from the perspective of scientific dialect, from the three aspects of aesthetic, and from symbolism and morality.

1. 审美
Aesthetic

所谓美，是指物象经过整理，在有统一感、秩序化的情况下，产生愉悦的精神感受。秩序是美的最重要条件，因而服饰配件设计要遵循形式美的规律，运用形式美的原理去创造出美的服饰品。❶考古学研究认为，距今30万年前，我们的先辈就开始使用动物的牙齿、骨头、贝壳、羽毛、石子来装饰自己身体的各个部分，并且还会在身上涂抹自然中萃取的颜料或文身，可见即使是茹毛饮血的时代，人类还没有穿着衣服，就开始尝试使用服饰品来修饰自己。从精神层面来说，修饰的目的是人类对美的追求。审美价值属性是艺术对象的本质属性，人及其社会生活的其他价值属性，都必须在一定条件下转化为审美价值属性，或者以审美的方式表现出来。❷美具有原始驱动力，古今中外，人类始终没有停止追寻美，服饰配件作为其中的一种艺术载体，承载着审美意识的精神内涵。

The so-called "beauty" refers to the fact that the imagery has been organized, and in the sense of unity and order, it produces a pleasant spiritual feeling. Order is the most important condition for beauty. Therefore, the design of apparel accessories should follow the rules of formal beauty and use the principles of formal beauty to create beauty apparel products [Chen Dongsheng, Wang Xiuyan. New clothing accessories research [M]. Beijing: China Light Industry Press, 2004.11]. Archaeological research believes that 300,000 years ago, our ancestors began to use animal teeth, bones, shells, Feathers and stones to decorate all parts of the body. Naturally extracted pigments or tattoos were also applied on the body. This shows that even in the rough era, humans had not yet wore clothes and began to try to use accessories to modify themselves. On the spiritual level, the purpose of the modification is the pursuit of beauty by human beings. The aesthetic value attribute is the essential attribute of the art object, and people and other value attributes of their social life must be transformed into aesthetic value attributes under certain conditions, or be expressed in an aesthetic way [Shi Hai-Bin. A Comparative Study of the Essence of the Aesthetic Essence of the Essence [J]. Hunan Social Sciences, 2005(5): 142-144.]. The beauty has its original driving force, from ancient and modern, both at home and abroad, humans have never ceased to pursue the pursuit of beauty. As one of these art carriers, apparel

---

❶ 陈东生，王秀彦. 新编服装配饰学［M］. 北京：中国轻工业出版社，2004.11.
❷ 石海滨. 从比较视角看艺术本质审美价值理性［J］. 湖南社会科学，2005(5):142-144.

accessories carry the spiritual connotation of aesthetic consciousness.

2. 象征
Symbol

象征作为一种强大的精神力量蕴含在服饰配件中，人们期望展示自身的美好以及获得赞美，设法从装饰品中获得快乐。美国人类学家格尔茨（又译作格尔兹）认为"象征符号是指作为观念载体的物、行为、性质或关系——观念是象征的'意义'"。❶怀特海德说："象征的目的就是要强化被象征物的重要性"❷例如，狩猎能手以猎物作为装饰品，随着文明的推演，凯旋的将士佩戴象征胜利的勋章，美丽的羽毛、稀有的宝石象征了力量和地位。古时甚至有一些部落为了铭记自己的荣耀，不惜损伤身体来装饰自己，来表现自己的年龄和社会地位。

As a powerful spiritual force, symbolism is embedded in fashion accessories. People expect to display their own beauty for praises, and to seek happiness from decorations. The American anthropologist Clifford Geertz believes that "a symbolic symbol refers to the object, behavior, nature, or relationship as a carrier of concept-concept is symbolic 'significance'." [Clifford Geertz. Cultural interpretation [M]. Nari Bilige et al., Wang Mingming. Shanghai: Shanghai People's Publishing House, 1999: 105]. Whitehead: "The purpose of the symbol is to strengthen the importance of being symbolized." [He xingliang. Types of Symbols [J]. Ethnic Studies, 2003(1): 39-47.]. For example, strong hunters use prey as ornaments. With civilizations progresses, medals worn by triumphant soldiers symbolize victory, beautiful feathers and rare gems symbolize strength and social status. In ancient times, some tribes even did not hesitate to damage the body to decorate themselves to show their age and social status in order to remember their own glory.

图腾作为象征的一种典型表现，最初来源于印第安语"Totem"，意思是"它的亲属"或"它的标记"。原始时代的人们把某种动物、植物或非生物当作自己的亲属、祖先、保护神，相信它们能够带来一种力量保护自己或者令自己变得强大。❸所以祖先们常常会将太阳、月亮、老虎、蛇、鸟等作为图腾，或是将动植物的一部分制作成服饰品佩戴在身上来象征其可能代表的一些意义。例如，狼牙、虎牙串成的项链象征男性的勇敢和力量，而贝壳制成的腰饰、植物果实或者种子做成的挂件象征女性繁衍后代的能力。

The totem is a typical representation of a symbol, originally from the Indian word "totem," which means "its relative" or "its mark." The people of the primitive era used animals, plants, or non-living creatures as their relatives, ancestors, and protectors, and believed that they could bring a kind of power to protect themselves or make themselves stronger [Zeng Qiang, Geng Yuan, Cai Xiaoyan. Fashion Accessories Design Tutorial [M]. Chongqing: Southwest Normal University Press, 2014.1: 4]. So ancestors often used the sun, moon, tigers, snakes, birds, etc. as totems, or a part of animals and plants as clothing items to be worn on the body to symbolize its possible representation. For example, a necklace strung together with spikes and tiger teeth symbolizes the bravery and strength of males. The pendants made of shells, plant fruits, or seeds are symbolic of female reproduction.

3. 寓意
Implication

寓意是指在表象事物中寄托或蕴含另一事物的意旨或意思。起初，服饰配件的寓意主要来源于服装起源论当中的保护说。早期人类在自然崇拜中，

---

❶ 克利福德·格尔兹. 文化的解释［M］. 纳日碧力戈等, 译. 王铭铭校. 上海：上海人民出版社, 1999:105.
❷ 何星亮. 象征的类型［J］. 民族研究, 2003(1):39–47.
❸ 曾强, 奚源, 蔡晓艳. 服饰品设计教程［M］. 重庆：西南师范大学出版社, 2014.1:4.

相信万物有灵，人的精神与躯体是分离的。灵魂有善恶，善灵带来幸福吉祥，恶灵带来疾病与灾祸。❶将自然界中的物质赋予超自然的力量即是寓意的由来，而后正向衍生为对美好愿望的向往和期盼，反向衍生为辟邪或者诅咒。在精神世界里，人们受到信仰的启示，选择为服饰配件赋予美好的寓意并且佩戴在身上，比如，中国喜欢为小孩子佩戴长命锁，期盼健康、长寿（图2-6）。

Implied implication refers to the meaning of sustenance of another thing in a representational thing. At first, the moral of clothing accessories mainly came from the theory of protection in the origin theory of clothing. In the worship of nature, early humans believed that everything had its spirits, that human spirit and body were separate. The soul can be good or evil: good spirits bring happiness and good fortune, while evil spirits bring diseases and evils [Zhang Jiaqiu, Che Yanxin. Furniture design [M]. Beijing: Communication University of China Press, 2012.7]. The supernatural power given to nature by nature is both the origin of the implication, and then the positive is derived from the desire for good wishes, and expectation; and the reverse is derived as evil or curse. In the spiritual world, people are inspired by beliefs and choose to wear the beautiful accessories with good implications. For example, in China, people like to wear long-life locks for children and expects health and longevity (Figure 2-6).

在民族风格服饰配件中，寓意尤为明显，一般由本身即富有寓意的传统图案担纲。由于人们社会交流频繁，服饰配件的形制或造型已大同小异，其外形寓意越见模糊。被赋予不同寓意的传统图案往往是民族文化的浓缩，自然成为人们借助服饰配件寄托某种意旨的重要载体。许多品牌也以此创造自己的品牌故事，比如丹麦珠宝品牌潘多拉，产品概念是赋予串饰不同个人价值观、意义和特殊品质，如智慧、勇气、爱情、信任。这种理念相当于一种消费者对商品"寓意"的定制（图2-7）。

In the national-style clothing accessories, the implication is particularly obvious. It is generally conveyed in the traditional pattern with rich meaning. Because people's frequent social exchanges, the shape of apparel accessories has been similar, and their meanings of appearances are increasingly blurred. The traditional patterns that have been given different meanings are often the essentials of national culture and naturally become an important carrier for people to rely on costume accessories to entrust certain intentions. Many brands also use this to create their own brand story. For example, the Danish jewelry brand PANDORA, the

图2-6 清代麒麟送子银如意形长命锁
Figure 2-6 Qing dynasty Kirin Sending Child Silver Shaped Longevity Lock

图2-7 潘多拉品牌产品每一颗串珠都包含独特的寓意，潘多拉官网
Figure 2-7 Pandora brand products contain unique symbols for each bead, Pandora official website

---

❶ 张嘉秋，车岩鑫. 服饰品设计［M］. 北京：中国传媒大学出版社，2012.7.

product concept is to give a series of different personal values, meanings and special qualities, such as wisdom, courage, love, and trust. This concept is equivalent to a consumer's "implications" of the product's customization (Figure 2-7).

第二节　服饰配件文化内涵
Section 2　Cultural connotation of costume accessories

　　服饰配件的文化内涵是其物质和精神内容的扩展，随着人类社会和文明的进步而共同变化，其本质是人类文明不可分割的一部分。要探索服饰配件的文化内涵，离不开对人类历史、文化、艺术、科学等方面的分析。结合设计学特征，从艺术要素和设计风格两个方面着手，领会服饰配件文化内涵的魅力所在。

　　The cultural connotation of apparel accessories is an extension of its material and spiritual content. With the changes and the civilization progress of human society , its essence is an inseparable part of human civilization. To explore the cultural connotation of apparel accessories can not be separated from the analysis of human history, culture, art, science and other aspects. Combining with the features of design, we will start from the two aspects of artistic elements and design styles to understand the charm of the cultural connotation of fashion accessories.

　　The cultural connotation of costume accessories is the expansion of its material and spiritual contents. With the progress of human society and civilization, it changes together. Its essence is an inseparable part of human civilization. In order to explore the cultural connotation of costume accessories, it is necessary to analyze the human history, culture, art and science. Combined with the characteristics of design, we should understand the charm of the cultural connotation of clothing accessories from the aspects of artistic elements and design style.

一、服饰配件中的艺术形态
A. Artistic form of clothing accessories

　　艺术形态是诉诸欣赏者感官的外部形式，这种形式由塑造艺术形象的各种媒介（色彩、线条、音乐、文字等）所决定，又可表现为抽象或具象，西方艺术形态和我国传统艺术形态等[1]。艺术作品首先应该被看作客观实在的物质结构存在表现在直觉者面前[2]。由此看来，服饰配件的艺术形态最主要的有三大元素——形、色、质。这三大元素组成了服饰配件的艺术语言。

　　Artistic form is an external one resorting to the viewers' senses. This form is determined by various media (colors, lines, music, words, etc.) that shaping the artistic image, and can also be expressed as abstract or figurative, Western art forms, and our country. Traditional art form and so on [Chen Jingsong. On the commodity form of operas [J]. Guangdong Arts, 2000(5):5-8.]. The works of art should first be regarded as the objective reality of the material structure present in front of the intuition [Yang Xiaoqing. The understanding and classification of the "non-objective" art morphology [J]. Literary Studies, 2001(4):41-47.]. From this point of view, the most important art form of apparel accessories carries three elements—shape, color and quality. These three elements build the artistic language of the costume accessories.

---

[1] 陈京松. 论戏曲的商品形态［J］. 广东艺术，2000(5):5-8.
[2] 杨小清. "非客观性"艺术形态学的理解与分类问题［J］. 文艺研究，2001(4):41-47.

1. 服饰配件的艺术形态元素之一——形
One of the artistic elements of apparel accessories —Shape

服饰配件作品是以空间立体的形式存在，通过复杂或简单的体、面关系来达到三维空间的艺术效果❶。所以在形与形之间就会出现主次、虚实、分合、交错、透叠等关系，从而塑造出更加丰富的视觉效果。从整体造型角度来说，服饰配件往往不是孤立存在的，所以服饰配件的形，往往需要与整体装扮和谐统一，起到提升审美价值的作用。服饰配件在整体造型中通常会以点、线、面的形式出现。

"点"在整体中表示一个位置。在服饰配件的实际应用十分普遍，可能是一个小图案、小色块，可能规则也可能不规则。❷独立的"点"具有收敛和集中的作用。理解和把握其中的奥妙，小小的一粒纽扣、胸针、勋章、首饰，通过适当地变化其颜色、材质、大小、位置、形状，都能够给人耳目一新的感觉。点的集散变化能够幻化出音乐的韵律感；点的大小变化、疏密变化是可以影响视觉的动向，产生美的节奏与旋律。两点成线，多点成面，"点"与"线""面"之间的融合变化也是分不开的。总之，点与点之间的排列能够产生运动的感受，形成渐变，甚至能够引导视觉的走向，制造情绪的起伏，比如平静、跳跃、活泼、柔和、硬朗。

"线"在服饰配件中主要起到三个作用：一是线的指向。线条和人体构成上下关系还是左右关系，在视觉上会有不同的暗示。竖直的配饰搭配使人拉长，比如手杖或者收起的伞，而左右指向则会拉宽整体效果。还有曲线和直线也有着不同的视觉暗示，曲线可以引导凹凸感，或者柔和的变化，而直线则更加强硬、挺拔、方正；二是线的分割。为了达到不同的视觉效果，线的运用也变得尤为重要。对于服装配饰，最典型的线的分割要数皮带或是腰带了。皮带和腰带暗示了人体腰线的位置，随着流行趋势的变化，将腰带缠绕在不同的位置，展现了不同时代、民族之间不同审美趣味。适当的提高腰线可以显得腿部更加修长，朝鲜族的服饰更是将腰线提到紧挨胸围线的位置；三是线的律动。水平线显示平静、沉着、安定、松弛、理性；螺旋线有腾跃感；工字线不断地改变方向；放射线富有动感、整体性强；波浪线让人产生紧张感，其中又分小而急的波浪线和大而缓的波浪线等。

"面"包含各种形状的概念，如多边形、圆形、不规则图形等。每一种图形带给人们不同的艺术感受，三角图形显得尖锐，三角结构有稳定、持久的特点，等边形让人觉得平静。造型中人们经常使服饰配件形成一个块面与服装相协调，比方说帽子、鞋子、包包、首饰，这些饰品从不同的视角形成了不同的块面，增加造型的层次。

Clothing accessories works stay in the form of three-dimensional space, through complex or simple relationships of lines and facets to achieve the three-dimensional space of artistic effects [Xu Xing. Clothing Accessories Art [M]. Beijing: China Textile Press, 2015.5]. Therefore, there will be relationships between the shapes, such as the order of the primary versus secondary, the actual versus the virtue, the separation versus integration, the interlacing versus the transposition, to create a richer visual effect. From the perspective of the overall model, clothing accessories often do not exist in isolation, so the shape of clothing accessories often needs to be in harmony with the overall dress and play a role in enhancing aesthetic value. Apparel accessories usually appear as dots, lines, and faces in the overall shape.

The "point" represents a position in the whole. Its practical application of clothing accessories is very extensive, it may be a small pattern, a small color patch with either regular or irregular shapes. [Chen Dongsheng, Wang Xiuyan. New clothing accessory studies [M]. Beijing: China Light Industry Press, 2004.11]. The independent "points" have a convergence and

---

❶ 许星. 服饰配件艺术[M]. 北京：中国纺织出版社，2015.5.
❷ 陈东生，王秀彦. 新编服装配饰学[M]. 北京：中国轻工业出版社，2004.11.

concentration effect. Understanding and mastering the techniques, a tiny button, brooches, medals, and jewellery could convey a fresh look by adjusting their color, texture, size, position, and shape. The distribution of points can be imagined as the music melody; the change of the size of dots and the change of density can affect the visual movement and produce the rhythm and melody of the beauty. Two points form a line, and more points form a facet. The fusion changes between "point" and "line" and "facet" are also inseparable. In short, the arrangement of points and points can produce an effect of movement in a gradual changing process, and can even guide the direction of the visual, creating emotional ups and downs, such as calm, jumping, lively, soft, and tough.

"Line" mainly plays three roles in clothing accessories: First, the line's direction. Whether the relationship between the lines and the human body is left or right, up or down would make the different visual implications. Vertical accessories could provide the stretching visual effect of people, such as walking sticks or umbrellas, while the left to right direction would widen the overall visual effect. There are also different visual cues for curves and straight lines. Curves can guide the sense of concavity and convexity, or softer changes, while straight lines are more rigid, tall and straight, and square; the second is the segmentation of lines. In order to achieve different visual effects, the use of lines has become particularly important. For clothing accessories, the most typical line is the number of belts or belts. Leather belts and waist belts suggest the location of the body's waist line. With the change of fashion trends, the belt is wrapped around different locations, showing different aesthetic tastes between different eras and ethnic groups. Appropriately lifting the waist line can make the legs more slender. Korean clothing even raises the waist line next to the bust line. Third, the line's rhythm. The horizontal line shows calmness, calmness, stability, relaxation, and reason; the spiral line has a leaping sense; the I-word line constantly changes direction; the radiation line is full of movement and overall integrity; the wavy line gives people a sense of tension, which is divided into small and urgent wavy. Lines, and large and gentle wavy lines.

The "facet" contains concepts of various shapes such as polygons, circles, irregular shapes, etc. Each kind of graphics brings people different artistic feelings. The triangular graphics appear sharp. The triangular structure has a stable and lasting characteristic. The equilateral shape makes people feel calm. In the modeling, people often make clothing fittings to form a face that matches with clothing, such as hats, shoes, bags, and jewelry. These ornaments form different blocks from different perspectives and increase the layers of modeling.

2. 服饰配件的艺术形态元素之二——色
The artistic elements of apparel accessories—Color

色彩史是一门值得深入研究的学问，从不同时代的色彩特点，可以得出有价值的人类学、社会学、心理学、美学、艺术学的结论。❶色彩的三要素是色相、纯度、明度。一般来说，服装有四种配色原则，这一点可以延伸到服饰配件当中来。第一，单色搭配，即不同部件采用相同或基本相同的色彩。第二，近似色搭配，即十二色环中的相邻的1~2各色之间搭配。第三，互补色搭配，即十二色环中的对角线色彩的搭配。第四，三色搭配，即十二色环中等距的三个颜色相搭配。其中第一、第二两种原则相容性强，偏向平静、柔和。第三、第四种配色原则产生分裂、冲突、动荡的效果。这些色彩原则不仅适用于服装配饰设计本身，更重要的是在配饰的搭配应用中使服饰配件与整体造型形成有机统一的整体。

---

❶ 彭永茂. 20世纪世界服装大师及品牌服饰［M］. 沈阳：辽宁美术出版社，2001: 9.

实际上，色彩并没有绝对的使用原则和规范，更没有美、丑的属性，要依靠设计者智慧的眼光和独到的审美观察，把色彩用在对的、合适的地方，给人带来整体统一的感受。

Color history is a study that is worthy of in-depth research. From the color characteristics of different eras, valuable conclusions about anthropology, sociology, psychology, aesthetics, and art can be drawn [Peng Yongmao. The 20th century world clothing masters and branded apparel [M]. Shen Yang: Liaoning Fine Arts Publishing House, 2001: 9]. The three elements of color are hue, purity, and brightness. In general, clothing has four color matching principles, which can be extended to clothing accessories. First, single-color matching means that different parts use the same or substantially the same color. Second, approximate color matching refers to the adjacent 1~2 colors in the twelve-color ring match. Third, complementary colors matching is, the matching of the diagonal colors in the twelve color ring. Fourth, the three-color matching requires the 12 colors of the twelve-color ring match the three colors. The first and second principles are highly compatible and tend to be calm and soft. The third and fourth clock coloring principles have the effect of splitting, conflict, and turbulence. These color principles not only apply to the garment accessories design itself, but more importantly, form an organic unity between the accessories and the overall shape in the matching application of the accessories.

In fact, there is no absolute principle and norm of color, and no beauty or ugly attributes. We must rely on the wisdom of the designer and the unique aesthetic observation to use colors in the right and appropriate places and bring people a whole, unity feeling.

3. 服饰配件的艺术形态元素之三——质

Artistic elements of clothing accessories—texture

"质"是作品的半成品，设计要沿着"质"规定的路线深化下去，使其成熟、完善和丰富，挖掘物质终端效果的过程，把质地所确定的审美框架填充起来，使材料风格定型为作品风格。❶质能够给人带来感情色彩，能够形容质的词汇有：粗糙、光滑、厚实、轻薄、坚硬、柔软、松脆、弹性、冰冷、温暖等。质所涉及的感官效应包括视觉上和触觉上的，是一种多维度的物质属性，当人们去欣赏一件艺术品的时候，总要尝试去触摸一下，或者产生触摸的欲望，观看和触摸共同完成了对质这一形态元素的感受。触摸使体验更加丰富，作为需要佩戴在身上的服饰配件，并且服饰配件的使用时间有可能比服装还要久。

相较于室内设计和建筑设计，触感是服饰配件设计中必不可少的因素，这取决于设计师选择的材质。这是设计师依照质来进行设计，是因为"任何一个天然或者人造的客体的设计，都隐藏在客体自身内部，必须让他从客体所要发挥的功能，从制造他的材料，从用以制造他的方法中必然地生发出来"。❷有一个故事是说，别人问起雕塑家怎样雕出这座完美的雕塑时，他答道"这座雕像原来就在那里，我只是将它多余的边边角角去掉而已"。这一特点在珠宝设计中尤其突出，由于宝石的珍贵，人们会选择尽可能地依照物体本身呈现出来的，原有质地和状态来进行设计和雕琢。

"Texture" is a semi-finished product. The design should be deepened along the route defined by "texture" to make it mature, complete, and rich. The process of material end effect should be excavated, and the aesthetic frame determined by texture should be filled into the finalized material style. For the style of the work [Peng Yongmao. 20th Century Master of World Clothing and Brand Apparel [M]. Shen Yang:

---

❶ 彭永茂. 20世纪世界服装大师及品牌服饰［M］. 沈阳：辽宁美术出版社，2001: 7.
❷ 涂途. 现代科学技术之花一技术美学［M］. 沈阳：辽宁人民出版社，1987: 50.

Liaoning Fine Arts Publishing House, 2001:7]. Texture can bring emotions to people. Words that can describe texture are: rough, smooth, thick, thin, hard, soft, crunchy, elastic, cold, warm, etc. The sensory effects involved in texture include both visual and tactile. It is a multi-dimensional material property. When people appreciate a piece of art work, they always try to touch it or the desire to touch arouses. Watching and Touching together completes the feeling of being in the form of confrontational elements. Touching makes the experience richer. As a clothing accessory that needs to be worn on the body, the clothing accessories may last longer than clothing.

Compared to interior design and architectural design, tactile sensation is an essential factor to consider in the design of apparel accessories. This depends on the materials chosen by the designer. It is necessary to design in terms of quality. "Any design of a natural or artificial object is hidden within the object itself, it must be allowed to function from the object, from the manufacture of his material, and from the method of making him born inevitably" [Tu Tu. Flowers of modern science and technology [M]. Shen Yang: Liaoning People's Publishing House, 1987:50]. There is a story that when people ask how the sculptor carved out this perfect sculpture, he replied, "The statue was there. I just removed it from the corners." This feature is particularly reflected in jewelry design. Because of the great value of the precious stone, people will choose to design and sculpt as much as possible according to the original texture and state of the objects themselves.

二、服饰配件中的设计风格
B. Design styles in apparel accessories

不同的艺术形态下的服饰配件，衍生出不同的艺术风格，划分不同艺术风格的角度很多，而不同的划分标准赋予了服饰配件风格不同的含义和称呼。服饰配件风格的形成往往没有非常固定的方式：地域可以形成一种风格，造型方式可以形成一种风格，历史上某一时期的潮流、风尚可以形成一种风格，一种流派可以形成一种风格，甚至是一些感受、用途等都可以形成一种风格。法国服装设计大师高缇耶意味深长地说过这样的话："风格，就像是爱。你深爱着某样东西，只要用心经营，并有信念，它就能永恒。"风格的形成难以捉摸，无可厚非的是，时间是检验风格的真理，在漫长的时光和文化中，渐渐沉淀出种种不朽的风格。正如香奈儿的名言："时尚来来去去，但风格永存。"❶ 随着人们对自身形象要求更为细致，对服饰配件与整体协调统一的要求也越来越高，服饰配件也随之有了更加细腻的风格倾向，也成为了设计师在设计过程中所需要考虑的一种感性因素，所以，在此从两个角度对特征鲜明的风格进行扩展：时尚类风格和历史类风格。

The costume accessories under different artistic forms derive different artistic styles, and there are many different perspectives for categorizing artistic styles. Different classification standards give different meanings and titles to clothing accessories. The formation of clothing accessories style is often not a very fixed way: geographical can form a style; modeling can form a style; in a historical period of time, fashion can form a style; a trend can form a style; even some feelings, uses, etc. can all form a style. French fashion designer Gaultier made a meaningful statement: "Style is like love. You love something deeply. If you manage it with your heart with strong faith, it will last forever." The formation of style is elusive. What is important is that time is a test of the truth of style, and in the long hours and cultures, various immortal styles have gradually emerged. As Chanel's saying goes: Fashion comes and goes, but the style is forever [Zheng Hui, Pan Li. Clothing

---

❶ 郑辉，潘力. 服装配件设计 [M]. 北京：辽宁科学技术出版社，2009.1: 29.

accessories design [M]. 北京：Liaoning Science and Technology Press, 2009.1: 29]. As people demand more nuanced images of themselves, more and more requirements are placed on the coordination and unity of apparel accessories. As a result, apparel accessories also have a more delicate style, which has become a kind of perceptual factor that designers need to consider in the design process. Therefore, the characteristics of distinctive styles are expanded from two perspectives:fashion style and historical style.

1. 时尚类风格
Fashion style

时尚类风格是比照或依附时尚潮流风格而存在的服饰配件风格。时尚类风格可分为七个典型的风格：经典风格、民族风格、华丽风格、浪漫风格、前卫风格、田园风格、简约风格。

Fashion style is the style of clothing accessory that contrasts or depends on fashion trends. Fashion style can be divided into seven typical styles: classic style, ethnic style, gorgeous style, romantic style, pioneer style, rustic style, simple style.

（1）经典风格。
Classic style.

经典风格的服饰配件通常设计比较保守，不易受到流行趋势的左右，偶尔流露出拘谨和耐人寻味的感觉，因此产品使用周期比较长，是很多消费者倾向的选择。经典风格追求严谨和优雅，文静而含蓄，是一种强调和谐统一的配饰风格，适合相对正式的场合（图2-8～图2-11）。

Classic-style clothing accessories are usually designed to be conservative, not easily affected by the popular trend, and occasionally showing a sense of restraint and intriguing, so the product life cycle is relatively long, and many consumers tend to choose. The pursuit of rigorous and elegant classic style, quiet and subtle, is an accent style that emphasizes harmony and unity, suitable for relatively formal occasions (Figure 2-8~ Figure 2-11).

（2）民族风格。
Ethnic style.

民族风格是在传承和借鉴传统民族服饰元素的基础上，结合现代生产、生活、社交等场合的需求而设计的兼具民族元素和现代服装设计元素的服饰配件。民族风格主要是对于民族元素的掌握和使用，其提炼的角度、范围很广，应用的手法也十分丰富。民族文化是艺术创作永不枯竭的灵感来源，民族风格的服饰配件常常包含一些特色的民族工艺，比如蜡染、刺绣、扎染、镶丝、钉珠（图2-12～图2-15）。

The ethnic style is an apparel accessory that combines both national elements and modern costume design elements based on the heritage and reference of traditional ethnic costume elements, combined with the needs of modern production, living, socializing and other occasions. The ethnic style is mainly for the mastery and use of ethnic elements, and its refining point of view, a wide range, and ap-

图2-8　经典的爱马仕伯金包，爱马仕官网
Figure 2-8　Classical Hermes Birkin Bag, Hermes official website

图2-9　巴黎世家机车包，巴黎世家官网
Figure 2-9　Balenciaga Motorcycle Bag, Balenciaga official website

图2-10　流氓扣帽，品牌官网
Figure 2-10　HOOLIGAN SNAP CAP, official website

图2-11　卡米亚，品牌官网
Figure 2-11　Carmina, official website

plication methods are also very rich. Ethnic culture is an inexhaustible source of inspiration for artistic creation. Ethnic-style costume accessories often include a number of unique ethnic techniques such as batik, embroidery, tie-dye, inlay, and beading (Figure 2-12~Figure 2-15).

（3）华丽风格。

Gorgeous style.

华丽风格源自人们对财富的炫耀和追求，服饰配件一般选用很多价格昂贵的珍稀材质，或者凝结了许多珍贵的技艺，有很高的品质感。华丽风格的服饰配件通常会带有一种高傲的气质，通过大胆的、夸张的表达方式来呈现。总的来说，华丽风格的服饰配件所表现出来的风格特点是华美的、大气的、夸张的、有气势的、醒目的综合印象（图2-16~图2-19）。

Gorgeous style is derived from people's display and pursuit of wealth. Costumes and accessories are generally made of expensive and rare materials, or have condensed many valuable skills and have a high sense of quality. Gorgeous style clothing accessories often have a proud temperament, presented through bold, exaggerated expressions. In general, the style characteristics of the gorgeous-style clothing accessories are gorgeous, high class, exaggerated, imposing, striking and comprehensive impression (Figure2-16~Figure 2-19).

图2-12　安娜苏 2018春/夏，品牌官网

Figure2-12　Anna sui 2018 summer and fall, official website

图2-13　多层项链，品牌官网

Figure2-13　Rosantica, official website

图2-14　红色华佗天奴2018春/夏，品牌官网

Figure2-14　Red Valentino 2018 spring and summer, official website

图2-15　雕刻2018春/夏，品牌官网

Figure2-15　Carven 2018 spring and summer official website

图2-16　杜嘉班纳品牌官网

Figure2-16　Dolce & Gabbana, official website

图2-17　古驰品牌官网

Figure2-17　Gucci, official website

图2-18　杜嘉班纳品牌官网

Figure2-18　Dolce & Gabbana, official website

图2-19　德赖斯·范诺顿2018春/夏，品牌官网

Figure2-19　Dries Van Noten 2018 summer and spring, official website

（4）浪漫风格。

Romantic style.

浪漫主义风格服饰配件在与其他风格相比，强调了对人类情感的表现，更加感性。此属性也决定了这类服饰配件更多地用于强调女性化风格的面貌，往往带有一种无拘无束的真诚和令人激情澎湃的热情。在形式法则上，浪漫主义风格服饰配件强调在遵循形式组合规律的基础上，打破僵硬的教条而追求幻想和戏剧化效果。造型夸张独特，线条柔美奔放，常见有不对称结构，色彩明亮多变，图案缤纷斑斓。典型的设计点包括流苏箱包、蕾丝饰品、刺绣、花边、抽褶、荷叶边、蝴蝶结、花结和花饰等（图2-20～图2-23）。

Compared with other styles, romantic style apparel accessories emphasize the expression of human emotions and are more sensible. This property also determines that this type of clothing accessories is more used to emphasize the feminine style, often with a kind of unfettered sincerity and passionate enthusiasm. According to the formation rules, romantic style apparel accessories emphasize the pursuit of fantasies and theatrical effects by breaking rigid dogmas on the basis of the rules of formation combination. The shape is exaggerated and unique, the lines are soft and unrestrained, and there are common asymmetrical structures, bright colors and colorful patterns. Typical design points include tassel bags, lace jewelry, embroidery, lace, shirring, ruffles, bows, rosettes, and floral decorations (Figure 2-20~ Figure 2-23).

（5）前卫风格。

Pioneer style.

凸显自我、张扬个性是前卫风格的首要设计理念。无常规的空间解构，大胆鲜明的色彩构成，刚柔并济的材质搭配，让人在突兀中寻求到一种超现实的平衡，而这种平衡也是对单一审美，单一理念的有力抨击。随着时代更替、推陈出新，前卫风格的服饰配件会在内容和形式上更加出人意料、夺人耳目。前卫风格乐于打破常规的形式美法则，结构夸张、大胆，多运用不对称、对比强烈、立体感和装饰感强的表现形式，突出了标新立异的特色。

材质方面倾向于保留部分传统材料，更多的选用人工后期再造、新潮、高科技或具有光泽感的材质，选取时更加注重材质的残旧、破损、刀割、破洞等后期再创作的外观效果。构成方面繁多无序，体现出杂乱无章、新颖独特的视觉冲击效果。设计师们将流行的目光从重视优雅的传统转向个性前卫的追求，并将这些观念注入他们的设计中，提醒人们也可以展现出在生活中常被忽略的非常态的美，作为另一种生活状态的转变，暂时摆脱现实生活中所承担的压力，达到心灵释放、身心愉悦的享受❶（图2-24～图2-27）。

图2-20 米兰街拍，全球时装网官网

Figure2-20 Street Shot, WGSN official website

图2-21 全球时装网官网

Figure2-21 WGSN official website

图2-22 全球时装网官网

Figure2-22 WGSN official website

图2-23 杜嘉班纳品牌官网

Figure2-23 Dolce & Gabbana official website

---

❶ 张珊，马颖. 前卫服装设计风格浅议［J］. 青年文学家，2015(20).

Emphasizing self and assertive personality is the primary design concept of pioneer style. There is no conventional spatial construction, bold and vivid color composition, and the combination of rigid and soft materials, people find a surreal balance in the abruptness, and this balance is also a powerful impact on a single aesthetic, single concept. As the times change and innovations are introduced, the pioneer fashion accessories will be more surprising and eye-catching in content and form. The pioneer style tends to break the conventional formal rules of beauty, exaggerated structure, boldness, and more use of asymmetrical, contrastive, three-dimensional and decorative sense of a strong expression, highlighting the characteristics of innovation.

The material tends to retain partly to be traditional, with more artificial reconstructed, trendy, high-tech or glossy materials. When it is selected, it pays more attention to the appearance of the material, such as old, damaged, knife-cut, hole, etc.effects, constituted in a variety of disorder, and reflecting the chaotic, novel and unique visual impact. Designers have turned their attention from the tradition of emphasizing elegance to the pursuit of individual pioneer, and infused these ideas into their designs, reminding people to show the uncommon beauty that is often ignored in life as another kind of the transition of the state of life is temporarily disengaged from the pressurcs that arc borne in real life to achieve the release of the mind and the enjoyment of body and mind [Zhang Shan, Ma Ying. Avant-garde fashion design style [J]. Young writer, 2015 (20)] (Figure 2-24~Figure 2-27).

（6）田园风格。

Rustic style.

田园风格的服饰配件追求一种原始的、纯朴自然的、非虚饰的美，是一种贴近自然，向往自然的风格。这种风格的服饰配件具有较强的活动机能，很适合人们郊游、散步和各种轻松的活动，迎合现代人的生活需求。田园风格倡导"回归自然"，在美学上推崇"自然美"，认为崇尚自然、结合自然，才能在当今高科技快节奏的社会生活中获取生理和心理的平衡。

田园风格力求表现悠闲、舒畅的田园生活情趣，最大的特点就是：朴实、亲切、实在，反对烦琐的装饰，反对表现浓墨重彩的华美，倾向于纯净自然的朴素，表现一种轻松恬淡的、超凡脱俗的情趣。粗糙和破损是被允许的，因为这更接近自然。服饰配件设计师从大自然中汲取设计灵感，常取材于树木、花朵、蓝天和大海，把触角时而放在高山雪原，时而放到大漠荒岳，虽不一定极近自然的风采，却定要褪尽都市的痕迹，表现大自然永恒的魅力。典型的配饰设计要素包括碎花图案、草帽、花边、纯棉质地、小方格、均匀条纹、木制品、编织大挎包

图2-24 普拉达2018春/夏，品牌官网

Figure 2-24 Prada2018 spring and summer, official website

图2-25 蒂维2018春/夏，全球时装网官网

Figure 2-25 Tibi 2018 spring and summer, WGSN official website

图2-26 伊莎贝尔·玛兰，全球时装网官网

Figure 2-26 Isabel Marant, WGSN official website

图2-27 德赖斯·范诺顿·德温2018春/夏全球时装网官网

Figure 2-27 Dries Van Noten dvn2018 spring and sumer, WGSN official website

等❶（图2-28～图2-31）。

The pursuit of a primitive, simple and natural, non-fake beauty is a style that is close to nature and yearning for nature. This style of clothing accessories has a strong activity, It is suitable for people outings, walks and a variety of relaxing activities to meet the needs of modern people's lives. The pastoral style advocates "returning to nature" and aesthetically promotes "natural beauty". It is believed that advocating nature and combining nature can achieve a balance between physical and psychological aspects in today's high-tech and fast-paced social life.

The rustic style seeks to express leisurely relaxing life. The greatest characteristics are: simple, friendly, practical, against cumbersome decoration, against the rich and colorful, and tend to purely natural simplicity, showing a relaxed and bleak, extraordinary sense. Roughness and breakage are allowed because it is closer to nature. Designers of fashion accessories draw design inspiration from nature, often drawing on trees, flowers, blue sky and sea, placing tentacles on alpine snowfields from time to time, and sometimes into desert wilderness. Although not necessarily close to the natural style, they are sure to discard the traces of the city and express the eternal charm of nature. Typical accessory design elements include floral patterns, straw hats, lace, cotton texture, small squares, uniform stripes, woodwork, and woven big bag [Long Fengmei. Discussion on the application of pastoral style in fashion design [J]. Art Science, 2012(4):56.] (Figure 2-28~ Figure 2-31).

（7）简约风格。

Simple style.

简约风格起源于现代派中的极简主义，现代派大师米斯·凡德罗提倡的"少即是多"，即在满足功能性作用的基础上做到最大程度的简洁。它以减少琐碎、去繁从简以获得设计最本质元素的再生，简洁明快的风格下往往隐藏着复杂精巧的构思。这些"简洁"并不意味着单纯的简化，它往往是丰富的集中统一化。这一理念符合了世界大战后各国经济萧条的困境，在当时得到人们的一致推崇。简约风格永远用一种低调的经典带来对时尚进行诠释的，使服饰配件设计的实用性和艺术性完美地结合起来，用以表达独特的着装风貌。

简约风格的服饰配件在造型上简化繁复的装饰，减少到几乎无以复加，关注简单造型的优雅，达到简约而不简单的艺术境界（图2-32～图2-35）。

图2-28 米索尼，全球时装网官网

Figure 2-28 Missoni, WGSN official website

图2-29 缪缪，品牌官网

Figure 2-29 MiuMiu, official website

图2-30 街拍图片，全球时装网官网

Figure 2-30 Street shot, WGSN official website

图2-31 巴杰利·米施卡2018春/夏

Figure 2-31 Badgley Mischka 2018 spring and summer

---

❶ 龙凤梅. 浅谈田园风格在服装设计中的应用［J］. 艺术科技，2012(4): 56.

  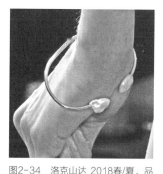 

图2-32　街拍图片，全球时装网官网
Figure 2-32　WGSN official website

图2-33　街拍图片，全球时装网官网
Figure 2-33　街拍图片，WGSN official website

图2-34　洛克山达 2018春/夏，品牌官网
Figure 2-34　Roksanda 2018 spring and summer, official website

图2-35　街拍图片，全球时装网官网
Figure 2-35　WGSN official website

The simple style originated in the comteporary minimalism, and the LESS IS MORE advocated by the modernist master Mess van der Rohe, that is, to achieve maximum simplicity on the basis of satisfying the functional role. It reduces the triviality and complexity of the design to obtain the most essential elements of the design. The simple and clear style often hides complex and delicate ideas. These "conciseness" do not mean just simplification. It is often rich and centralized. This concept was in line with the predicament of economic depression in various countries after the World War, and it was unanimously praised by people at that time. The minimalist style always brings an interpretation of fashion with a low-key classic, which perfectly combines the practicality and artistry of the costume accessories design to express a unique style of dress.

The simple style of the clothing accessories simplifies the complicated decoration in the shape, reduces to maximun extent, pays attention to the elegance of the simple styling, and achieves the simple but not simple artistic realm (Figure 2-32~ Figure 2-35).

2. 历史类风格
Historical style

历史服饰有着长期的文化积淀，带着一个时代或民族的深深烙印，是取之不尽的设计灵感源泉。细分下来，历史类风格可分为七个比较典型的风格：古埃及、巴比伦及亚述风格、古希腊及古罗马风格、拜占庭风格、哥特风格、文艺复兴风格、巴洛克风格、洛可可风格。❶

Historical costumes have long-term cultural accumulation, with a deep imprint of an era or ethnicity, and are inexhaustible sources of design inspiration. By careful classification, the historical style can be divided into seven typical styles: Ancient Egypt, Babylonian and Assyrian, Ancient Greek and Roman, Byzantine, Gothic, Renaissance, Baroque, and Rococo [Zhang Jiaqiu, Che Yanxin. Apparel design [M]. Beijing: Communication University of China Press, 2012.7:11].

（1）古埃及、巴比伦及亚述风格。

Ancient Egyptian, Babylonian, and Assyrian styles.

埃及由于气候炎热干燥，衣物简单，甚至裸露，人们依靠佩戴华美的服饰配件来装饰自己，其中包括各种图腾崇拜，比如鹰、太阳、甲虫、莲花。其中材料常用各种金属以及宝石，这些材料稳定性高，可以长存于世，使得古埃及、巴比伦及亚述风格呈现出绚丽的色彩。另外古埃及、巴比伦及亚述风格中有一些十分独特的服饰配件，包括国王的权杖和连枷、王后的秃鹫头饰、王冠、项圈、流苏"坎迪

---

❶ 张嘉秋，车岩鑫. 服饰品设计［M］. 北京：中国传媒大学出版社，2012.7: 11.

斯"等（图2-36~图2-39）。

Due to the hot and dry climate in Egypt, clothing is simple and even bare, people rely on wearing gorgeous accessories to decorate themselves, including various totem worship, such as eagle, sun, beetle, lotus. The materials used are various kinds of metals and precious stones. These materials have high stability and can survive for a long time, making the ancient Egyptian, Babylonian, and Assyrian styles show brilliant colors. In addition, ancient Egyptian, Babylonian, and Assyrian styles have some very unique costume accessories, including the king's scepter and chain, the queen's bald headdress, crown, collar, tassel "Candace" and so on (Figure 2-36~ Figure 2-39).

（2）古希腊及古罗马风格。

Ancient Greek and Roman style.

古希腊及古罗马风格（公元前1500~公元前100年）服饰配件整体感觉舒适慵懒，宽松的设计加上褶皱、垂坠和立体花卉成了希腊式服装的经典搭配。古希腊的服装多采用不经裁剪、缝合的矩形面料，通过在人体上的披挂、缠绕、别饰针、束带等基本方式，形成了"无形之形"的特殊服装风貌。随意、自然、富于变化是这类服饰配件的重要特点。具有代表性的包括裹束头发的缎带、头绳、发环、鞋子及喜欢系带的凉鞋，选用牛皮绳、丝带、藤草等材质编织而成，另外加之金子、银子或者宝石装饰（图2-40~图2-43）。

The overall feel of Ancient Greek and Roman style (1500 B.C.~100 B.C.) accessories is comfortable and lazy. The loose design combined with folds, drape and three-dimensional flowers has become the classic combination of Greek clothing. The ancient Greek dresses used rectangular fabrics that were not cut and stitched, and formed the "invisible shape" special clothing styles through the basic methods of draping, winding, affixing pins, and straps on the human body. Random, natural, and rich in change are important features of this type of clothing accessories. Typical examples include ribbons, headbands, and hair loops that wrap hair, and shoes that are laced with sandals. They are made from woven ropes, ribbons, rattan, and other materials, plus gold, silver, or precious stones (Figure 2-40~ Figure 2-43).

（3）拜占庭风格。

Byzantine style.

拜占庭时期（公元300~公元1400年），随着基督教文化的普及，服饰配件把表现的重点转移到了质地、色彩和装饰纹样的变化上。丝绸之路的开通使得中国的丝绸流入，拜占庭风格也体现了希腊、罗马以及东方对其的影响。随着染织业的不断发达，产生了珍贵华美绣着花纹的丝织品，这更加增添了服饰配件的绚丽色彩，体现了材料丰赡华美、极尽奢华的特点。色彩变得绚丽、美艳，装饰性的流

图2-36　古驰 2017秋/冬，胡狼面具，品牌官网
Figure 2-36　Gucci 2017 autumn and winter, wof mask, official website

图2-37　古驰2017秋/冬品牌官网
Figure 2-37　Gucci 2017 autumn and winter. official website

图2-38　阿妮欧娜2015春/夏，品牌官网
Figure 2-38　Agnona 2015 spring and summer, official website

图2-39　阿妮欧娜2015春/夏，品牌官网
Figure 2-39　Agnona 2015 spring and summer, official website

苏、绲边以及宝石非常普遍。在男女宫廷服的大斗篷、帽饰以及鞋饰上都出现了光彩夺目的珠宝和华丽图案的刺绣，营造出一种充满隆重感的服饰装饰美。同时，服饰配件造型对称、呆板、僵硬，材料方面多用珍珠、宝石、金线。男装中以封闭式鞋取代了以前的凉鞋，披肩遮住身体的大部分，前面和后面有方形或长方形的绣饰，给人一种强烈的宗教性（图2-44～图2-47）。

During the Byzantine Period (AD 300—AD 1400) with the spread of Christian culture, costume accessories shifted the focus of performance to changes in texture, color and decorative patterns. The opening of the Silk Road led to the influx of Chinese silk. Byzantine style also reflected the influence of Greece, Rome and the East. With the continuous development of the dyeing and weaving industry, the precious and silky embroidered pattern of silk fabrics have been created, which added to the splendid colors of fashion accessories, and the material has been beautifully presented and extravagant. More beautiful, glamorous colors, decorative tassels, piping and gems are very common. In the large royal cloaks, hats, and shoes of the men's and women's courts, embroidered with dazzling jewelry and ornate designs appeared, creating a sense of grandeur and decorative apparel. At the same time, the accessories are symmetrical, boring and stiff, and the materials are often made of pearls, precious stones, and gold threads. In the men's wear, closed sandals were used to replace the previous sandals. The shawl covered most of the body. Square or rectangular embroidery was placed on the front and back, displaying strong religious elements (Figure 2-44~ Figure 2-47).

图2-40 瓦伦蒂诺2014秋/冬高级定制，Vogue官网
Figure 2-40 Valentino 2014 fall and winter haute couture, Vogue official website

图2-41 瓦伦蒂诺2014秋/冬高级定制，Vogue官网
Figure 2-41 Valentino 2014 fall and winter haute couture, Vogue official website

图2-42 坦尼娅·泰勒2018，全球时装网官网
Figure 2-42 Tanya Taylor 2018, WGSN official website

图2-43 JW安德森2018春/夏，Vogue官网
Figure 2-43 JW Anderson 2018 spring and summer, Vogue official website

图2-44 香奈儿秋季前2011，网站官网
Figure 2-44 Chanel Pre Fall 2011, Popsugar official website

图2-45 香奈儿秋季前2011，Popsugar官网
Figure 2-45 Chanle 2011 Pre Fall, Popsugar official website

图2-46 香奈儿秋季前2011，网站官网
Figure 2-46 Chanle 2011 Pre Fall, Popsugar official website

图2-47 香奈儿秋季前2011 1stdibs官网
Figure 2-47 Chanle 2011 Pre Fall 1stdibs official website

（4）哥特风格。

Gothic style.

最早来源于欧洲早期的一个叫作西哥特的部族，这个以破坏和掠夺为乐的部族以无知和缺少艺术品位而著称。哥特的词义本身意味着粗野、倒退、不开化，造型特点包括很多锐角，腰带高至胸部，视觉上强调高耸感，堆砌出的层叠效果突出哥特风格的繁复，增加奢华感。在头饰中出现了锥状帽，帽子的高度还象征了身份的高低。尖头鞋也是其中非常具有代表性的饰品。14世纪的尖头鞋长度达到了极致，最长的有1米左右，多余部分以苔藓之类的填塞，由于宗教原因，鞋子的长度为6或者6的倍数。在时间的沉淀中，哥特风格呈现出神秘、阴郁、妖冶、华丽的倾向。色彩多以黑色为主，配饰均呈现锐三角形造型，这源于基督教精神。此外还有长头巾、披肩、斗篷、长手套、面纱等。现代哥特风格的服饰配件常常出现十字章、太阳神之眼、五角星、十字架、基督像这种体现神秘感的元素以及锁链、铁钉、领带、铆钉、项圈、渔网、丝绒绳等元素（图2-48～图2-51）。

Gothic originated from an early Western European tribe called Visigoth. The tribe enjoys destruction and plunder and known for its ignorance and lack of artistic status. The Gothic meaning itself means rough, backward, and non-open. The styling features include many sharp corners, belts up to the chest, visual emphasis on towering feelings, and the stacking effect of up highlights the complexity of the Gothic style and increases the sense of luxury. Cone caps appear in the headwear, and the height of the hat also symbolizes the level of identity. Pointed-toe shoes are also very representative of the 14th-century pointed shoes. The length of the pointed shoes has reached the extreme. The longest one is about 1 meter. The extra part is filled with moss and the like. For religious reasons, the length of shoes is 6 or 6 the multiples. In the precipitation of time, the Gothic style tends to be mysterious, gloomy, flirtatious and gorgeous. The colors are mostly black, and the accessories are all sharp triangles. This is due to the Christian spirit. There are also long turbans, shawls, cloaks, long gloves, veils and so on. Modern Gothic-style clothing accessories often appear in cross chapters, the eyes of the sun god, pentagram, crosses, Christ like this embodiment of the elements of mystery and chains, nails, ties, rivets, collars, fishing nets, velvet rope and other elements (Figure 2-48~ Figure 2-51).

图2-48 纪梵希2012秋/冬，Vogue官网-1
Figure 2-48 Givenchy 2012 fall and winter, Vogue official website-1

图2-49 纪梵希2012秋/冬，Vogue官网-2
Figure 2-49 Givenchy 2012 fall and winter, Vogue official website-2

图2-50 亚历山大·麦昆2016年秋/冬，GQ官网-1
Figure 2-50 Alexander McQueen 2016 fall and winter, GQ official website-1

图2-51 亚历山大·麦昆2016年秋/冬，GQ官网-2
Figure 2-51 Alexander McQueen 2016 fall and winter, GQ official website-2

（5）文艺复兴风格。

Renaissance style.

在14～17世纪时人们认为，文艺在希腊、罗马古典时代曾高度繁荣，但在中世纪"黑暗时代"却衰败湮没，直到14世纪后才获得"再生"与"复兴"，因此称为"文艺复兴"。文艺复兴时期的服饰主要是以欧洲各国王权为中心发展起来的服饰文化，主要的服饰特征是鲜明的构筑性。主要的服饰品有拉夫领，整体呈车轮造型，主要是波浪形的褶皱以"8"字形连续的褶裥构成，有时使用金属丝作为支架，由于造型越来越大、越来越挺，以至于最后形成了一个独立的部分（图2-52～图2-55）。

In the 14th and 17th centuries, people believed that literature and art were highly prosperous in the classical Greek and Roman times, but in the Middle Ages, were lost and annihilated in "dark age". It was not until the 14th century that it gained "regeneration" and "rejuvenation," hence it was called "literary revival". The costumes of the Renaissance period were mainly costume cultures developed around the kingship of various European countries. The main features of clothing are distinct architectural features. The main clothing products are Ralph's collar. The overall shape of the wheel is mainly wavy folds made up of "8"-shaped continuous pleats. Sometimes metal wire is used as a bracket. As the shape becomes bigger and bigger, it becomes more and more striking, leading to the final formation of a separate part (Figure 2-52~Figure 2-55).

（6）巴洛克风格。

Baroque style.

巴洛克艺术风格原本是指17～18世纪上半叶强调炫耀财富、大量使用贵重材料的建筑风格，也影响了当时艺术的全面性变革。"巴洛克"的字义源自葡萄牙语，意指"变了形的珍珠"，也被引用作为脱离规范的形容词。巴洛克风格气势宏伟、色彩艳丽、生机勃勃、线条优美、富丽豪华。在服饰配件上体现为夸张的廓型，注重装饰，常使用蕾丝、绣花和缎带作为装饰元素，强调华丽、繁复的核心思想，色彩方面厚重饱和、以暖色为主，常用贵族服饰的金色、黑色、红色，色彩明度偏低，巴洛克风格具有一种性别混淆的视觉效果，具有中性化气质，既不同于文艺复兴时期风格迥异的两性风格，也不同于女性化的洛可可风格。❶常见配饰有三角帽、花卉

图2-52 瓦伦蒂诺 2016秋/冬高定秀，Vogue官网-1

Figure 2-52 Valentino 2016 fall and winter haute couture, Vogue official website-1

图2-53 瓦伦蒂诺2016秋/冬高定秀，Vogue官网-2

Figure 2-53 Valentino 2016 fall and winter haute couture, Vogue official website-2

图2-54 瓦伦蒂诺 2016秋/冬高定秀，Vogue官网-1

Figure 2-54 Valentino 2016 fall and winter haute couture, Vogue official website-1

图2-55 瓦伦蒂诺2016秋/冬高定秀，Vogue官网-2

Figure 2-55 Valentino 2016 fall and winter haute couture, Vogue official website-2

---

❶ 董怡. 巴洛克风格在现代男装中的运用状况研究［J］. 装饰，2009(4):114-115.

等（图2-56～图2-59）。

Baroque art style originally referred to the architectural style that emphasizes wealth display and extensive use of valuable materials from the 17th century to the first half of the 18th century, and thus influenced the comprehensive reform of art at that time. The word "baroque" is derived from Portuguese, meaning "a pearl that has changed shape," and it has also been cited as an adjective deviating from the norm. Baroque style is magnificent, colorful, vibrant, colorful, beautiful lines, rich and luxurious. Embodied in the costume accessories for the exaggerated profile, the style pays attention to the decoration, often uses lace, embroidery and ribbons as decorative elements to emphasize the gorgeous, complex core ideas, heavy color saturation, warm color, commonly used noble dress gold, black and red is low, and thus has a gender-blurred visual effect. It has a neutral temperament, which is different from the bisexual style of the Renaissance style and the feminine Rococo style [Dong Yi. Baroque style in the modern men's wear study [J]. Decoration, 2009 (4): 114-115.]. common accessories are triangular caps, flowers and so on (Figure 2-56~ Figure 2-59).

（7）洛可可风格。

Rococo style.

"洛可可"一词由法国词汇"Rocaille"演变而来，它具有"螺贝"的意思，故用以形容造型艺术中那种善用蜷曲的线条和繁复装饰的风格，是一种强调C型的漩涡状花纹及反曲线的装饰风格。洛可可艺术更趋向一种精制而优雅，具有装饰性的特色。洛可可风格表现在服饰配件上，是以追求小巧的花纹图案，色彩趋于明快淡雅，强调优美的曲线造型，轻柔而富于动感的视觉效果，比如，各种绸带、花边、褶皱、烦琐的假发、头纱、面具、扇子等小巧精致的饰品，带着纤巧而富丽的光芒。❶香奈儿2013早春珠宝配饰就是典型的洛可可宫廷风格，造型精致繁复而色彩轻盈通透，除了常见的项链、耳环、手镯、领口、发缘、腰际都不忘巧心装点。模特们头上的各色蝴蝶结贯穿始终，烘托出洛可可风格的精髓，成为当季的热门配饰（图2-60～图2-63）。

The word "Rococo" originated from the French word "Rocaille". It has the meaning of "spike shell". Therefore, it is used to describe the use of swirling lines and complicated decorative styles in the shaping arts. It is an emphasis. C-shaped swirl pattern and anti-curve decorative style. Rococo art is more refined and elegant, with decorative features. Expressed in

图2-56　2012秋/冬杜嘉班纳，Vogue官网-1
Figure 2-56　2012 fall and winter Dolce&Gabbana, Vogue official website-1

图2-57　2012秋/冬杜嘉班纳，Vogue官网-2
Figure 2-57　2012 fall and winter Dolce&Gabbana, Vogue official website-2

图2-58　2012秋/冬杜嘉班纳，Vogue官网-3
Figure 2-58　2012 fall and witner Dolce&Gabbana, Vogue official website-3

图2-59　2012秋/冬杜嘉班纳，Vogue官网-4
Figure 2-59　2012 fall and winter Dolce&Gabbana, Vogue official website-4

---

❶ 李芃，杨贤春. 洛可可风格在中西方服饰上的体现［J］. 湖南工业大学学报，2001, 15(4): 43-45.

第二章 服饰配件的设计基础

图2-60 香奈儿2013早春珠宝配饰，style官网
Figure 2-60 Chanel 2013 spring, style official website

图2-61 香奈儿2013早春珠宝配饰
Figure 2-61 Chanel 2013 spring

图2-62 洛可可风格面具
Figure 2-62 Rococo styled mask

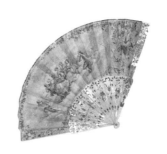

图2-63 洛可可风格折扇
Figure 2-63 Rococo styled folding fan

the clothing accessories, the Rococo style pursues small patterns, bright and elegant colors, beautiful curve shapes, soft and dynamic visual effects, such as a variety of ribbons, lace, pleats. The wigs, veils, masks, fans and other small and exquisite ornaments, with a delicate and rich light [Li Jie, Yang Xian-chun. Rococo style in the expression of Chinese and Western clothing [J]. Journal of Hunan University of Technology, 2001, 15 (4) :43-45.]. Chanel2013 early spring jewelry is a typical rococo court style, modeling exquisite and complex and light and transparent colors, in addition to common necklaces, earrings, bracelets, collar, hair edge, waist with delicate decoration. The colored bows on the models' heads are always there, highlighting the essence of Rococo style and becoming a popular accessory of the season (Figure 2-60~Figure 2-63).

三、服饰配件中的技术美感
C. Technical beauty in apparel accessories

1. 基本概念
Basic concepts

技术从生产劳动中来，它体现了人类制造工具的生产活动的根本特征。因此，用技术生产活动创造出来的产品所具有的美称为技术美。❶这里的技术，是泛指意义上的应用于服饰配件设计制造的现代技术。与服饰设计相同，现代技术的广泛使用，不仅使服饰及服饰配件的生产变得迅速，同样，在其美学设计中也起到了重要作用。服饰配件中的技术美，可以说是艺术与技术的交融体。技术不仅仅为艺术创作提供了实现审美情趣的多种方式，同时设计理念的表达也置入了现代技术之中。技术美是一种潜藏着物质功能的美，但是它的审美价值并不是直接来自产品的物质功能和效用本身。技术美是由形式所表现的科学技术文明的历史内容，它反映了产品的合规律性与合目的性的统一以及人在创造历史中所取得的自由。人通过产品的结构形式把自然规律纳入了自身目的的轨道，使它与人的社会生活环境和人的活动相和谐。❷服饰配件技术美的审美体验，只有在具象与抽象思维的相互转换中才得以完整。

技术美的生成更多地与理性联系在一起。现代技术不断地更新着服饰配件美的艺术表现，从而生成区别于传统的设计美感。与服饰设计中的技术美相同，服饰配件中的技术美感，同样体现在精确的秩序感、视觉图式的层次感与技术性痕迹等几个方面。❸

Technology comes from production labor, which embodies the essential characteristics of the produc-

---

❶ 李超德. 设计美学 [M]. 合肥：安徽美术出版社，2004.8: 120.
❷ 李超德. 设计美学 [M]. 合肥：安徽美术出版社，2004.8: 121.
❸ 黄思华. 服饰技术美及其艺术表现的研究 [J]. 服饰导刊，2017, 6(1): 43–46.

tion activities of human manufacturing tools. Therefore, the beauty of products created with technological production activities is called technical beauty [Li Chaode. Design Aesthetics [M]. Hefei: Anhui Fine Arts Publishing House, 2004.8:120]. The technology here refers to a wide range of modern technologies applied to the design and manufacture of apparel accessories. As with apparel design, the widespread use of modern technologies not only makes the production of apparel and accessories faster, but also plays an important role in its aesthetic design. The technical beauty in fashion accessories can be said to be a blend of art and technology. Technology not only provides a variety of ways for artistic creation to realize aesthetic tastes, but at the same time, the expression of design concepts is also incorporated into modern technologies. Technical beauty is a kind of beauty with hidden physical functions, but its aesthetic value is not directly from the material function and utility of products. Technical beauty is the historical content of the scientific and technological civilization represented by the form. It reflects the unity of the product's regularity and purpose and the freedom people have achieved in creating history. Through the structural form of the product, man incorporates the laws of nature into the track of his own purpose, making it in harmony with the human social environment and human activities [Li Chaode. Design Aesthetics [M]. Hefei: Anhui Fine Arts Publishing House, 2004.8:121]. The aesthetic experience of the beauty of fashion accessories can only be completed in the mutual conversion of figurative and abstract thinking.

The generation of technical beauty is more associated with logic reasoning. Modern technology constantly updates the artistic expression of the beauty of fashion accessories, thereby creating a design aesthetic that is different from the traditional design. Similar to the technical beauty in apparel design, the technical beauty in apparel accessories is also reflected in the precise sense of order, the level of visual schemata, and technical traces. [Huang Sihua. Research on Costume Beauty and Its Artistic Performance[J]. Clothes & Accessories, 2017, 6(1):43-46.].

2. 技术美的艺术表现
Artistic Expression of Technical Beauty

服饰配件中的技术美，依托于服饰配件所采用的材料、设计制造过程中的工具、工艺等。与此同时，高新科技的引入，更使得技术美得到了新的阐释。

The technical beauty in apparel accessories depends on the materials used in apparel accessories, tools and processes in the design and manufacturing process, At the same time, the introduction of hi-tech technology has given new interpretation to the beauty of technology.

（1）材料。
Material.

服饰配件美的基本依托是其所采用的材料，虽然设计师可以自由地构思其设计，但无法摆脱所采用的材料对设计的限定。服饰配件的审美质量与科学技术的发展水平密切相关。随着人们的审美需求变化，促使服装配饰材料的科研与生产更加注重艺术附加值。

The basic reliance on apparel accessories is the material it uses. Although designers can freely conceive their designs, they can't get rid of the restrictions on the design of materials used. The aesthetic quality of apparel accessories is closely related to the level of development of science and technology. As people's aesthetic needs change, research and production of clothing accessories materials will be more value-added.

（2）工具。
Tools.

工具是生产劳动时所采用的器具，作为生产力基本要素之一，生产劳动工具可顺畅地转化为艺术创造工具。而不同工具对服饰配件的生产所产生的影响，直接体现在服饰配件的设计和生产的结果上。

Tools are used in production labor. As one of the basic elements of productivity, production and labor tools can be smoothly transformed into tools for artistic creation. The impact of different tools on the production of apparel accessories is directly reflected in the design and production of apparel accessories.

（3）工艺。

Craftsmanship.

工艺是利用工具作用于材料之上，达到设计目的的操作过程，表现为一种实践形态。工具与材料是服饰配件美的基础要素，而工艺是产品美的核心要素。在材料与工具高度标准化的今天，工艺则是体现个性、表达差异化的最好途径，服饰配件的高附加值通常通过工艺得到最完美的体现。

Craftsmanship is the use of tools on the material to achieve the purpose of the design of the operation, as a form of practice. Tools and materials are the basic elements of beauty accessories, and craftsmanship is the core element of product beauty. In the highly standardized materials and tools today, craftsmanship is the best way to reflect individuality and express differences. The high added value of apparel accessories is usually best reflected in the process.

（4）高新科技。

High and new science and technology.

高新技术的概念较为宽泛，泛指近代发展起来的、依托于自然科学进步的科学内容。高新技术的发展对服饰配件设计起到了非常大的影响，着重表现在新材料的应用、新设计方法的选择、新功能的发明以及新生产方式的组织等方面。新材料的开发及使用，使得服饰配件的艺术价值得到更好的依托，从根本上提升其艺术附加值；新技术的使用，例如，计算机辅助设计、大规模建模分析等，使得服饰配件的设计生产得到进一步的增强；而依托于新技术的工艺方法，使得服饰配件的设计目的得到更好的表现，提升了其艺术附加值；新的生产方式，以一种更加环保、更加高效的方式，使得所生产产品的美学价值得到进一步的推广和传播。

The concept of high and new technology is relatively broad, referring to the scientific content developed in modern times and based on the progress of natural science. The development of high technology has exerted a great influence on the design of apparel accessories, with emphasis on the application of new materials, the selection of new design methods, the invention of new functions, and the organization of new production methods. The development and use of new materials will enable the artistic value of apparel accessories to be better supported and fundamentally enhance its artistic added value; the use of new technologies, such as computer-aided design, large-scale modeling and analysis, will enable the design of apparel accessories. Production has been further enhanced; and relying on new technological processes, the design objectives of apparel accessories have been better demonstrated, and their artistic added value has been enhanced; new production methods have become more environmentally friendly and more efficient. Make the aesthetic value of the products produced to be further promoted and spread.

四、服饰配件设计的底蕴
D. The backgrounds of apparel accessories design

1. 知识结构

Knowledge structure

服饰配件设计是一项充分发挥主观能动性的思维过程，要求大量的知识储备和专业认知作为基础。甚至可以说，服饰配件设计不仅要求设计师能够掌握服饰配件专业知识，更要熟悉服装相关的知识、造型搭配相关的知识。优秀的服装配件设计应该能够与服装呈现和谐统一的氛围，还要能够将自然和生活中的灵感提炼到服饰配件作品中去，并且升华为高级的艺术作品，除此之外，更要对服装材料、服饰配件的材料和工艺有充分的了解和经验。

Clothing accessories design is a thinking process that gives full play to subjective initiative. It requires a large amount of knowledge and expertise as the basis. It can even be said that the design of apparel accessories not only requires designers to master the professional knowledge of apparel accessories, but also to be familiar with the relevant knowledge of clothing and related knowledge of fashion combination. Excellent clothing accessories design should be able to present a harmonious and unified atmosphere with the clothing, but also be able to extract the inspiration of nature and life into the clothing accessories works, and sublimate into advanced works of art, in addition to clothing. Material and clothing accessories have a good understanding and experience.

2. 平面构成
Plane composition

平面构成是服装配饰设计的基础知识之一。首先服饰配件要能够与整体服装造型相称，其次服饰配件本身作为一件独立的艺术品也需要具备有秩序的、经过设计呈现出艺术的美感。平面构成知识能够很好地安排局部和整体，使之达到点、线、面、体的审美平衡。

Plane composition is one of the basic knowledge of clothing accessories design. First of all, clothing accessories should be able to be commensurate with the overall clothing style. Secondly, as an independent artwork, apparel accessories themselves also need to have an orderly and artistic design. The knowledge of plane composition can well arrange the local and the whole, so that it can reach the aesthetic balance of point, line, face, and body.

3. 色彩构成
Color composition

色彩构成和平面构成均属于设计最根本的知识范畴，但凡涉及设计，都是对这两项基础掌控力的考验。色彩是配饰品的第一印象和外在质量的重要内容，也是取得附加经济价值的捷径。很大程度上，色彩决定了服饰配件的优劣。❶色彩构成知识能够使服饰配件更加生动且富有表现力。

Both the color composition and the plane composition belong to the most basic knowledge category of design. When it comes to design, it is the test of these two basic control abilities. Color is the first impression of accessories and the important content of external quality. It is also a shortcut to obtain additional economic value. To a large extent, color determines the pros and cons of apparel accessories [Zheng Hui, Pan Li. Clothing accessories design [M]. Beijing: Liaoning Science and Technology Press, 2009.1:10]. Color composition knowledge can make costume accessories more vivid and expressive.

4. 材料应用
Application of materials

服饰配件有着广泛的材料选择空间，不同的服饰配件倾向于不同材料选择，比如制作首饰的各种金属、宝石、蕾丝、缎带，制作鞋子的材料有橡胶、PVC、PV、动物皮革、棉麻竹木等。这些材料的基础属性、呈现的质感、处理的技巧都有一定的理论知识，设计过程需要掌握好材质本身的表现力。

Apparel accessories have a wide range of material choices, and different apparel accessories tend to be selected for different materials, such as various metals, gems, laces, ribbons for making jewelry, rubber, PVC, PV, animal leather, cotton, and linen for making shoes. Bamboo and other wood. The basic properties of these materials, the texture they present, and the skills they handle have some theoretical knowledge. The design process needs to master the expressiveness of the material itself.

5. 行业知识
Industry knowledge

服饰配件分布的行业十分宽泛，每个行业有着

---

❶ 郑辉，潘力. 服装配件设计 [M]. 北京：辽宁科学技术出版社，2009.1:10.

各自的行业知识，每一个种类的服饰配件又有自身的名词术语，比如，鞋子就包括鞋尖、鞋头、鞋口背、鞋舌、口条、鞋跟帮、鞋跟加强皮、鞋跟、马蹄跟、底上围、底面、侧底等。除此之外，在历史的变迁中留下了许多风格独特的有特定名称的专业词汇，仍然以鞋为例有乐福鞋、牛津鞋、莫卡辛鞋、木屐等。箱包有肩包、钱包、手提包、挎包、肩包、书包、箱型包等。而深入到相关的制作过程中还有特定的制作技法术语，比如，鞋子的制作过程中就有钉中底、绑帮、帮革底面起毛、勾芯、黏外底等。❶ 可以说很多种类的服饰配件已经形成了非常完善的专门知识，需要相关设计人员深入了解设计品类的相关知识体系，才能够更好地将设计与生产结合起来。

The distribution of apparel accessories industry is very broad, each industry has its own industry knowledge, and each type of clothing accessories have their own terms, such as shoes, including the tip of the shoe, shoe toe, shoe back, shoe tongue, heel help, heel reinforced leather, shoe heel, horseshoe heel, bottom upper, bottom, side bottom and so on. In addition to this, many special vocabulary words with specific names have been left behind in the historical changes. For example, there are shoes such as Loafers, Oxford shoes, Moccasins shoes and clogs. Bags include backpacks, purses, handbags, messenger bags, shoulder bags, school bags, suitcases, etc. As deep into the relevant production process there are specific manufacturing techniques and terminologies, such as nailing the middle bottom in the shoes production process, tying, the bottom surface of the leather raising, hook core, sticky outsole and so on [Ma Rong. Apparel Accessories Design [M]. Chongqing: Southwest China Normal University Press, 2002.1:30]. Many types of clothing accessories have formed a very complete expertise, the need for related designers to understand the design category related knowledge base can better integrate design and production.

6. 前沿科技
Pioneer Technology

科技的进步是审美趋势转变的巨大动力，服饰配件的实用性被越来越多地关注，带有先进功能的服饰配件也越来越多地受到消费者的青睐，所以服饰配件设计师应该实时把握科技进展，将有益的新技术新材料转化为服饰配件产品造福人类。

The progress of science and technology is a huge driving force for the transformation of aesthetic trends. The practicality of apparel accessories is being paid more and more attention. Clothing accessories with advanced functions are also increasingly favored by consumers, so clothing accessories designers should be real-time. Mastering the progress of science and technology, transform useful new technologies and new materials into apparel accessories products for the benefit of mankind.

7. 商业资讯
Business Information

服饰配件具有商品属性，要求服饰配件的设计需要考虑商业价值，能够取悦消费者，赢得市场和收益。在服饰配件设计过程中，要时刻把握商业动态和流行趋势，了解商品的价值规律和使用周期、流行周期，考虑系列化的产品开发。只有这样才能获得更多的市场资源和话语权。

Due to the commercial attributes apparel accessories are required to be designed with commercial value in mind to please consumers, and to win market and revenue. In the process of apparel accessories design, we must always grasp the business trends and trends, understand the value of goods and the use of cycles, epidemic cycle, consider serialized product development. Only in this way can we gain more market resources and the right to speak.

---

❶ 马蓉．服饰配件设计［M］．重庆：西南师范大学出版社，2002.1:30.

8. 工艺技能
Craft skills

工艺是指服饰配件的制作流程与规则，技能是指在服饰配件设计过程中所掌握的技巧和能力。服饰配件设计是一项综合的、复杂的设计任务，在设计的过程中需要考虑选材、制作、环境影响等诸多因素。服饰配件是一件空间艺术品，工艺技能是实现设计创意的必要支持。历代的设计师、工匠早已将许多服饰配件分门别类地总结出了一套行之有效的工艺技能，这些工艺技能需要通过实践去掌握和学习。在创作过程中，也有许多环节是值得重视的。比如，腕表的工艺就相当的复杂，许多零件小到需要用到放大镜来组装，有的表盘设计更是极尽奢华，不得不赞叹其中所凝结的工艺技能。正如手表的工艺方法一样，每一种服饰配件的工艺技法都各有不同，需要分门别类地了解和学习。

Process refers to the production process and rules of apparel accessories. Skill refers to the skills and abilities that are mastered during the design of apparel accessories. Clothing accessories design is a comprehensive and complex design task. In the design process, many factors such as material selection, production, and environmental impact need to be considered. Costume accessories are a space artwork, and craft skills are the necessary support to achieve design creativity. Designers and craftsmen of the ages have already categorized many costume accessories into a set of effective craft skills. These craft skills need to be mastered and learned through practice. In the creative process, there are many aspects that are worth paying attention to. For example, the craft of the wrist watch is quite complicated. Many parts are so small that they need to be assembled using a magnifying glass. Some dial designs are so extravagant that we have to admire the process skills they condense. Just like the craftsmanship of the watch, the craftsmanship of each dress accessory is different, and you need to understand and learn in different ways.

服饰配件设计涉及许多与具体品类相关的工艺技能，所以在具体的制作实践中，设计者需从总体上把握工艺技能的共性，从原材料的选择到最终产品的诞生，在原料配方、工具设备、流程步骤、操作要点等各个方面进行系统的学习，并根据具体的设计目标选择不同的工艺技能，比如，一顶帽子的制作材料可能会涉及表布、里布、黏合衬、帽内带、帽檐管线等，设备需要使用平缝机、包缝机、套口机、熨烫机等，工具则是纸、笔、尺子、针线、剪刀、滚轮等，这些都会依据不同的服饰配件品类需要而有所不同。一般的服饰配件制作流程和工艺方法也是多种多样的，比如，包袋制作的工艺以及流程需要烫边、锤边、上胶、压麦、定位、打针孔、车缝等；鞋履制作需要裁剪、裁片刮边、折边、鞋帮缝合、钉住鞋尖、钉住两侧、钉住鞋跟、放入材料、接底面、用机器压紧、装鞋跟、钉鞋跟等。设计师只有清楚相关设计品类的一般制作流程和制作技巧，才能在设计实践的过程中有效发挥主观能动性，将自己的审美意识通过工艺技能体现出来，根据具体的设计，选择合适的制作步骤，比如，包袋的各个零部件怎么安排分布，制作的先后顺序等。

The design of apparel accessories involves many process skills related to specific categories. Therefore, in the specific production practice, the designer needs systematical study to grasp the general characteristics of the craft skills from the selection of raw materials to the birth of the final products, in the raw material formulation, tools and equipment, processes, operation points, and to select different process skills according to specific design goals. For example, the production materials of a hat may involve surface cloths, linings, fusible interlining, and caps. Along the pipeline, the equipment needs to use sewing machine, overlock machine, looping machine, ironing machine, etc. The tools are paper, pen, ruler, sewing thread, scissors, scroll wheel, etc. These will be based on the needs of different apparel accessories category. General apparel accessories are also varied in production process and process methods, such as bag making process and process needing hot edge, hammer edge, glue, pressure wheat, positioning, needle hole, sewing, etc.; shoe production needs Cutting, trimming, scraping, folding, sewing of the upper, nailing the toes, prying the sides, pegging the heel, inserting the material, connecting the underside, compressing with the machine, mounting the heel, and spiked heel, etc. Designers need to be clear about the general production process and production skills of related design categories in order to effectively exert their subjective initiative in the process of design practice, embody their aesthetic consciousness through technological skills, and select appropriate production steps according to the specific design, such as how to arrange the distribution of various parts of the bag, the order of production, etc.

充分学习工艺技能方面的知识、培养实践能力能够使设计师更加全面地了解服饰配件设计的各个方面和相关环节，一方面使得书本上学到的知识能够学以致用，转化为实际的作品。另一方面也可以通过完成作品，完善并积累自身的设计经验。总的来说，掌握服饰配件设计的工艺技能，应具备以下能力：

Fully learning the skills and knowledge of process skills and cultivating practical abilities enable designers to understand all aspects and links of apparel accessories design more comprehensively. On the one hand, the knowledge in books can be learned to use and transform into practical works. On the other hand, It can also improve and enrich its own design experience through the realization of the design. In general, mastering the technical skills of apparel accessories design should have the following capabilities:

①准确了解和掌握服饰配件具体品类设计的一般制作流程和方法。

Accurately understand and master the general production processes and methods for the specific category design of apparel accessories.

②根据设计图准备服饰配件具体品类设计的相关材料、设备、工具。

Prepare relevant material, equipment, and tools for the design of apparel accessories according to the design drawings.

③利用相关材料、设备、工具制订具体品类的制作步骤和工艺标准。

Formulate the producing process and processing standard for the specific categories by taking the advantage of the relative materials, equipment and tools.

④根据制作步骤和工艺标准进行具体品类的加工过程和操作实践。

Process the producing process and operating practice according to the production steps and processing standard.

9. 美学素养
Aesthetic literacy

人的需要是多层次的，其中包含着物质和精神两个不同层面。产品的功能可划分为实用的、认知的和审美的，服饰配件常常既作为消费品又作为工艺品而存在。在长期的实践过程中，人们从追求美的意图出发，不断地将层层提炼的美通过设计蕴藏于服饰配件当中，这就要求设计师能够掌握美学的基本规律和感性理念，也就是所谓的美学素养。黑格尔在《美学》中指出，艺术美是理念发展到精神阶段的产物，艺术美高于自然美。人的审美需要融合和渗透在人的日常生活当中，墨子说："食必常饱，然后求美；衣必常暖，然后求丽；居必常安，然后求乐"（《墨子·佚文》）。人类的审美需求使人的日常生活转化成一个审美的世界，只有能够很好地感知美，才能更好地创造美。波特兄弟在《市场要素设计》一书中列举了五种不同类型的产品，分别给出了它们对使用性能、人机特性、审美要素和工艺性的侧重顺序。其中手工电锯的人机性能排在第一位，其次是使用性能、工艺性，最后才是审美性能。而花瓶则是审美要素排在第一位，人机性能则排在最后。由此推断，作为装饰品的服饰配件，审美性能应当是设计者首要考量的要素，这就要求设计师具有良好的审美素养，能够进行良好的审美创造。

The needs of people are multi-layered, and they contain two levels of material and spiritual. The function of a product can be divided into practical, cognitive and aesthetic, and apparel accessories often exist as both consumer goods and handicrafts. In the long-term practice process, people start from the intention of pursuing beauty, and continuously apply the refined beauty collected into fashion accessories. This requires designers to master the basic laws and concepts of aesthetics, that is, the so-called aesthetics. Literacy. Hegel pointed out in "*Aesthetics*" that artistic beauty is the product of the development of ideas to the spiritual stage, and the beauty of art is higher than that of natural beauty. The aesthetic needs of people are integrated and penetrated in people's daily lives. Mo Zi said: "Only when we have enough food will we seek daintiness; only when we have warm clothing will we seek beauty; only when we have secure housing, will we seek happiness ("Mo Zi · YiWen"). The human aesthetic needs transform people's daily life into an aesthetic world. Only by being able to perceive beauty well can we better create beauty. The Porter Brothers listed five different types of products in the book "*Design of Market Elements*", and gave them the order of emphasis on performance, human-machine characteristics, aesthetic elements, and craftsmanship. Among them, the human-machine performance of electric saws comes first, followed by the use of performance, craftsmanship, and finally aesthetic performance. The vase is the aesthetic element in the first place, human performance is ranked last. From this, it is inferred that the aesthetic performance of apparel accessories as an accessory should be the primary consideration of the designer, which requires the designer to have a good aesthetic quality and be able to perform good aesthetic creation.

美学素养是从事服饰配件设计的前提。它是一种伴随着情感体验的感知觉过程，也是一种感情色彩浓烈的认知活动。审美感受力的高低，不仅表现在感觉能力的细致和精微，即视觉灵敏度的高低，而且与理解力想象力和对整体的统摄力相关。一个审美感受力强的人，善于捕捉和发现事物中美的因素和形式以及所蕴含的美的意义，从而为自己的创造提供营养和参照。德国艺术评论家莱辛说："凡是我们在艺术作品里发现美的东西，并不是直接由眼睛，而是由想象力通过眼睛去发现其为美的。"设计师既要善于挖掘形象的意义蕴含，又要从整体的联系中去把握形象。不论是在形象特征的借鉴上，还是在细部形态的构成上，都要依靠审美感受力来做出判断和选择。❶西班牙艺术家萨尔瓦多·达利专门为妻子加拉设计了一枚独一无二的心形胸针，著名的"皇家之心"，黄金质地的心形底座上镶嵌着红宝石，象征着血液和血管，以此来表达他那疯狂的爱情，这是一枚机械胸针，内部安装了精微复杂的动力系统，每走一步，它就跳动一下，如同一颗人类心脏一般的真正的金属心脏，如此诗意的设计正是美学素养在配饰设计中的体现（图2-64）。

Aesthetic literacy is a prerequisite for engaging in apparel accessories design. It is a process of perceptual sensations accompanied by emotional experiences, and it is also a cognitive activity with intense emotions. The level of aesthetic sensibility is not only reflected in the subtlety of sensory abilities, ie, the level of visual acuity, but also related to the understanding of the imagination and the overall resilience of the whole. A person with a strong aesthetic sense is good at capturing and discovering the factors and forms of things and the meaning of the beauty they contain, so as to provide nutrition and reference for their own creation. German art critic Lessing said: "Everything we find in the works of art is not directly from the eyes but from the imagination through the eyes to find it beautiful." Designers must

---

❶ 徐恒醇. 设计美学［M］. 北京：清华大学出版社，2006.(7): 80.

be good at excavating the meaning of the image. It is also necessary to grasp the image from the overall connection. Whether it is in the reference of image characteristics or the composition of detailed forms, we must rely on aesthetic sensibility to make judgments and choices [Xu Heng Chun. Design aesthetics [M]. Beijing: Tsinghua University Press, 2006.7:80]. Spanish artist Salvador Dali designed a unique heart-shaped brooch for his wife Gala, the famous "Royal Heart "The heart-shaped base of gold texture is inlaid with rubies, symbolizing blood and blood vessels, to express his crazy love. This is a mechanical brooch with subtle and complex power systems installed inside. Every step is taken. Just like a human heart, a real metal heart, this poetic design is the embodiment of aesthetic literacy in accessory design (Figure 2-64).

图2-64 皇家之心
Figure 2-64 The Royal Heart

美学素养是升华服饰配件设计的基础。在我国古代艺术创作理念中，把"感物"作为艺术创作的源头。"感物"是指艺术家对客观现实的体察和感受，它是创作的来源和基础。对于服饰配件设计来说，只有产品创意构思是不够的。一个好的构想不一定能够产出好的设计作品，还需要进行艰苦细致的设计操作。这时要求设计师能够从兴奋的心态转入沉静安定的心境，把艺术构思和理性的思考结合到一起进行产品造型的细节处理。服饰配件的细节不仅具有画龙点睛的作用，还能够产生引人入胜的魅力，设计师需要在平和的心境中反复地思考和推敲，对每一个细节进行紧密把控。服饰配件的设计能否经得住时间的考验往往取决于设计师对细节非凡的感悟力。杰罗·查尔斯·尊达在1970年以"设计一款前所未有的运动表"为宗旨，设计出了举世闻名的皇家橡树系列手表（图2-65、图2-66）。皇家橡树原本是英国皇家海军一艘于1830年下水启用的战舰，在当时它的负载武器种类及吨位数均为全球之最，杰罗·查尔斯·尊达就是从此艘被命名为皇家橡树号的战舰上找到的设计灵感。战舰上的八角形舷窗象征了力量和防水，是独特的八角形表盘的由来。特殊的八角形舷窗表圈用8颗6角螺钉与表壳紧密地结合在一起，再配合一体式的链带，不但奢华、大胆、前卫，而且能结合复杂的机芯、表壳结构、新型材质和优良制表传统，使其在当时超凡脱俗。这些细节的把控无不体现了设计师的美学素养，而经过如此这般精心推敲的设计作品也永远留在人们的心中，成为具有极高艺术价值的艺术品。此表一经推出，立刻受到了人们的喜爱和市场的认可。

Aesthetic literacy is the basis of sublimation apparel accessories design. In the concept of ancient Chinese art creation, "sensibility" is the source of artistic creation. "Sensibility" refers to the artist's perception and feeling of objective reality. It is the source and foundation of creation. For apparel accessories design, only product creative ideas are not enough. A good idea may not necessarily produce good design work. It also requires meticulous and meticulous design operations. At this time, designers are required to move from an excited mindset to a calm and steady state of mind, and combine artistic conception and rational thinking into the details of product modeling. The details of the clothing accessories not only have the finishing touch, but also can produce fascinating appeal. The designer needs to think and scrutinize repeatedly in peaceful mind and keep close control of each detail. The design of clothing accessories can withstand the test of time or not often depends on the designer's extraordinary feelings of the details. In 1970, Gerald Charles Genta designed the world-famous Royal Oak watch with the aim of "designing an unprecedented sports watch" (Figure 2-65、Figure 2-66). The Royal Oak was originally a battleship launched by the British Royal Navy in 1830. At that time, the type and tonnage of the loaded weapons were the highest in the world. Gerald Charles Genta's ship design inspiration was found on the battleship named Royal Oak. The octagonal porthole on the battleship symbolizes strength and water resistance and is the origin of the unique octagonal dial. The special octagonal porthole bezel is tightly integrated with the case by means of eight hexagonal screws. Together with the integrated chain belt, it is not only luxurious, bold and avant-garde, but also combines complex movements, case structures, and new materials. And the tradition of fine watchmaking made it extravagant at the time. The control of these details all reflect the designer's aesthetic qualities, and after such meticulously elaborate design works will remain in the hearts of people forever and become highly artistic works of art. Once this form was introduced, it immediately became popular with people and recognized by the market.

图2-66　由杰罗·查尔斯·尊达设计的爱彼表皇家橡树1972年款
Figure 2-66　Audemars Piguet Royal Oak, designed by Gerald Charles Genta, 1972

五、服饰配件的发展趋势
E. Development trends of fashion accessories

服饰配件的发展趋势依附于人类社会发展趋势之中，是人类社会文化发展的一个缩影，未来社会的文化艺术、生活方式、思想潮流等都影响着服饰配件的发展，要准确把握其发展趋势就需要了解人类社会的发展趋势。

在服饰配件的发展上，人类社会的发展趋势直接表现在多种多样的造型要求、实用功能、独特个性等方面，其设计往往打破了固有的界限，产生了文化融合与领域跨越，更加追求实用性和装饰性的和谐统一，呈现出以下发展趋势。

The development trend of costume accessories is attached to the development trend of human society, which is a miniature of the development of human society and culture. The culture, art, lifestyle and ideological trend of the future society all influence the development of costume accessories.

In the development of apparel accessories, directly shows the trend of the development of human society in various aspects such as design requirements, functional features, unique personality, its design often broke the boundary of the inherent, produced across cultures and areas, more pursuit of harmony, practicality and decoration, presents the following trends.

图2-65　由杰罗·查尔斯·尊达设计的爱彼第一款皇家橡树
Figure 2-65　Audemars Piguet first Royal Oak designed by Gerald Charles Genta

1. 设计概念多元化
Development trends of fashion accessories

多元化的内涵既是静态的也是动态的。静态的设计多元化就是由于不同的民族、不同的地域、不同的文化、不同的生活习性而形成的不同的审美情趣和产品要求，它是地域的、民族的、静态的；动态的多元化是指在现实生活条件下，为满足不同生活和兴趣爱好的人群，对传统的不同地域、民族和审美的设计产品的继承、发展和再利用。侧重于动态的多元，是指对人类静态多元的设计历史和文化资源的挖掘、继承、发展和再创造。设计的多元化不但符合设计本身的发展规律，也满足人类生活审美的实际要求，在现实社会生活中有着重要的指导作用。

中国是一个既重视传统与现代交织又容纳东西方文化融合的文明古国，服饰配件也向着新的形式和目标发展。东西方艺术的相互借鉴使许多服饰配件富有迷人的情调和韵味，本民族与外来民族的灵感碰撞将各自的特征加以提炼和创新，设计出更具深层次意义的新形式。

The content of diversification is both static and dynamic. Static design diversification is the result of different ethnic tastes, different regions, different cultures, and different living habits that form different aesthetic tastes and product requirements. It is geographical, ethnic, and static; dynamic diversity refers to the inheritance, development, and reuse of traditional, geographical, ethnic, and aesthetic design products for people living in different lives and hobbies under real-life conditions. Emphasis on dynamic pluralism refers to the exploration, inheritance, development, and re-creation of the static and diverse design history and cultural resources of humanity. The diversification of the design not only conforms to the law of development of the design itself, but also satisfies the practical requirements of the human life aesthetics, and plays an important guiding role in the actual social life.

China is an ancient civilization that not only values the integration of tradition and modernity but also embraces the fusion of East and West cultures. Apparel and accessories are also developing toward new forms and goals. Mutual use of Eastern and Western arts has brought many charming accessories and charms to the atmosphere. The inspirations of the nation and foreign nationals have refined and innovated their respective characteristics and designed new forms with deeper meanings.

2. 设计面向未来化
Design for the future

设计从本质上看是属于未来的。面向未来的设计，是对未来理想的适应性设计，设计是人类实现未来理想的工具，通过设计，将人们对未来的理想具体化、现实化。未来化设计也许在当代不能被接受，但其必然是对未来社会愿景的一种实践方式，这种设计具有明显的实验性、前卫性和不确定性。❶服饰配件设计在未来化的趋势中往往表现得光怪陆离，拥有不可思议的功能性，或是不能够被理解的装饰效果，诸如此类的设计往往先呈现在电影、戏剧中，再慢慢靠现实中的产品演变来实现。由此可见，在高速发展的现代社会，服饰配件设计将会成为面向未来的创造性设计。

Design essentially belongs to the future. The future-oriented design is an ideal adaptive design for the future. Design is a tool for mankind to realize its future ideals. Through the design, people's ideals for the future will be concrete and realistic. Future design may not be accepted in the contemporary era, but it must be a practical approach to the future social vision. This design has obvious experimental, pioneer and uncertainty [Li Zhaozu. Introduction to Art Design [M]. Wuhan: Hubei Fine Arts Publishing House, 2009.3:151]. Fashion accessories design tends to be bizarre in the trend of future, has incredible function-

---

❶ 李砚祖. 艺术设计概论［M］. 武汉：湖北美术出版社，2009.3: 151.

ality, or the effects can not be understood decoration, such as this kind of design often appears in the film, the drama first, then gradually realizes by the product development in the reality. It can be seen that in the rapidly developing modern society, the design of apparel accessories will become a creative design for the future.

3. 设计诉求人性化
Design appeal humanization

"为人的设计"体现了设计的全部价值或者说真正的价值,设计的出发点和根本目的也正是如此。设计本身是一种社会化行为,设计师从事的设计,不是为自己的设计,而是为他人、为众人的设计,因而这种设计体现了人类共同的愿望。❶对于服饰配件设计,为市场竞争目的而进行的设计,其价值在于物,而立足于人,以人为本的设计具有更崇高的道德价值,是人性最终的向往和归属。以人为本,体现了人特有的互助心和责任感,关照了社会发展和人类文明等更高层次的追求。

"Design for Humanity" embodies the full value or real value of the design. The starting point and the fundamental purpose of the design are also the same. Design itself is a kind of social behavior. Designers are engaged in the design, not for their own design, but for the design of others for everyone, so this design reflects the common desire of humanity [Li Zhaozu. Introduction to Art Design [M]. Wuhan: Hubei Fine Arts Publishing House, 2009.3:153]. For apparel accessories design, design for the purpose of market competition, the value lies in the matter, while people-oriented design has the higher moral values, leading to the ultimate yearning and ownership of human nature. People-oriented design reflects the unique spirit of mutual help and responsibility of people, and takes care of the higher levels of social development and human civilization.

4. 设计体验感性化
Design experience sensuality

感性化设计是指从人的感觉特性出发的设计。"感"是人第一信号系统对外界的感知,"感者,动人心也",所以感就是心有所动;性主要指本能、欲望、情感。所以感性是感觉、情绪、渴望对事物的一种内心判断。如今商品的销售越来越倾向于出售某种"体验"。"感觉"即是一种体验,而这种感觉带来的刺激越多越丰富,则其价值就越大。

设计进入感性层面,即形成了所谓的"感官设计",人的感觉包括视觉、嗅觉、听觉、味觉、触觉等。服饰配件设计也越发倾向于带给人感性化的体验,比如,根据脚掌着力的方式进行鞋底软硬的细微调整,使得穿着者有更加服帖舒适的触觉;或是在箱包和帽子上设计反光条,增加特定环境中的视觉警示等。这些微妙的变化本质上都是出于满足人们对感官的体验。而在未来的服饰配件设计中,设计者将会更加敏感地体察到人的感官体验,使"为人的设计"不只停留在实用功能、造型、尺度等方面,而是进入到一个更复杂、更深入的感觉层面。

Perceptual design refers to a design that starts from the person's sensory characteristics. "Feeling" is the first person's perception of the outside world by the first signal system. "Senses can touch heart". So the feeling is that the heart is touched; the sex mainly refers to instinct, desire, and emotion. So sensibility is an inner judgment of feelings, emotions, and desires for things. Today's sales of goods are more and more inclined to sell an "experience." "Feeling" is an experience, and the more exciting and stimulating this feeling brings, the greater its value.

The design enters the perceptual level, which forms the so-called "sensory design". The human senses include sight, smell, hearing, taste, touch and so on. Apparel accessories design is more and more inclined to bring emotional experience to people. For example, minor adjusting the hardness of the shoe

---

❶ 李砚祖. 艺术设计概论 [M]. 武汉:湖北美术出版社, 2009.3: 153.

bottoms through the feet stepping ways, to achieve more comfortable experiences for wearers. or the design of reflective strips on bags and hats, to increase visual warnings in specific environments. These subtle changes are essentially to satisfy people's experience of the senses. In the future design of apparel accessories, designers will be more sensitive to the perception of human experience, so that "human design" not only lingers in terms of practical functions, modeling, scale and other aspects, but moves into a more complex, Deeper feeling level.

5. 材质运用丰富化
Enrich the material applications

饰品材质的演化随着科技进步、生活习俗、流行时尚、文化艺术思潮的变化而改变。进入当代，先进的科学技术导致社会的飞速发展，各种新思维和新观念层出不穷，对于饰品的艺术性、独创性及个性化的需求越来越高。部分艺术家已不再满足于传统的工具材料，他们正在寻求更广泛的材料来表现自己的激情和观念。

随着服饰配件设计概念的多元化和材料科学的进步，大量新材料相继出现，如现代合金材料、纳米材料、超导材料、相变材料等，拓宽了材质之间的互相搭配与融合的界限，材料语义变得更加丰富，情感、绿色、可循环、材料的二次开发都成了未来服饰配件材质运用的新焦点。新型服饰配件品类的出现得益于科技的进步和人类的想象力，被赋予了新的情感意义，使用新时代产生的材料和技术而制作出的服饰配件有着更独特的时代烙印。

The evolution of jewelry materials changes with advances in science and technology, lifestyle, fashion, and cultural and artistic trends. Into the modern era, advanced science and technology have led to the rapid development of the society. Various new ideas and concepts have emerged in an endless stream. The demand for the artistry, originality and individuality of jewelry has become higher and higher. Some artists are no longer satisfied with the traditional tool materials. They seek more extensive materials to express their passion and ideas.

With the diversification of design concepts of clothing accessories and advances in materials science, a large number of new materials have appeared one after another, such as modern alloy materials, nanomaterials, superconducting materials, phase change materials, etc., which broaden the boundaries of mutual compatibility and fusion between materials. Material semantics have become more abundant, emotions, green, recyclable, and secondary development of materials have become the new focus of the future use of apparel accessories materials. The emergence of new clothing accessories category benefits from the advancement of science and technology and human imagination. It has been given a new emotional meaning. The clothing accessories made by the materials and technologies produced in the new era have a more unique era mark.

6. 服饰组合配套化
Clothing combination and matching

服饰品之间的组合配套化主要表现在服装与服饰配件之间的配套设计和服饰配件与服饰配件的配套设计。现代人的消费理念逐渐倾向于"生活方式"化的消费，单一的产品往往很难具备吸引力，服饰配件往往需要搭配其使用场合、系统搭配效果、产品设计创意以及佩戴者的精神向往等因素。设计师在新系列的开发时也越来越重视服装与服饰配件的系列化效果。

随着人们生活方式的多样化，服饰的应用场合也随之增多，人们对各种不同场景下的要求也丰富起来，时装开始以更多的组合方式呈现给消费者，例如，腰带、项链、帽子、鞋子、箱包等多样化组合，这种配套化展示和销售成了一种生动的商家与顾客的沟通渠道，一方面佩戴效果的展示拓展了顾客对服装搭配的考量。另一方面使产品以及设计师理念得到了更加充分的表达，也是品牌个性的一种表达渠道。

The combination of apparel products is mainly reflected in the matching design between apparel and apparel accessories and the matching design of apparel accessories. The concept of modern people's consumption gradually tends to "lifestyle" consumption. A single product is often difficult to be attractive. Clothing accessories often need to match their use occasions, system effects, product design ideas and wearer's spiritual aspiration and other factors. Designers are paying more and more attention to the serialization effect of clothing and accessories during the development of new series.

With the diversification of people's lifestyles, the application of apparel has also increased, people's requirements for various occasions have also increased, and fashion has begun to be presented to consumers in more combinations, such as belts, necklaces, hats, shoes, bags, and luggage. That combination has turned to be a lively channel of communication between merchants and customers. On one hand, the display of wearing effects expands the consideration of costumes for fashions, and on the other, it makes products and designs' concept of fully expressed, as well as an expression channel for brand specification.

7. 实用功能智能化
Practical intelligence

随着科技的发展，服饰配件行业也在不断创新不断走向智能化，服饰配件可以随时监测我们的身体状态，给予人们更加科学的保护。未来智能服装将是一个覆盖全球数十亿人、数百万种应用产品的庞大市场。照此发展下去，相信在不久的将来，服饰配件行业将会大量投入科技创新元素，以崭新的姿态屹立于世界时尚产业。

在信息化的今天，尽管当前还面临着种种难题，但智能化已是必然趋势，技术的进步将使服饰配件产生质变，人们对于这种趋势和流行的接纳，将使智能服装成为一个主流的选择。目前，智能服饰配件正处于一个初步的探索阶段，没有任何一家公司能够完全驾驭这个领域。智能服饰配件无论是在硬件还是在软件方面，都还存在很大的提升空间。近几年，小猫头鹰公司推出了一款针对婴儿的智能袜子，这款袜子能通过红光和红外线，在无侵扰的情况下测量宝宝的心率、含氧量和皮肤温度。这款基于无线操作的袜子未使用任何黏合剂，过敏性低，所有电子部件都藏在防水、绝缘的硅胶套里面，确保安全。通过蓝牙模块，它还能将记录到的信息传输到配套APP上。

With the development of science and technology, the apparel accessories industry is constantly innovating and constantly moving toward intelligence. Clothing accessories can monitor our physical status at any time and give people more scientific protection. Smart clothing in the future will be a huge market covering billions of people and millions of applications worldwide. With this development trend, it is believed that in the near future, the apparel accessories industry will invest heavily in technological innovation elements and stand in the world fashion industry in a brand-new perspective.

In our current information era, despite the various difficulties we have, intelligentization is an inevitable trend. Advances in technology will change the quality of apparel accessories. People's acceptance of this trend and popularity will make smart clothing a mainstream selection. At present, smart clothing accessories are in an initial stage of exploration. No company can completely control this area. There is still much room for improvement in smart apparel accessories, both in hardware and software. In recent years, Owlet launched a type smart socks for babies that can measure baby's heart rate, oxygen content, and skin temperature without intrusion through red light and infrared light. The wirelessly-operated socks use no adhesive and low in allergy. All electronic components are hidden inside a waterproof, insulating silicone sleeve to ensure safety. Through the Bluetooth module, it can also transmit the recorded information to the related APP.

8. 设计制造一体化
Design and manufacturing integration

科学技术的进步为包括服饰配件在内的制造业提供了设计与制造一体化方案。当前的3D打印技术、CNC数控技术、数码印刷技术、激光成型技术、粉末冶金技术、纳米涂装技术、液体缝制技术、物联网技术及自动化生产线等新型生产技术自身正在不断进步和成熟，未来的某些服饰配件行业可以充分利用这些技术，不仅可以使服饰配件新增前所未有的功能，还能加快产品从设计开发到批量制造的速度，满足快速、精准、安全的服饰配件个性化定制，打通了设计、试制、评价、量产、物流、推广、销售的全产业链环节。

比如生物饰品，就是在新技术突破下产生的。生物饰品又称基因饰品，是利用生物技术通过基因物质的提取、分离、扩增和固化作用，将生命体基因物质包埋于不同的饰品材料中，再经过雕饰而制备的一种生化饰品。

Advances in science and technology have provided integrated design and manufacturing solutions for the manufacturing industry, including apparel accessories. The current 3D printing technology, CNC numerical control technology, digital printing technology, laser forming technology, powder metallurgy technology, nano-coating technology, liquid sewing technology, Logistic network and automated production lines and other new production technologies are constantly improving and maturing. Some apparel accessories industry can make full use of these technologies, not only can add unprecedented features of apparel accessories, but also speed up the product from design and development to mass production speed, to meet the fast, accurate and safe clothing accessories personalized customization, open up The whole industrial chain link of design, trial production, evaluation, mass production, logistics, promotion and sales.

For example, biological jewelry is produced under the breakthrough of new technologies. Biological jewelry, also known as genetic jewelry, is the use of biotechnology to extract, separate, amplify, and solidify genetic material, embed the genetic material of living things in different jewelry materials, and then prepares a biochemical jewelry.

人类社会处在一个生产力快速上升期，新技术的产生和突破无时无刻不在影响着人类，现在社会学家普遍认为未来人类社会是信息化社会、大工业社会，世界各国文化在信息和交通技术的发展下不断互相碰撞、影响及融合。虽然复杂多样，但大体上还是可以从中把握到未来人类社会发展趋势的一些主要脉搏：

首先是文化多元化。各国文化交流日益丰富是不可逆转的趋势，文化的多元化发展是必然的，这对服饰配件设计者来说既是机遇也是挑战，设计者要放眼全局把握关键，学会找到准确的设计立足点，成就自己的设计作品。

其次是更加注重生态文明，可持续发展成为人类普遍认可的理念，这与人类的生存息息相关，设计者需要有其设计使命感和社会责任感。

最后是以人为本的思维深入人心。人们越来越重视自身的情感体验，设计者的作品是服务于人类本身的，拥有优秀的人、物互动体验是服饰配件设计成熟与否的一个关键标志。

Human society is in a period of rapid productivity growth, and the emergence and breakthrough of new technologies are affecting human beings all the time. Now sociologists generally believe that the future human society is an information society and a large industrial society. Although complex and diverse, it is generally possible to grasp some of the main pulse of the future development trend of human society:

The first is cultural diversity. It is an irreversible trend that the cultural exchanges of various countries become increasingly rich, and the diversified development of culture is inevitable, which is both an opportunity and a challenge for the designers of clothing accessories. Designers should keep an eye on the overall situation, grasp the key, learn to find an accurate design foothold and achieve their own design works.

Secondly, more attention should be paid to ecological civilization, and sustainable development has become a universally recognized concept, which is closely related to human survival. Designers should have a sense of design mission and social responsibility.

Finally, people-oriented thinking is deeply rooted in people's hearts. People pay more and more attention to their own emotional experience, and designers' works serve human beings themselves. Having excellent interaction experience between people and things is a key sign of the maturity of costume accessories design.

第三章　常用服饰配件设计
Chapter 3　The design for Daily fashion and jewelries

课程名称：常用服饰配件设计

教学目的：通过学习能使学生掌握首饰的历史沿革；首饰的种类与仪态个性的搭配；首饰与服装的搭配艺术。

教学重点：本章学习以掌握首饰与服装的搭配、创意设计为重点。

Course Name: Commonly Used Fashion Accessories Design

Teaching objectives: Through learning, students can understand the history of jewelry; master the combination of types, and personality of jewelry; master the art of matching jewelry and clothing.

Teaching focus: This chapter is designed to master the combination of jewelry and clothing, and creative design.

学习重点：本章学习的重点是在首饰的创意设计与服装搭配方法上。理论部分以示范讲述为主，加强直观教学，以利于知识的传授。技能练习通过作品、示范、个别辅导、作业讲评等教学方法，最终达到让学生掌握知识，掌握技能的目的。

Learning focus: The focus of this chapter is on the creative design of jewelry and the way it is paired with clothing. The theoretical part focuses on demonstration and strengthens intuitive teaching to facilitate the transfer of knowledge. Skills practice through the teaching, demonstration, individual counseling, homework evaluation and other teaching methods, and ultimately achieve the purpose of helping students master the knowledge and skills.

课时参考：32课时。

Lesson period reference: 32 class hours.

首饰的产生可以追溯到远古的史前文化，国内传统定义"首饰"一词始于明清时期，主要是指佩戴于头上的饰物。那时又将首饰称"头面"，如梳、钗、冠等。其后由于戒指的使用与发展超过了其他饰品的品种发展，又因"手"与"首"同音，因而将戒指等"手饰"也被统称为"首饰"。

The generation of jewelries could be traced back to the ancient prehistoric culture. The traditional definition of 'jewelries' started from Ming Qing dynasty. It mainly refers to the jewelries for the hair. At that time it was also called 'Tou Mian', like comb, fork, crown. Later, due to the rapid development and appliance of rings, far exceeding other jewelries, and with the identical pronunciation of 'hand' and 'head' in Chinese, all the jewelries for hands are categorized into 'head jewelries'.

现代对首饰的定义又称广义首饰，是指用各种金属材料，宝玉石材料，有机材料以及"仿制品材料"，制成装饰人体及环境的装饰品，与服装搭配使用的饰品总称。由于首饰较多地使用稀贵金属和珠宝，因而它们的价值较高。而现在的服饰配件多以织物为主，同时也会使用多种非贵金属的价格低的材料，这样新材料，新观念不断进入首饰领域，从而首饰的品种、范围就不断扩大，这也是当今首饰业的发展方向。

The contemporary definition for jewelry is generalized. It refers all the jewelries for decorating human bodies, environments, and fashions, made by gems, jade, organic materials and synthetic ones. Because quite a few of the jewelries' materials are rare metals and jewels, they are quite valuable. Currently the jewelries for fashion are mainly made by fabrics, also by various non-precious metals and low priced materials. With the continuous new materials and new ideas applied in the jewelries area, the categories and scope keep expanding. It is also the future development direction for contemporary jewelries industry.

首饰虽然体积不大，却清晰印有不同时代的烙印。不同时代的首饰体现了不同时代人们的文化习俗及审美情趣，也受到社会礼制的约束和传统习俗的影响，同时还与制作工艺水平和设计理念有关。

Athough the sizes of jewelries are not big, they have the clear marks of different eras. The specific culture and taste of that certain time were reflected by them. Meanwhile, they are influenced by the social regulations and customs, related with design ideas and manufacturing capability.

从人类开始意识到用装饰来美化自身的时候起，人类也就与首饰结下了"不解之缘"。人类最原始的首饰，距今已有7000年的历史，我们可以追溯到遥远的石器时代，当时由于生产力非常低下，人类在适应自然的过程中，面对无限神秘的自然世界，希望得到大自然强大力量的庇佑与恩赐，从而这种原始本能的神灵崇敬，通过各种艺术形式表现了出来，从而形成了独特的艺术形式——原始图腾，并在首饰造型或图案上表现出来。

Since the human beings had the awareness to decorate their bodies, the unbreakable relationships have been formed with jewelries. The most original jewelries enjoy the 7000 years' history, which could be traced back to the stone age. At that time, the production capacity was quite low. During the process of natural evolving, dealing with the enormously mysterious nature, people wished they would be blessed and gifted by the great power of nature, thus, these primitive worship emotions had been exerted through various artistic ways. Then the unique artistic totem was originally formed, reflected on the patterns and shapes of jewelries.

第一节　西方首饰发展简史
Section 1　The brief history of western jewelries

早期的首饰主要有几个时期的首饰发展为代表：

The early jewelries are categorized by the these particular periods below :

一、石器时代
A. The stone age

在旧石器时代原始部落的史料中，记载了早期原始"体饰"，主要形式为：项饰、腰饰、臂饰、腕饰、头饰等几种，而在这些形式中，尤以项饰和腰饰为主。它们在很大程度上是围绕人体生殖区而装饰的。究其原因，除了这些部位有支持佩戴物的能力外，有研究表明是有原因的。如动物身上的色彩和图案也是一种体饰，是自然的体饰。如雄鸟的头饰、项饰、胸饰、尾饰等，往往在繁殖季节呈规律性的变化，而且这些装饰对吸引异性颇有功效，大量生物学材料已证实动物的漂亮装饰在性选择过程中具有很大的优势。

In Paleolithic time, from the historic records about primitive tribes, we find the forms of 'body jewelries' were classified as: neck jewelries, waist jewelries, jewelries, wrist jewelries, head jewelries. Among those categories, the majority is neck jewelries and waist jewelries. They are mostly for the decoration around the genital areas. Apart from the reason that these body parts could be decorated by these jewelries, other purposes are obvious with the support of abundant researches. The colors and patterns on the animals are also defined as body decoration. These natural body decorations, like the head, neck, breast, tail decorations on the male birds would vary regularly in the reproduction seasons. These decorations work quite well on attracting females. Enormous biological records indicate these pretty decorations enjoy great advantages in sex selection.

世界各地发现的旧石器时代晚期的最早的"人体饰物"，无论是动物牙齿、羽毛，还是石珠等它们均有一个十分显著的特点：光滑、规则、小巧、美观。而这一特点充分说明了"体饰"产生的装点装扮、自我炫示、吸引异性的重要心理动因。而由这种起源动因衍变而来的"人体美化"功能是首饰最原始最根本的功能。

The earliest 'body jewelries' found in the late Paleolithic period around the world, regardless of the materials of animal teeth, feather, or stone beads, they share one distinguished feature:smooth, regular shaped, small, pleasing appearance. This feature clearly demonstrates the important psychological motives for decorating, self-presenting, attracting sex. The function for 'beautifying bodies' of jewelries developed from the primitive motives is ultimate and primitive.

据近年《西伯利亚时报》报道，考古学家在俄罗斯西伯利亚地区的丹尼索瓦洞穴发掘了一批50000年前埋藏的"珠宝"，年代可以推至旧石器时代早期，比之前在南美发现的相似珠宝要更早。它们是人类祖先用鸵鸟蛋制成的，这些做工精细的珠子当时可能被缝在衣服上，也可能做成手镯和项链，体现了先人们非凡的审美和技艺（图3-1）。

From the reports of *Siberia Times* In recent years, archaeologists in Siberia area found a batch of 'jewels' buried 50000 years ago in Denisova cave. They can be traced back to the early Paleolithic period, even earlier than the similar ones found in South America. They were made into refined beads by ostrich eggs, possibly sewed on the clothes, or made to bracelet and necklace, reflecting the extraordinary taste and techniques of our ancestors (Figure 3-1).

图3-1 距今50000年，旧石器时代早期手镯
Figure 3-1 50000 years ago the bracelets of early Paleolithic period

二、苏美尔时期
B. The Sumer period

根据记载，在公元前大约4000年前，西方人首次使用金属首饰，而在西方最早制作黄金首饰的是苏美尔人。苏美尔人是波斯高原的游牧民族，他们迁徙到美索不达米亚地区，造就了古代苏美尔文明。苏美尔人将黄金敲打成极薄的金箔，用金箔制成的树叶花瓣，如同流苏一样可以悬挂。其中，最具代表性的是公元前2600年的舒巴德皇后所陪葬的头饰，它采用天青石、光玉髓、玻璃和黄金制成，与头饰一起的还有金耳环和以黄金、天青石和光玉髓串成的项链（图3-2、图3-3）。苏美尔首饰在当时，是为逝去的人做的，因他们相信死后会去另外一个世界，所以在许多古代帝王贵胄的陵墓里都发现有大量的殉葬品。

Upon the recordings, the metal jewelries were applied by westerners around 4000 B.C. The Sumerians were the first to manufacture the gold jewelries. Sumerians were the nomads of Persian highland. They migrated to Mesopotamia area, created the Sumerian culture. They hammered gold into very thin fold, made them into the shapes of leaves and pedals, hanging like tassels. The most distinguished one is the head jewelries buried with Queen Shubad 2600 B.C. It was made by celestite, camelian, glass and gold. Gold earrings and necklace of celestite and camelian were found along with it(Figure 3-2、Figure 3-3). At that time, the Sumerian jewelries were made for the persons passing away, as they believed another world for the deads. Thus, large amount of goods were found in many tombs of ancient kings and nobles.

图3-2　苏美尔普阿比女王（"Pu-abi nin"）头饰　公元前2550~公元前2400年乌尔（伊拉克南部）美国费城宾夕法尼亚大学考古与人类学博物馆

Figure3-2　The head accessory for Queen Shubad Sumer, about 2550 B.C.~2400 B.C. South Iraq. University of Pennsylvania Museum of Archaeology and Anthropology

图3-3　苏美尔普阿比女王花圈头饰　早王朝三期乌尔美国费城宾夕法尼亚大学考古与人类学博物馆

Figure 3-3　The floral head accessory for Queen Shubad Sumer, University of Pennsylvania Museum of Archaeology and Anthropology

三、古埃及时期
C. The Ancient Egypt era

古埃及文明的遗物中，首饰的使用相当广泛。社会上的各个阶层，上至法老，下到平民，生者、死者，都佩戴首饰，甚至连神兽也不例外。

古埃及法老、贵族的首饰多用贵金属（主要指金银的合金，因为埃及国内盛产黄金，而白银则十分稀有，且金银合金较为耐用）和半宝石（是指介于宝石和石头之间的各种色彩斑斓的矿石，如绿长石、绿松石、孔雀石、石榴石、玉髓、青金石等）制成，二者都是当时人们喜爱的饰物材料。平常百姓所戴首饰一般用釉料制成，通常以石英砂为坯，再饰以玻璃状的碱性釉料，也可在石子上涂釉彩而成。

In the remains of ancient Egyptian culture, jewelries were vastly applied in every social status. From Pharaoh to civilians, everyone wore jewelries alive and dead, even the mythical animals.

Ancient Egyptian pharaohs and nobles used precious metals (mainly gold and silver alloys, because Egypt is rich in gold, while silver is very rare, and gold and silver alloys are more durable) and semi-precious stones (which are between gemstones and stones). All kinds of colorful ores, such as green feldspar, turquoise, malachite, garnet, chalcedony, lapis lazuli, etc., are the favorite materials at that time. Usually, the jewelry worn by ordinary people is usually made of glaze, usually with quartz sand as the blank, and then decorated with glassy alkaline glaze, or glazed on the stone.

古埃及制作首饰的材料多具有仿天然色彩，取其蕴含的象征意义。金是太阳的颜色，而太阳是生命的源泉；银代表月亮，也是制造神像骨骼的材料；天青石仿似保护世人的深蓝色夜空，这种材料均从阿富汗运来；来自西奈半岛的绿松石和孔雀石象征尼罗河带来的生命之水，也可用利比亚沙漠的长石甚至绿釉料代替；尼罗河东边沙漠出产的墨绿色碧玉像新鲜蔬菜的颜色，代表再生；红玉髓及红色碧玉的颜色像血，象征生命。

The materials of ancient Egyptian jewelries had the similar colors of nature. They were applied for it. Gold is the color of sun, the origins of lives; silver represents moon, like the materials to create bones by Gods; celestite, which shipped from Afghan, is alike the dark blue night sky,blessing human beings; the turquoise and malachite from Sinai Pen stand for the water of lives brought by the Nile, they could be replaced by feldspar or even green porcelain of Libyan

Desert; the dark green colored jade produced in east Nile desert is alike the color of fresh vegetable, representing revival; the red camelian and red jade is alike blood, representing lives.

古埃及首饰的种类主要有项饰、耳环、头冠、手镯、手链、指环、腰带、护身符及项饰平衡坠子等，制作精美，装饰复杂，并带有特定含义。

现存最古老的首饰，是在古埃及的中期（公元前1991～公元前1778年），它早期的首饰是偶像崇拜及宗教生活的具体表现。这些首饰以各种象征性图案及纹样出现，如太阳纹、鹰、蛇、圣甲虫（即屎壳郎）等十大偶像为主体，佩戴这些首饰以寻求神灵庇佑及保护。如埃及人尤其对蛇（古蛇）特别崇拜，因此在法老的面具及头饰上带有蛇的纹样（图3-4）。

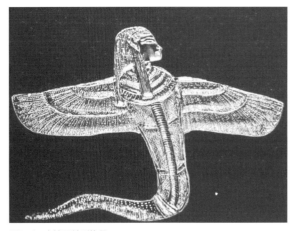

图3-4 古埃及蛇形饰品
Figure 3-4 Snake shaped jewelry in ancient Egypt

The main categories of Egyptian jewelries are necklace, earring, crown, bracelace, bracelet, ring, waist belt, amulet, pendant, etc. They are refined with complicated decoration for certain meanings.

The oldest existing jewelries are from the mid Egyptian times (1991 B.C.~1778 B.C.), the early jewelries are the specific reflection for worship and religious life. These jewelries are with the symbolic patterns of ten idols like sun, eagle, snake, scarab (beetle) for the blessing and protection of Gods. For instance, the Egyptians especially worship the snakes, thus the mask and head accessories bear the snake patterns (Figure 3-4).

图3-5 青金石戒指 约公元前6世纪
Figure 3-5 Lapis lazuli rings, 6 centuries B.C.

图3-6 古埃及圣甲虫护身符
Figure 3-6 The ancient Egyptian scarab amulet

公元前1600年前，埃及的首饰工艺达到辉煌的巅峰。古埃及人把圆柱形印记当作首饰佩戴，成为贵族阶级的身份象征。大约公元前6世纪，底部平坦的图章形印记逐渐取代了圆柱形印记，而圆柱形印记的流行又促成戒指的发展（图3-5、图3-6）。

In 1600 B.C., the Egyptian jewelry manufacture techniques reached at the glorious peak. The Egyptians wore the cylinder-shaped stamps as jewelries, as the symbols of nobles. About 6 centuries B.C., the flat-bottomed stamps gradually replaced the cylinder-shaped stamps. The cylinder-shaped promoted the development of rings (Figure 3-5、Figure 3-6).

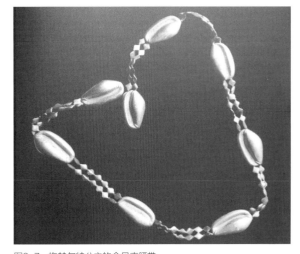

图3-7 梅赫尔特公主的金贝壳腰带
Figure 3-7 Golden shell belt of princess mehherdt

第十二王朝（公元前1900～公元前1800年）的腰带由贝壳、护身符和珠饰组成，材料包括天然银金、银、光玉髓、紫晶、青金石、长石和玻璃等

（图3-7）。在贝壳的中空，由银金制成。古埃及人认为贝壳与女性生殖器的形状相似，可起保护作用。两个以天然银金制成的中空护符，代表一束继承王位者的发式。两个鱼形护身符，用以祈求佩戴者免被水淹。跪坐的人像是长寿之神一赫，他通常手执棕榈枝。其余都是圆形珠子，包括中空的银金珠子和光玉髓、紫晶、青金石、绿色长石制成的珠子。这件腰带体现了古埃及人有将珍贵金属与色彩艳丽的半宝石串连的爱好，代表着当时珠宝佩饰工艺的高度成就。项饰平衡坠子是当胸前佩戴的项饰过重时，用以平衡其重量的坠子，常佩于背后两肩胛骨的中间。平衡坠子的基本形状像个钟摆，和项饰一起原为供奉天空之神哈托尔的女祭司们专用，后来所有贵族妇女都佩戴它，以示对神的尊敬。项饰象征着死后的复生和多子，而平衡坠子佩子背后则有庇佑之意。

The waist belt (Figure 3-7) in 12th dynasty (1900 B.C.~1800 B.C.) was composed of shell, amulet and beads, by natural silver gold, silver, camelian, amethyst, lapis lazul, feldspar and glass, etc. The shell shape is hollow, made by silver gold, as the Egyptians thought the shell shape could exert the protection function for resembling the female genital area. Two hollow amulets by natural silver gold represent the heir to the throne's hairstyle. Two fish shaped amulet is for praying not being drowned by flood. The one sitting on his knees is alike the God of longevity, usually holding the palm tree branch. The rest are the round beads, including the hollow beads of silver gold, camelian, amethyst, lapis lazuli,, green feldspar. This waist belt indicates the Egyptians' preference about stringing precious metal and colorful semi-gems. It demonstrates the high achievement of the jewelry techniques of that time. The pendant is for balancing the heavy weight of the necklace, normally being worn between the two scapulars behind. The basic shape of the pendant is like the clock pendulum, at first particularly for the priestesses of Hathor, later all the female nobles wore it to express the worship for the Gods. Necklace represents the revival and many decedents after death, while the pendant at the back means protection and blessing.

代表古埃及首饰最高成就的当然是法老的首饰。法老墓中曾随葬着大量精美无比的珍宝首饰，但历经多年的盗掘，大多流失殆尽。在为数不多的未被盗掘的法老墓中，第十八王朝法老图唐卡门墓的首饰最为有名。除上述实物外，古埃及雕像、浮雕及图画上人物所佩戴的首饰，也以其逼真的刻画，向今天的我们展示着这个文明古国在首饰工艺上的辉煌成就。

The highest level of the ancient Egyptian jewelry is certainly the ones for Pharaoh. There were large amount of refined precious jewelries in the Pharaohs' tombs. Most of them are lost after years of digging and stealing. Among the limited Pharaohs' tombs untouched, the eighteenth dynasty's Pharaoh Tutankhamen's jewelries are the most well known. Apart from the real ones, the jewelries appearing in the ancient Egyptian Statues, cameo and drawings demonstrate the glorious achievement in jewelry techniques once in that civilized nation.

四、古希腊时期
D. The Ancient Greek age

古希腊有着丰富的首饰遗产。早期的希腊金属工艺，明显受到古埃及的影响，形式单纯、表现有力，且多用圆、三角形等几何纹样装饰。古希腊的首饰主要品种有王冠、花环、耳环和项链。花环（图3-8）是用于宗教行列，也用于奖励某位功臣。希腊工匠的最主要的贡献是直接用金子铸造人和动物的形象的首饰（图3-9）。在公元前3世纪希腊首饰，很少用宝石，色彩效果是靠珐琅工艺获得。但在公元前3世纪马其顿国王亚历山大大帝征服希腊后，东方宝石渐渐用于希腊首饰。

The jewelry remains of ancient Greece are rich. The early Greek metal techniques were obviously influenced by ancient Egyptians: simple forms, effective presentation, geometric shapes like round

图3-8 金花冠 公元前3~公元前4世纪 希腊国立考古博物馆
Figure 3-8 Gold floral crown 3~4 centuries B. C. National Archaeological Museum, Greece

图3-9 金发罩 公元前3世纪 希腊国立考古博物馆
Figure 3-9 Gold hair cover 3 centuries B.C., National Archaeological Museum, Greece

and triangle were often applied. The main jewelry categories of that time were: crown, floral crown, necklace, earrings. The floral crowns were for the religious purpose, also for rewarding the meritorious royal officers. The most significant contribution of Greek jewelry technicians is to molding the shapes of human beings and animals directly by gold. Before 3 centuries B.C., gems were rarely applied. The colorful effects were achieved by enamel techniques. But after 3 centuries B.C., Greece was conquered by Alexande, the Great of Macedon, the oriental gems were gradually applied on Greek jewelries.

五、古罗马时期
E. The ancient Roman age

公元前2世纪至公元2世纪的古罗马首饰继承了希腊的传统，用宝石和玻璃人造宝石的首饰渐渐增多（图3-10），耳环的款式亦有着各种造型各异的品种，如一种两耳酒瓶形状垂坠，就显示了当时古罗马首饰受埃及和叙利亚的文化影响较大。还有不少耳饰配有S型的钩子（图3-11），项链则是8字链结构，扣上具有护身符意义的挂坠。手镯则变得粗大，常有盘曲的蛇（图3-12）作为装饰。

From the first two centuries B.C. to the second century AD, ancient Roman jewelry inherited the tradition of Greece. The jewelry made of gemstones and glass artificial gemstones gradually increased (Figure 3-10), and the styles of earrings also had various types of shapes. Many of the earrings come with s-shaped hooks (Figure 3-11), while the necklace is an 8-character chain with a talisman pendant. The bracelets are thickened and decorated with coiled snakes (Figure 3-12).

图3-10 古罗马时期耳环——带绿玻璃珠子
Figure 3-10 Ancient Roman earrings with green glass beads

图3-11 S形耳钩
Figure 3-11 S-shaped ear hook

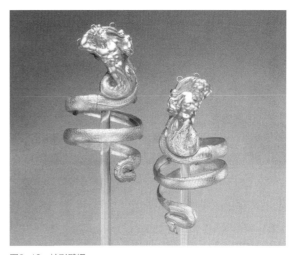

图3-12 蛇形臂镯
Figure 3-12 Snake bracelet

古罗马最重要的首饰是戒指，相传是古罗马人率先将戒指当作订婚和结婚的标志（图3-13）。镶嵌硬币的戒指（图3-14）是当时整个帝国时代最流行的首饰。

同时古罗马还盛行一种浮雕宝石饰品，称为"卡米奥"，是用玛瑙琢成的椭圆形片，利用宝石上下不同层面的不同色彩，将浮雕部分与底层子部分区分开来，例如，在黑色或棕色底层上刻有白色的浮雕（图3-15）。罗马人凭借其政治、军事上的天才，创建了庞大的帝国。作为希腊文明的继承者和传播者，罗马人的浮雕艺术延续着东方和希腊的古典法则。在处理空间和深度上罗马艺术家比希腊古典艺术家又有所进步，在处理平面化人物背景的柔软性上，罗马浮雕在技术上要更精进娴熟，这为浮雕首饰的发展奠定了基础。

公元2世纪古罗马人的首饰采用了两项新技术：一是浮雕细工技术；二是乌银镶嵌技术。古希腊人用这种技术来装饰兵器，而古罗马人则用以首饰制作。

The most important piece of jewelry in ancient Rome was the ring, which is said to have been the first symbol of engagement and marriage in ancient Rome (Figure 3-13). The coin ring (Figure 3-14) was the most popular piece of jewelry throughout the empire.

At the same time, ancient Rome also prevails a kind of relievo jewelry, called "camio", which is an oval piece cut with agate, and uses the different colors of the upper and lower layers of the gemstone to separate the relief part from the lower part, such as the white relief carved on the black or brown bottom (Figure 3-15). With their political and military genius, the Romans created a vast empire. As the successor and disseminator of Greek civilization, Roman relief art continues the classical rules of the east and Greece. In terms of processing space and depth, Roman artists have also made progress compared with Greek classical artists. In dealing with the softness of the flat background of characters, Roman relief sculpture is more sophisticated in technology, which laid a foundation for the development of relief jewelry.

In the 2nd century AD, the ancient Romans adopted two new techniques for jewelry. One was relief carving technology, the other was the black and silver Mosaic technology. The ancient greeks used this technique to decorate weapons, while the ancient Romans used it to make jewelry.

六、拜占庭时期（8世纪）、14～15世纪时期
F. The Byzantine period (8th century)、14th~15th century

图3-13　古罗马订婚戒指
Figure 3-13　An ancient Roman engagement ring

图3-14　古罗马硬币戒指
Figure 3-14　An ancient Roman coin ring

图3-15　古罗马武士族徽戒指
Figure 3-15　Ancient Roman knight's signet ring

拜占庭服饰时期可谓是"交会期"的服饰，它是和希腊罗马的古典风格，东方文化的神秘色彩，宗教文化的禁欲精神三种遗址文化相融合的。当人们从古埃及、希腊、古罗马那种简朴形态，单调色彩的服装世界走向中世纪的拜占庭时代时，突然看到了一个五彩缤纷令人目眩的服装世界。奢华五彩缤纷的服饰有赖于拜占庭发达的织染工业和中国丝绸的输入，还有赖于出色的饰品镶嵌工艺。

拜占庭时期是基督教发展时期，也是宗教艺术的典型发展时期。基督圣像是拜占庭艺术的主体，因此拜占庭的首饰设计中广泛运用十字架以及基督圣像等图案（图3-16）。拜占庭首饰一反罗马帝国首饰的简朴风格，造型图案极其华丽。如许多雕透工艺极其繁复。而他们的主要贡献是掐丝珐琅技术。图案的轮廓是由金属细条围在金属胎边，匕、焊在表面，在中间填上釉彩（图3-17）。

15世纪宝石工艺的发展表现在宝石切割越来越精妙。设计形式受到火焰式和垂直式哥特风格建筑的影响，壁龛和窗花格微型化为首饰式样（图3-18）。在当时低敞胸裙子的流行，增加了对项链和垂饰的需求，服装的宽袖衣给予了手镯露面的机会，男子佩着镶宝石的皮带扣，对于宝石材料的需求就多了，女子项链挂着可开合的空心小盒，对于小盒内外的"视觉传达"有了许多新的内容。如盒面外有着珐琅的宗教画如耶稣受难、受胎告知、圣母像、天使等，盒里装一爱人的头发。

在15世纪，还流行帽饰，而且还用黄金或其他金属制成。

The fashion in Byzantine period is the collection period. It combined the remained cultures of classical Greek Roman styles, mysterious oriental styles, and the religious Ascetic spirit. From the ancient Greek and Roman simple styled, few colored fashion to the mid-century Byzantine period, a colorfully glamorous fashion world was explored. These luxurious and colorful fashion relied on the advanced weaving and dying industry, the distinguished jewelries setting techniques, the import of Chinese silk at Byzantine period.

Byzantine period is the period for Christian religion development, also the typical development for religious art. The Saint Christ is the main theme for the Byzantine art, the Cross and the Saint Christ patterns were widely used in jewelries design Figure 3-16. Quite opposite to the simple style of jewelries of Rome, the Byzantine jewelries' pattern and shape are extremely luxurious. For example, the carving technique is extremely complicated. Their main contribution is the cloisonne technique. The pattern outline was enclosed by metal string besides the metal mold, with the enamel filled in the middle, welded on the surface.

The development of gem technique of 15th century was presented on the refined gem cutting. The design was influenced by the flame shaped and vertical were influenced by the Gothic architecture. The wall end and the window grid were minimized into jewelry design. The popularity of the low neck gown promoted the demand for necklace and pendant. The wide sleeved clothes got bracelets revealed. The belt buckle men wore were set with gem, which increased the needs for gem materials. The hollow accessible little box hung in the necklace for women were added the "visual expression" from inside and outside: like the Saint Chris in enamel, the Virgin of Guadalupe, angels, arrival information of baby outside, a bunch

图3-16 拜占庭时期油画
Figure 3-16 Byzantine oil painting

图3-17 十字架图案首饰
Figure 3-17 Crucifixion jewelry

图3-18 拜占庭时期吊坠
Figure 3-18 Byzantine pendant

of hair from lover.

In 15th century, the hat accessories were popular too, they were even made by gold and other metal.

七、文艺复兴时期
G. The Renaissance period

在黑暗的中世纪时期，人们生活在被统治阶级扭曲的宗教律法里，当时的作品带有强烈的宗教色彩。经过文艺复兴运动，随着启蒙运动和人文主义兴起，围绕以人为中心的首饰艺术创作主题盛行，将文艺复兴的艺术风格渗透在珠宝首饰中，从而人物塑像的图案出现在珠宝首饰上（图3-19）。

同时，经典传说也成为创作的主题，真实的和虚构的人物形象开始交替出现，最为精美的典范，要数维多利亚和阿拉伯特博物馆的"罐头宝石"，这是一件诞生于16世纪后期的意大利吊坠。是文艺复兴时期的意大利艺术家用涂上珐琅的黄金做的微型雕塑。

In the dark midcentury age, people were governed under the tortured religious laws and regulations. The works that time carried a strong sense of religion. After the Renaissance, humanism was aroused, thus, the theme for human was greatly applied in jewelry art design, as well as the Renaissance style, and thus the pattern of humans had been appearing on the jewelries (Figure 3-19).

Meanwhile, the classical legends were adopted as the theme for jewelries design, with the real and hypothetical figures appearing. The best one is the "Canning Jewel" in Victoria and Albert Museum. It is an Italian pendant of 16th century, a mini gold statue with enamel painted by Italian artists.

从16世纪中期开始，浮雕在首饰上的表现让位于镌刻。首饰中心也从意大利、法国移向奥地利和德国。首饰越来越被看作是妇女的专用装饰品而且越发奢华精巧。妇女佩戴一整套首饰，包括手镯、衣领角、有垂饰的项链和头饰。垂饰有龙形、海马、魔鬼、动物、圣经和其他题材，都是用涂上珐琅的黄金制成，有时还有方形的宝石镶边。

From the mid 16th century, sculpturing had been replaced by craving.

The center of jewelries had been moved from Italy to Austria and Germany. The jewelry had been regarded as for the women specifically, and it turned to be more luxurious and refined. Women wore a complete set of jewelries, including bracelets, the accessories for collar, the necklace with pendant and head accessory. The themes of the pendants are dragon, sea horse, devils, animals, bible, etc. They were made by the gold covered with enamel, sometimes framed by square gem.

八、17世纪首饰
H. The 17th Century jewelry

17世纪的首饰设计以花卉图案为基调。服装款式的改变引导了首饰的潮流。随着切割技术高度发展，出现一大批各式各样的切割方式，人们也把对涂珐琅的黄金的制作更多地转向了运用宝石材料。但从1650年后，圆顶宝石很少再用。托架从自然花朵形态转向叶形。

The base theme for jewelries in 17th century is floral pattern. The change of fashion led the jewelries' style. After the highly developed cutting techniques,

图3-19　文艺复兴时期首饰
Figure 3-19　Renaissance jewelry

numerous various cutting methods had emerged. People turned to applying more gem materials from covering gold with enamel. From 1650, round top gems were rarely used, the shape of the holder turned from flower to leaf.

17世纪末至18世纪，人们的兴趣转向欣赏首饰的光芒。当时的切割技术更为完美，如钻石切割技术发明于1700年，到18世纪初一颗钻石最多可以切割出58个面。钻石在宝石中占了"统治地位"，其次是水晶，最次的是白铁矿石。一套首饰包括耳环、项链、手镯、胸饰以及鞋扣可以用不同的材料组成。

From the late 17th century to 18th century, the lust of the jewelries were appreciated more. At that time, the cutting techniques turned to be sophisticated. For example, the diamond cutting were invented in 1700. In 18th century, 58 facets could be cut at most on a diamond. The diamond dominated the gems, the next was crystal, the very next was the marcasite. A complete set of jewelries was composed of earrings, necklace, bracelet, brooch and shoe buckle by different materials.

首饰一般分成在日光下使用和在晚间烛光下使用。日间（日光下使用）配饰有妇女的衣带串饰，戴在妇女的手腕，系一串钥匙，钱袋或小装饰品，鼻烟盒，镶金的石英的怀表，烛光（晚间烛光下使用）首饰是闪光的钻石、祖母绿、黄玉或红宝石。

Normally jewelries were categorized into daily sunlight use and candle light use at night. Daily used jewelries were string accessories on the clothes belt, bracelets on wrists with a bunch of keys, purses and other small accessories, snuffbox, the pocket watch in quartz and set with gold. The candle light used jewelries were glittering diamond, emerald, topaz and ruby.

九、18～19世纪的首饰
I. The jewelries from 18th century to 19th century

19世纪前期的首饰在各种美术流派的影响下，产生了烦琐华丽的具有代表性的洛可可风格，出现在古代造型要素上，创新了新古典主义款式。同时又受到了浪漫主义和自然主义美术思潮的冲击而形成了相类似风格的首饰款式。洛可可式首饰采用不对称图案和鲜艳的颜色，运用了彩色宝石和珐琅彩釉，尽显了首饰的富贵华丽。

Influenced by various art styles of 19th centuries, the Rococo style emerged representing the flourished luxury. Neoclassical style was innovated on the base of classical elements, impacted by the artistic ideology of Romanticism and naturalism, thus the similar styled jewelries emerged. The asymmetric patterns and bright colors were applied in Rococo jewelries with colorful gems and enamel, presenting the luxury of the jewelries.

俄国的费伯杰是珠宝首饰历史中最伟大的艺术家之一。他曾为俄罗斯皇室成员设计制作珠宝装饰的复活节彩蛋而闻名。

Faberge, one of the most distinguished artists from Russia in jewelry design is recognized for designing the eggs in Easter day for Russian royal members.

在17世纪中期，有了制造人造宝石的行业，18世纪，人造宝石有了合法交易市场，这种新的材料能凸显首饰的各种艺术造型，因此人造宝石是珠宝首饰历史上最重大的革新。1800～1820年，一种新的合金材料问世，它是由17%的锌和83%的铜合成的金属铜，它是黄金代用品并很快成为贵族的新宠。同时，造就了如蒂凡尼、宝格丽、卡地亚等，这些品牌一直延续至今并使传统手工艺开始向现代艺术设计过渡。18世纪末，工业革命爆发，工业机器的使用不仅仅导致生产方式的改变，同时也对社会关系、生活方式及文化等各个方面带来巨大冲击。首饰不再是皇室及贵族的特权，社会的上层阶级和中产阶级都有能力拥有属于自己的首饰。

In the mid 17th century, synthetic gem manufacturing emerged. In 18th century, artificial gem trading had the legal market. These new materials could promote the various art forms for jewelries. It was recognized as the greatest innovation in jewelry history. From 1800 to 1820, a new synthetic metal material was generated. It was made by 17% zinc and 83% copper, replacing the gold and soon appreciated by the nobles. At that time, Tiffany, Bvlgari, Cartier were founded. They turned to migrate to the contemporary art design from traditional hand making technique. These brands are still active nowadays. At the end of 18th century, industrial revolution took place, the applying of industrial machines changed not only the production methods, but brought great impact on the social relationships, living styles and cultures, etc. Jewelries were no longer the priority of the nobles and royal members, the upper and middle classes had the capability to own their jewelries.

十、英国"手工艺运动"对首饰的影响
J. The impact of english 'Hand making activity'

"手工艺运动"是"新艺术运动"的先导。1851年英国举办了第一次万国博览会，展示了工业革命的成果，同时也让一些先觉的知识分子发现，工业化批量生产导致家具、室内产品、建筑设计的水准明显下降。相对于手工制品，机器生产的产品似乎千篇一律的造型。于是，他们发起了改变这一现状的"手工艺运动"，这时珠宝又开始回归手工为本的方向。

"Arts and crafts movement" led the "Art Nouveau activity". In 1851, the Expo was first held in England, for demonstrating the achievement of industrial revolution. Meanwhile some sensitive intellectuals noticed that the quality of furniture, interior design, architecture design seriously fell due to the standardized industrial manufacture. Comparing to hand made products, these industrially manufactured ones appeared to be stereotyped. Thus, they held the "hand making activity" to change that situation. Then jewelries had returned to the hand making.

十一、新艺术主义与装饰主义时期
K. The Art Nouveau and Art Deco Time

新艺术主义运动是19世纪80年代初在"手工艺运动"作用下，影响整个欧洲乃至美国等许多国家的一次大规模艺术运动。

这一时期的艺术家们，从自然形态中吸取灵感，以蜿蜒的纤柔曲线作为设计创作的主要设计语言。藤蔓、花卉、蜻蜓、甲虫、女性、神话等等成为艺术家常用的主题。当时在他们的作品中，表现出一种清新的、自然的、有机的、感性的艺术风格，被称为"新艺术风格"。此时，贵重宝石使用较少，钻石也只是起到辅助性的作用，而玻璃、牛角和象牙等被广泛使用。这是新艺术主义时期首饰作品的重要特征之一。艺术家对于自然生动而别具情趣的刻画，加上工艺极其精湛，使首饰作品的装饰效果不仅在视觉上呈现出华丽的美感，且更具有意境和传情的气质。

Art Nouveauis is a large scale art movement led by "Arts and crafts movement" impacting the whole Europe and many nations like U.S.A. in early 1980s'.

The artists at that time, sought the inspiration from nature. The delicate and elegant curved lines are the mainly used in design. Ivy, flowers, dragonfly, beetle, female, legends are the commonly used themes by artists. A fresh, natural, organic sensible art style was presented, called "Art Nouveauis". At that time, precious gems were rarely use. Diamond was only for support, while glass, ivory, horn were widely used. That was one of the significant features of "Art Nouveauis". The vivid and natural presentation by artists, plus the extremely refined techniques, delivered a splendid beauty, as well as the touching and emotional atmosphere by the jewelry.

十二、现当代首饰（1910年至今）
L. The contemporary jewelry (1910 till now)

20世纪以后首饰没有统一的造型和风格。是因

为以材料探索的形式来突破，使其归于当代首饰艺术范畴；另一条表现主线则是时装首饰。时装首饰一般是在首饰金属架上镶嵌玻璃、白铁矿石或其他材料制成。它与时装配套，流行周期短暂，不具有保值价值，只具有装饰价值。现代首饰创作在很大程度上摆脱了传统首饰严密、繁复的工艺程序，使首饰变得更为自由、简洁，使其创作主题、材料选择等都很有时代感。在西方，现代首饰逐渐成为现代艺术的一部分，因而现代首饰作品中，多样的材料和技法的创新为首饰设计带来更多的灵感。伦敦和巴黎是当今世界主要的首饰生产中心。

After 20th century, no standardized design or style could be applied on jewelry, for the breakthrough of the exploration of the materials, is categorized into jewelry art. The other main category is the fashion jewelry. Usually glass, marcasite, or other materials are set on the metal frames. They are set with fashion, with short life cycle, and only with decoration function without value for preservation. The contemporary jewelry design is exempt from the complicated technical protocols, which enables the simpler and more free-form jewelry design. The theme and material selection bears the contemporary sense. In western, contemporary jewelry is embedded into part of the contemporary art. Various innovations in materials and techniques promote more inspiration in design. London and Paris are the main jewelry manufacture center nowadays.

第二节　中国珠宝首饰发展简史
Section 2　The brief overview of Chinese jewelry

　　在原始社会，就有中国玉首饰文化出现在人们的生活中。而金银饰品的制造，是从商代开始。在唐代，已开始出现金、银、珍珠、宝石等材料，结合工艺使首饰文化达到鼎盛时期。明清年间，中国的珠宝首饰的制作达到了一个新的高度，尤其是皇室用的金银饰品所用的材料、设计、制作等，形成了完善的体系。进入了现代，中外珠宝文化的交融，使中国珠宝首饰由最初的巫术与偶像崇拜，发展为人们生活中的时尚点缀。

From the primitive society, jade jewelry has been staying in Chinese culture. The manufacture of gold and silver jewelry started from Shang dynasty. In Tang dynasty, the technique of combining gold, silver, pearl, gem had emerged, reaching the peak time of jewelry culture. In Ming Qing dynasty, another great standard of jewelry manufacture had been achieved, especially the royal gold, silver jewelry. The material, design, and manufacturing elaborates into a optimal system. Currently, oriental and western jewelry cultures get blended to make the Chinese jewelries turn to be the fashion accessories from the original witchcraft and idol worship function.

一、商周时期：小巧简约
A. Shang and Zhou time: simple and refined

　　商朝是中国迄今在考古发掘中发现最早的黄金制品的年代，距今已有3000余年的历史。商朝金器的分布范围，主要是以商文化为中心的中原地区，以及商王朝北部、西北部和偏西南的少数民族地区（在今天的河南、河北、山东、内蒙古、甘肃、青海及四川等地）都曾发现过这一时期的金器。这个时期的金器，形制工艺比较简单，器形小巧，纹饰少见，大多为装饰品。如为金箔、金叶和金片等，主要用于器物装饰。在商王朝北部和西北部地区的金饰品，是人们身上佩戴的黄金首饰。最令人瞩目的是四川广汉三星堆早期蜀文化遗址出土的商朝金器。不仅数量多，而且形制别具一格。其中颇为独特的是金面罩、金杖和各种金饰件。金银器早期的发展情况，也反映了中国早期文明、文化发展的多元性和不平衡性。

Shang dynasty is the earliest time with more than 3000 years' history recorded in the Chinese archaeology history for the discovery of gold wares. The distribution area of these gold wares is in the central, which is the center of the Shang Culture and north , north west and south west tribal area (now Henan, Hebei, Shandong, Interior Mongolia, Gansu, Qinghai and Sichuan, etc). The gold wares at that time—were small sized with simple molding technique and rare patterns. Most of them were for decoration, like gold film, gold leaf and gold slice. The gold wares found in north and north west part, Shang dynasty were for wearing. The most distinguished ones are from Sanxingdui in Sichuan Guanghan area. The amount is huge and the style is unique like gold mask, gold stick and various gold accessories. The early situation of gold an silver wares demonstrates the multi-cultures and unbalance in the ancient Chinese civilization.

二、魏晋南北朝时期：清新活泼
B. The Wei Jin, Northern and Southern dynasties: Fresh and lively

魏晋南北朝时期妇女的首饰以假髻、步摇为多，俗称珠松。当时手镯的使用已比较普遍，有条脱、跳脱、腕阑、臂钗等多种名称，造型也很丰富。

魏晋南北朝，各类金饰逐渐增多，如河北磁县东魏茹茹公主墓、太原北齐东安王娄睿墓所出土金饰，都是制作精良，式样新颖。指环在魏晋南北朝时期流行已较普遍，江苏宜兴晋墓和辽宁北票房身村晋墓出土的金指环，环面一头狭一头宽，在宽的环面上凿出点纹，既可装饰，又可在缝衣时作顶针之用。

Females in Wei Jin, Northern and Southern dynasties tended to wear fake hair knot, step swings also called Zhu Song. Bracelets were popular at that time with various names and shapes.

At that time, various gold jewelries grew common. For example, the gold jewelries discovered from the princess Ruru of East Wei's tomb in Ci county Hebei, Dongan King Lou Rui of North Qi's tomb in Taiyuan were all finely made with novel style. The rings were quite popular in that time. The gold rings discovered from Jin tomb of Yixing Jiangsu and Beipiaofangsheng Village Liaoning were wide one side and thin on the other side. The dots patterns were made on the wide ring side for both decoration and sewing purposes.

三、隋唐时期：绚丽多彩
C. The Sui Tang dynasty: vividly colorful

隋唐妇女的发髻式样比较丰富，因此在妇髻上配有各种的首饰，常见的有梳、篦、簪、钗、步摇、翠翘、珠翠金银宝钿、搔头等。插载的钗梳多至十数种，除了金银镶玉的髻钗外，名贵的象牙也被用于制翩钗之用。当时，金粒镶嵌工艺已从黑海沿岸的希腊地区传到中国：用细小的金颗粒镶嵌在光滑或浮雕金属的表面，以形成各种图案的装饰艺术，这种工艺被广泛地应用在唐代的首饰制作中。

The hair knot styles of women in Sui Tang dynasty were variously rich, so the jewelries for hair were diverse such as comb, step swings comb, grate, hairpin, step shake, green forsythia, gold and silver prohibition, titillation. There were more than 10 categories of forks. Apart from gold and silver, ivory was also applied. At that time, the gold dot setting technique had been introduced to China from Greece along the Black Sea: tiny gold sands were set on the smooth surface of the metal to form various pattern for decoration. This technique had been widely applied in Tang's jewelry manufacturing.

隋唐时期项链的运用也很普遍，以金、玉等材料为多，较为典型的是在西安出土的隋大业四年的一条金项链，项链的"链条"由28颗金珠构成，上有镶刻着鹿纹的蓝色宝石搭扣，下有双层项坠，制作极其精致，在金玉宝石的搭配上具有良好的效果，高贵富丽，体现出隋代的细金属工艺的高超水平。

Necklace was very popular in Sui Tang dynasty made by gold and jade commonly. One typical necklace was the gold necklace of 4th year after Sui dynasty, discovered in Xi'an. The necklace was composed of 28 gold beads with blue gem buckle carved with dear pattern. Double pendant was added with extremely refined manufacturing. The noble and luxurious effect of combining jade and gold was fully achieved to demonstrate the high level of thin metal manufacturing in Sui dynasty.

四、宋元时期：清丽典雅
D. The Song Yuan dynasty: elegant

宋元时期，受"程朱理学"的影响，首饰从唐代的富贵繁华风格锐减为清冷消瘦的宋代文人风格，植物纹样饰品比较常见，尤以松、竹、梅等象征气节的植物为多。

In Song Yuan dynasty, effected by 'Cheng Zhu philosophy', the style of jewelries turned sharply from luxurious to simple and thin intellectual style. The plants pattern was usual, especially common in seasonal plants like pine, bamboo, plum blossoms.

宋元女子金银首饰的基本构成，为冠梳、钗簪、耳环及钏镯、戒指、帔坠。金银首饰的纹样构成，除传统的龙凤和螭虎之外，所取用的多是清新俊丽并且很生活化的物象，如牡丹、莲花、蝴蝶、鸳鸯等，以此来表现丰盈和谐的情致。其实这些素材在唐代和辽代的艺术品中已经出现，宋元时代则以新的造型将其重新组织为各种图式，且灵活自然的运用，使之成为流畅的艺术新样式。

The silver and gold jewelries for women in that time were crown comb, Hairpin, earrings, bracelets, rings, frame droppings. Apart from the traditional patterns like dragon, phoenix, tiger, the fresh and life styled patterns were commonly used, like peony, lotus, butterfly, mandarin duck, to present the harmonious and enriched atmosphere. In fact, these themes had been appeared in Tang and Liao dynasty, in Song Yuan dynasty, they had been reorganized into various patterns with flexible application into smoother new art styles.

五、明清时期：华丽繁缛
E. The Ming Qing dynasty: luxuriously complicated

明清以后的首饰，多为大家所熟悉，出土的首饰以及流传下来的首饰很多。明清时期的项圈、项链、缕络等首饰，制造得都很讲究，有金制、以金包玉、在金上镶嵌宝石等方式。有的还在金项圈上附加一些长穗和垂件，妇女、男子均有佩戴。

The jewelries after Ming Qing dynasty were familiarized. The amount of the jewelries left or discovered was huge. The manufacturing of necklace, necklet, complex ray is refined: gold, gold covered with jade, gold set with jade. Some gold necklet were added with long droppings and pendants for both women and men.

从制作及艺术风格上看，明清时期的首饰有两个相反的特点：一是复杂烦琐，集各种名贵材料于一体，加以金为件，在其上镶嵌珠宝;有的以玉为针，包金镶银，精雕细刻，还附加复杂的垂饰;二是极为简朴，不在金、银坯上加饰任何纹样和装饰，金银圈或玉环由本身材料的质地展示出自身美感。

From the art styles and manufacturing, there were two opposite features for Ming Qing dynasty: one was extremely complicated with various precious materials based on gold. Some were set with jewels. Some were needles in jade, covered by gold or set with silver with refined carving and sculpturing. Besides, complex pendants were attached. The other feature was extremely simple. No pattern or decoration was added on silver or gold mold. Gold and silver or jade rings were presenting their beauty by their own textures.

明代以后，玉石在首饰中的作用更加重要，特别是白玉，一直是人们欣赏的首饰材料。明清时期的玉石首饰通常采用"深浮雕"的方法，充分利用了各种玉石的温润的特殊视觉效果，再装饰上各种动物和花卉图案，给人以风雅得体的感觉。

After Ming dynasty, the jade enjoyed even more importance in jewelries, particularly the white jade, always the preferable jewelries materials. The 'deep sculpture' technique was applied on jade jewelries in Ming Qing dynasty fully exerting the special gentle visual effects of jade. The appropriately elegant atmosphere was achieved with the addition of various animal and floral patterns.

总之，清代的金银首饰形式丰富多彩，技艺精湛。其制作工艺包括了浇铸、锤鍱、炸珠、焊接、镌镂、掐丝、镶嵌、点翠等，并综合了起突、隐起、阴线、阳线、镂空等各种手法。应该说，清代金银工艺的繁荣，不仅继承了中国传统工艺技法而又有所发展，并且为今天金银工艺的发展创新奠定了扎实的基础。

After all, the forms of gold and silver jewelries in Qing dynasty were various with sophisticated techniques. These techniques were casting, plating, beads cracking, welding, engraving, setting, dot coloring, etc. The methods applied were bulging, hiding, inside or outside lining, hollowing, etc. The prosperity of the techniques in Qing dynasty was developed based on the inherited traditional craftsmanship, set the solid base for the future development.

第三节　中国古代首饰分类
Section 3　The classification of Chinese ancient jewelries

一、笄【jī】
A. Ji

笄是古人用来簪发和连冠的饰物，是簪、钗的鼻祖。古代男女均留长发，笄为古代男女用来插定绾起的头发或弁冕的。"弁"为古代冠名。"冕"即皇冠。固定冠帽的笄称为"衡笄"，周代设"追师"的官来进行管理。用来固定发髻的笄称为"鬠笄"。从周代起，女子年满十五岁便算成人，可以许嫁，谓之及笄。如果没有许嫁，到二十岁时也要举行笄礼，由一个妇人给适龄女子梳一个发髻，插上一支笄，礼后再取下（图3-20、图3-21）。

Ji is the accessories for hair bum and hat, the original form of hair pin, hair clasp. Men and women all grew long hair in ancient time. Ji is for holding and inserting into the hair bum. Bian' is for the ancient crown while 'Mian' is for royal crown. The hair pin for fastening the crown is called 'Heng Ji'. In Zhou dynasty, certain officers were arranged to manage that. From Zhou dynasty, women aged 15 years were legal to marry, it was called 'Ji reach'. If she was not married, the Ji ceremony would be held at her 20 year age: an appropriately aged woman combed a hair bum for her, inserting a Ji, and took off Ji after the ceremony (Figure 3-20、Figure 3-21).

图3-20　商代骨笄
Figure 3-20　The Ji of bones in Shang dynasty

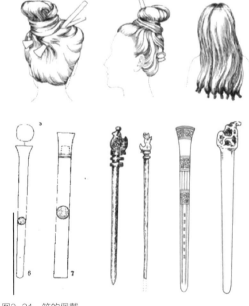

图3-21　笄的佩戴
Figure 3-21　The wearring of Ji

二、簪【zan】
B. Zan (hairpin)

簪是笄的发展，同样起固定发髻之用，在前端加以纹饰，雕刻成植物（花草）、动物（凤凰孔雀）、吉祥器物（如意）等形，并可用金、玉、象牙、玳瑁等贵重材料制作，工艺也愈发丰富，有錾花、镂花及盘花等（图3-22～图3-25）。

Zan develops from Ji, also used for fastening the hair bum. The rear part is decorated with patterns or sculptured into plants (grass and flowers), animals (Phoenix, peacock), lucky wares (Ruyi,), in precious materials like gold, jade, ivory, turtle shells, etc. The techniques are richer, like carving, engraving etc (Figure 3-22~Figure 3-25).

三、钗【chai】
C. Chai (Hair Fork)

钗由两股簪子交叉组合成的一种首饰。钗用来绾住头发，也有用它把帽子别在头发上。发钗的安插有多种方法，有的横插，有的竖插，有的斜插，也有自下而上倒插的。所插数量也不尽一致，既可安插两支，左右各一支；也可插上数支，视发髻需要而定，最多的在两鬓各插六支，合为十二支（图3-26、图3-27）。

Chai is composed of two crossing hairpins for holding the hair, sometimes for clipping the hat on the hair. The wearing ways are multiple: vertical wearing horizontal wearing oblique wearing, or reversed wearing. The number is not uniformed, two can be worn while one on each side, multiple forks

图3-22 永泰公主墓出土鎏金银钗
Figure 3-22 The gold silver fork from Princess YongTai's tomb

图3-23 敦煌壁画中晚唐妇女头饰
Figure 3-23 The hair accessories of late dyny from Dunhuang fresco

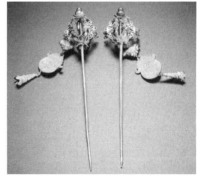

图3-24 清 金制杖形发簪一对
Figure 3-24 A pair of gold stick shaped hair pins, Qing dynasty

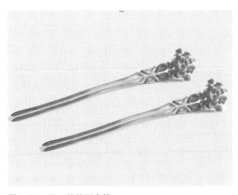

图3-25 元 莲花形金簪
Figure 3-25 Lotus shaped gold hair pin, Yuan dynasty

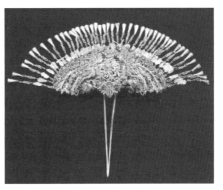

图3-26 苗族银凤钗
Figure 3-26 Silver Phoenix hair fork of Miao Tribe

图3-27 银点翠双鱼蝶恋花头钗
Figure 3-27 Butterfly on flowers hair forks

can be worn, at most six forks can be worn on each side of the head, total twelve forks (Figure3-26、Figure 3-27).

四、步摇【buyao】
D. Buyao (step swings)

步摇是在顶部挂珠玉垂饰的簪或钗，是古代妇女插于鬓发之侧以作装饰之物，同时也有固定发髻的作用。一般形式为凤凰、蝴蝶、带有翅膀类的，或垂有旒苏或坠子，走路的时候，金饰会随走路的摆动而动，栩栩如生。取其行步则动摇获名（图3–28）。

The step swings are the hair pin or hair forks with the pearl, jade droppings on the neck part, as the decoration on the hair side, also for fastening the hair bun. The forms are usually the phoenix butterfly, or the animals with wings, or with the droppings. So it will vividly swing when walking, thus the name is given (Figure 3-28).

五、扁方
E. Bian Fang

扁方是满族妇女梳旗头时所插饰的特殊大簪，形制与作用与汉人妇女髻上的扁簪类似。清代贵族妇女梳"两把头"或"大拉翅"，都使用扁方，起到连接真、假发髻之中"梁"的作用。装饰的同时更重要的是控制发髻使其不散落下来（图3-29、图3-30）。

Bian Fang is the special hair pin for women of Man Tribe to comb Qi style hairstyle. The form and function is similar to the women of Han's flat hair pin. It serves as the bridge to connect the real and fake hair bum, to hold the hair not getting loose and dropping (Figure 3-29、Figure 3-30).

图3-28 唐 金镶玉步摇
Figure 3-28 Jade step swings set with gold,Tang dynasty

图3-29 清代金镂空镶珠扁方 故宫博物院
Figure 3-29 The gold hollowBian Fang of Qing dynasty set with pearl. Forbidden Palace Museum

图3-30 扁方佩戴图
Figure 3-30 The wearring of Bian Fang

六、梳篦【shū bì】
F. Shu Bi(Comb)

古代简称"栉"，与簪、髻、钗、步摇等并称为中国古代八大发饰之一，昔为宫廷御用珍品，故有"宫梳名篦"之誉。梳篦在古时是人手必备的发饰，尤其是妇女，几乎梳不离身，时间一久，便形成插梳的风气（图3-31～图3-33）。

Shu Bi was also named "zhi", one of eight ancient hair decorations like Zan, Ji, Chai, and Bu Yao. It used to be recognized a treasure in Palace, thus enjoyed the fame of "Palace Comb". Shu Bi was a daily necessity for women in ancient time. With time passing by, the style of inserting combs in hair was formed (Figure3-31~Figure 3-33).

七、华胜
G. Hua Sheng

华胜是古代妇女的一种花形首饰。为华丽的首饰，华丽的头饰（图3-34、图3-35）。

Hua Sheng is a floral shaped luxurious jewelry wearing on head for ancient women (Figure 3-34、Figure 3-35).

图3-31　朴素的宋元梳篦
Figure 3-31　The simple styled comb in Song Yuan dynasty

图3-32　宋代玉梳篦
Figure 3-32　Jade comb in Song dynasty

图3-34　华胜
Figure 3-34　Hua Sheng

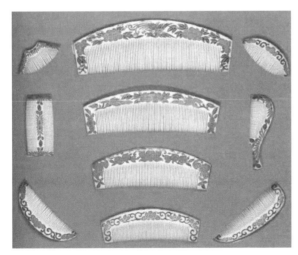

图3-33　清代梳篦
Figure 3-33　Combs of Qing dynasty

图3-35　华胜佩戴图
Figure 3-34　The wearring of Hua Sheng

八、抹额
H. MO E (Frehead Bnd)

抹额，也称额带、头箍、发箍、眉勒、脑包，汉族服饰，明代较盛行。妇女包于头额，束在额前的巾饰，一般多饰以刺绣或珠玉（图3-36～图3-39）。

Mo E, also called forehead band, hair band, eyebrow band, is the Han styled fashion, popular in Ming dynasty, for women covering forehead. It is a cloth accessories set with embroidery, jade or pearls (Figure 3-36~Figure 3-39).

图3-36 唐代女子戴透额罗敦煌壁画
Figure 3-36 The women with transparent Mo E

图3-37 清代女子画
Figure 3-37 Qing dynast

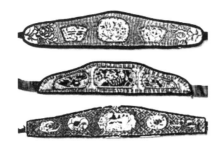

图3-38 抹额形式
Figure 3-38 Erasure form

图3-39 清代点翠抹额
Figure3-39 Mo E in Qing dynasty

九、花钿【diàn】
I. Dian

花钿是古时妇女脸上的一种花饰。花钿有红、绿、黄三种颜色，以红色为最多，以金、银制成花形，蔽于脸上，是唐代比较流行的一种首饰。花钿的形状除梅花状外，还有各式小鸟、小鱼、小鸭等，十分美妙新颖（图3-40、图3-41）。

Dian is a kind of decoration on female's face in ancient time. There are three colors for it: red, green and yellow while. red is the most popular. It is made of gold or silver into floral shape, pasted on face, popular in Tang dynasty. Besides the shape as plum blossoms, the other appealing shapes are like birds, fish, and ducklings (Figure 3-40、Figure 3-41).

图3-40 《侍女图》唐 花钿
Figure 3-40 *portraits of ladies*, Tang dynasty, Dian

图3-41 唐朝流行的花钿款式，消失于元朝
Figure3-41 The popular styles of Dian in Tang dynasty, disappeared in Yuan dynasty

十、珥珰【ěr dāng】
J. Er Dang(Earring)

耳坠，俗称耳环，古代称它为"珥""珰"等。

耳坠是国人佩戴历史最悠久、最普及的一种饰物，至今长盛不衰（图3-42）。

Earrings are called 'Er', 'Dang' in ancient time. It has been the most preferable jewelries ever since with longest history for Chinese people (Figure 3-42).

十一、玉玦【yù jué】
K. Yu Jue

是我国最古老的玉制装饰品，为环形形状，有一缺口。在古代主要是被用作耳饰和佩饰。据说古人饰玉有两个含义：一是表示有决断性；二是用玉玦表示断绝之意（图3-43）。

It is the oldest jade accessories, ring-shaped with a gap, and used as the accessories for ear or wearing. Two meanings were given by the ancient people to this: one indicated the ability of decision, the other meant breaking off (Figure 3-43).

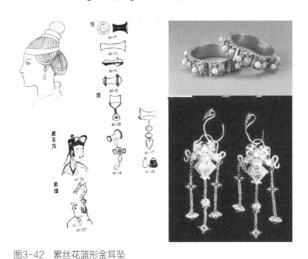

图3-42　累丝花篮形金耳坠
Figure 3-42　Flower basket shaped earrings

图3-43　汉代玉玦故宫博物院
Figure 3-43　Yu Jue of Han dynasty, the Palace Museum

十二、项圈
L. Necklace

颈饰是原始社会就很普遍的装饰，春秋战国时代的颈饰出土不少，河南三门峡上村岭虢国墓地出土颈饰，其中一件系由许多不同形状的=组成，其余为=形器及一小系璧（图3-44）。

Necklace has been popular since primitive society. Quite a few necklaces of Spring and Autumn dynasty were discovered. The one was discovered of Guo nation tomb from Shangcun Ridge in Sanmenxia, Henan was composed of different = shaped connected by a small piece of jade (Figure 3-44).

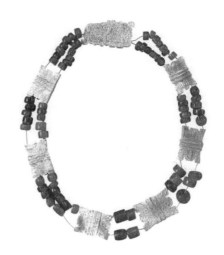

图3-44　虢国墓地出土颈饰 商
Figure 3-44　Necklace from Guo nation tomb, Shang dynasty

十三、璎珞【yīng luò】
M. Ying Luo

古代用珠玉串成的装饰品，多用为颈饰，璎珞原为古代印度佛像颈间的一种装饰，后来随着佛教一起传入我国，唐代时，被爱美求新的女性所模仿和改进，变成了项饰。它形制比较大，在项饰中最显华贵（图3-45）。

It is the accessories threaded in jade and pearl in ancient time, common for necklaces, It used to be the accessories for ancient Indian Buddha, and later was introduced to China with Buddhism. In Tang dynasty, it was adjusted by the fashionable women into the neck accessories. The size is large. It is the most noble one in neck jewelries (Figure 3-45).

十四、胸饰
N. Brooch

佩戴一枚胸针，常可产生画龙点睛的效果，尤其当衣服的设计比较简单或颜色比较朴素时，别上一枚色彩鲜艳的胸针，就会立即使整套装饰活泼起来，并具有动感（图3-46）。

Wearing a brooch could achieve a dramatically vivid effect, especially when the design and color of the clothes are simple. A brightly colorful brooch would dynamically lighten the whole set of clothes fashion (Figure 3-46).

十五、腰饰
O. The waist jewelries

主要包括玉佩、带钩、带环、带板及其他腰间携挂物。材料一般以贵金属镶宝石或玉石居多。我国早期的腰饰主要是玉佩，即挂系腰间的玉石装饰物。玉佩在古代是贵族或做官之人的必佩之物（图3-47）。

Jade plate, belt hook, belt ring and the others are the main accessories for waist. The materials normally are precious metal set with jade or jem. In the early time, the jade plate was hanging on the waist. It is the main waist jewelry in China. It is the necessary for ancient nobles and authorities (Figure 3-47).

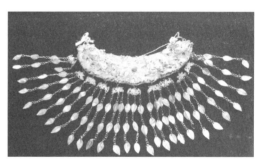

图3-46 水族龙凤戏珠坠须银胸饰
Figure 3-46 Dragon and phoenix with pearl dropping brooch

图3-45 璎珞示图
Figure 3-45 Ying Luo

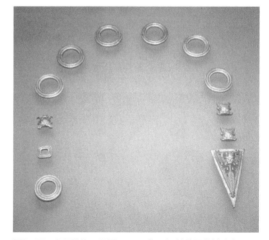

图3-47 金腰带饰 西周 1990年三门峡虢国墓地出土
Figure 3-47 Gold belt accessories West Zhou dynasty discovered in 1990 Guo nation tomb in Sanmenxia

十六、禁步
P. Jin Bu (step cautious)

古代的一种饰品。将各种不同形状玉佩，以彩线穿组合成一串系在腰间，最初用于压住裙摆。佩戴行步之时，发出的声音缓急有度，轻重得当。如果节奏杂乱，会被认为是失礼，古人对此是十分注重的（图3-48）。

It is an ancient jewelry, threaded with various shaped jade in a colorful string, bound on the waist. At the early time, it was for holding the skirt. It would sound hasty or slow with appropriate volume when walking. If the rhythm sounded messy, it would be regarded as lack of manner, as the ancient people cared much about it (Figure 3-48).

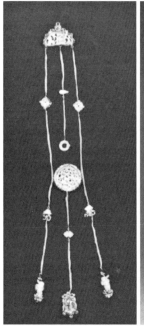
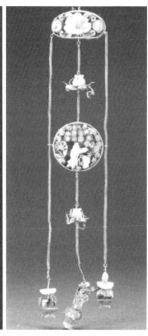

图3-48 明代 金镶玉禁步
Figure 3-48 Gold set with jade Jin Bu, Ming dynasty

十七、臂钏【bì chuàn】
Q. Bi Chuan (Armlet)

臂钏，一种套在上臂的环形首饰，特别适合于上臂滚圆修长的女性，能够表现女性上臂丰满浑圆的魅力。西汉以后，佩戴臂环之风盛行，臂环的样式很多，有自由伸缩型的，这种臂环可以根据手臂的粗细调节环的大小。臂钏是最女性化的首饰，只能被女子并且是大臂粗的女子佩戴（图3-49、图3-50）。

Armlet is a ring shaped accessory for upper arm, especially for the round long armed woman to highlight her round arms' charm. After West Han dynasty, wearing armlet turned popular. The styles are various, extendable and flexible for adjusting the arms' thickness. It is only for thick armed woman (Figure 3-49、Figure 3-50).

图3-49 镶金镶宝玉臂钏
Figure 3-49 Jade armlet set with gold

十八、手镯
R. Bracelet

手镯是一种套在手腕上的环形饰品。按结构，一般可分为两种：一是封闭形圆环，以玉石材料为多；二是有端口或数个链片，以金属材料居多。按

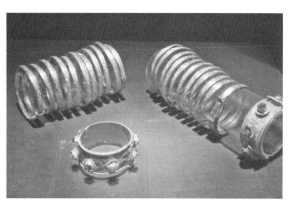

图3-50 唐代臂钏
Figure 3-50 Armlet of Tang dynasty

制作材料，可分为金手镯、银手镯、玉手镯、镶宝石手镯等（图3-51）。

Bracelet is a ring-shaped accessory for wrist. Classified by structure, bracelet usually can fall into two categories: one is the enclosed ring, mainly using jade as materials, and the other is a ring with connectors and pieces, chiefly using metal as materials. Classified by materials, there are gold bracelet, silver bracelet, jade bracelet, jewelry bracelet etc (Figure 3-51).

十九、戒指
S. Ring

我国大约在距今4000多年前就已有人佩戴戒指，据现存文献记载，它有"手记""约指""驱环""代指""指环"等诸多异名。而这些异名中数"指环"使用的频率最高，使用的时间最长。直到明代以后，戒指的称呼才渐渐多起来（图3-52）。

Ring wearing was recorded 4000 years ago in China. There are various names for it in the documents recorded. The name 'ring' is the most frequently used till now (Figure 3-52).

二十、指甲套
T. Nail shield

清代服饰贵族女子留长指甲，清代的指甲套，也称"护指"。

以凤仙花或指甲花染指甲的风气由来已久。清代用金银做成指甲套，纹饰极为精美华丽，种类丰富。后宫嫔妃用的就非常普遍了，样式也增加了许多。中国古代妇女蓄指甲及饰戴护指套的传统由来已久，现在能看到的最早的护指套是吉林省老河深地区出土的汉代金护指。它是由薄薄的金片卷曲而成，呈螺旋状向上延伸，粗细长短可任意调节，简练而实用（图3-53、图3-54）。

Noble women of Qing dynasty grew long nails. The tradition of dyeing the nails by nail flowers had been lasted for a very long time. In Qing dynasty, the nail shield was made by gold and silver with refined pattern in various styles. It was also called nail shield in Qing dynasty.

It is very popular among the concubines with many additions in styles. The tradition is lasted for a very long for growing nails and put on nail shieldby Chinese ancient women. The earliest gold nail shield of Han dynasty discovered is from deep river valley in Ji Lin province. It is made by thin gold film folding, extending upwards spirally. The length and thickness are adjustable, simple and applicable (Figure 3-53、Figure 3-54).

图3-51　珊瑚手镯　台湾故宫博物院
Figure 3-51　Coral bracelet, Palace Museum Taiwan

图3-52　镶嵌戒指　清　台湾故宫博物院
Figure 3-52　Ring with settings of Qing dynasty, Palace Museum Taiwan

图3-53　清　点翠镶宝石与米珠镂空指甲套
Figure 3-53　Nail shield with gem and tiny pearls of Qing dynastythe simple styled comb in Song Yuan dynasty

图3-54　清　竹叶纹镂空指甲套
Figure 3-54　Nail shield in bamboo leave pattern of Qing dynasty

第四节　首饰设计
Section 4　Jewelry Design

一、常用首饰的分类
A. Classification of Commaonly used jewelry

根据首饰的应用部位、工艺、应用，可以做以下的分类：

According to the parts, processes and applications of jewelry, the following classifications can be made:

1. 按人体装饰部位分类
Classification by the body parts

（1）发饰：包括发卡，钗等。

Hair accessories: including cards, hairpin, etc.

（2）冠饰：冠，帽微。

Crown ornaments: crown, cap micro.

（3）耳饰：耳钉，耳环，耳线，耳坠。

Earrings: Ear studs, earrings, ear lines, earrings.

（4）脸饰：包括鼻部在内的饰物（多见印度饰物）。

Face decoration: ornaments including the nose (mostly Indian ornaments).

（5）颈饰：包括项链，项圈。

Neck ornaments: including necklaces and collars.

（6）胸饰：吊坠，链牌，胸针，领带夹。

Chest ornaments: pendants, chain cards, brooches, tie clips.

（7）手饰：包括戒指，手镯，手链，袖扣。

Jewelry: including rings, bangle, bracelets, cufflinks.

（8）腰饰：腰带，皮带头。

Waist ornament: belt, belt head.

（9）脚饰：脚链，脚镯等。

Foot ornaments: Anklets, ankle bracelets, etc.

2. 按工艺手段分类
Classification by process

（1）镶嵌类。

Mosaics.

①高档宝玉石类：钻石，翡翠，红蓝宝，祖母绿，猫眼，珍珠。

High class gemstones: diamonds, jade, red saphires, emeralds, cat's eyes, pearls

②中档宝玉石类：海蓝宝石，碧玺，丹泉石，天地然锆石，尖晶石等。

Mid-grade gemstones: aquamarine, tourmaline, tanzanite, zircon, spinel, etc.

③低档宝玉石类：石榴石，黄玉，水晶，橄榄石，青金石，绿松石等。

Low-grade gemstones: Garnet, Topaz, Crystal, Peridot, Lapis, Turquoise, etc.

（2）非镶嵌类：素金首饰。

Non-inlaid: gold jewelry.

3. 按用途分类
Classification by purpose

（1）流行首饰。

Fashion Jewelry.

①大众流行：普遍追求商品化的首饰。

Popular pop: The ubiquitous pursuit of commercial jewelry.

②个性流行：个体追求艺术性，个性化的首饰。

Personality Popularity: Individuals pursue artistic and personalized jewelry.

（2）艺术首饰。

Fine jewelry.

①收藏：出之大师的作品，不易佩戴，具有藏用价值。

Collection: The master's works are not easy to wear and have hidden value.

②摆件：供摆设陈列之用。

Decoration: For display purposes.

③佩戴：实用化的艺术造型首饰。

Wear: Practical art jewelry.

二、首饰设计元素
B. Jewelry design elements

1. 灵感来源与素材搜集整理
Collecting and sorting inspiration sources and materials

设计灵感是任何一名设计师在创作初期，会不断寻找且不可或缺的要素。它或许是一种情感、一段经历、某个特殊的地方，或者大自然中存在的某种颜色、物品的肌理、质感，甚至是一种特殊的味道或声音等。这些都能给设计师带来创作的艺术灵感。

Design inspiration is an element that any designer will keep looking for in the initial stage of creation. It may be an emotion, an experience, a special place, or a color, texture, or even a special taste or sound that exists in nature. These all can bring the artistic inspiration of creation to the designer.

如果仔细分析，灵感来源并非空穴来风，而是有其产生的思想轨迹。通常，灵感来源来自生活中的观察与思考，是对周围事物耳闻目睹的反映。在首饰艺术家和设计师形成自己的创作风格过程中，能够认识到这种轨迹非常重要，如果不能明白灵感的来源，就不能很好地运用形、色、体等设计手段很好地表达设计想法。每个物体都具有无限的可能性和很多种观察和认知的方法，关键是怎样明白它所承载的"语言"，也就是如何能恰如其分地表达作者的意念和思想（图3-55）。

If analyzed carefully, the source of inspiration is not groundless, but has its thought track. Usually, inspiration comes from observation and thinking in life, which is a reflection of the things around us. In the process of forming their own creation style, jewelry artists and designers can realize that this kind of trajectory is very important. If they cannot understand the source of inspiration, they cannot use shape, color, body and other design methods to express design ideas well. Every object has infinite possibilities and many ways of observation and cognition. The key is how to understand the "language" it carries, that is, how to properly express the author's ideas and thoughts (Figure 3-55).

在明白灵感来源的基础上，才能正确地使用设计语言。对造型、颜色、肌理、材料的把握会恰到好处，经过日积月累，设计师们就会在创作中逐步形成自己的创作风格。因此，设计师要做的就是成为一个有心人，拥有一双训练有素的眼睛，在日常生活中有意识地观察和搜集有用的素材，并进行及时的归纳和整理。比如，在自然界，通过搜集摄影或实物来寻找具有形、色、肌理、组织结构等特征的果荚、花朵、种子等；在生活中观察周围的景观、事物、人物、新闻、搜集废旧的有纪念意义的材料和物品；从人生经历和记忆中搜寻可供叙述的转化为视觉形态的故事等；从观看的影片、杂志、书籍中搜集可供使用的花样、图案、色彩样本等（图3-56）。

Only by understanding the source of inspiration can the design language be used correctly. Holding to modelling, color, flesh texture, material can be just right. Accumulating over time, stylist people can form his creation style gradually in the creation. Therefore, what the designer should do is to become a conscientious person, with a pair of well-trained eyes, consciously observe and collect useful materials in daily life, and timely summarize and organize them. For example, in nature, by collecting photographs

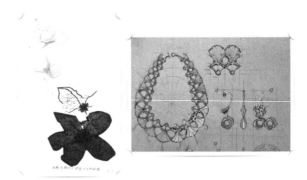

图3-55　上海视觉艺术学院　珠宝首饰设计专业学生　灵感来源图
Figure 3-55　Source of inspiration for svisual arts colleges tudents majoring in jewelry design in Shanghai

图3-56　上海视觉艺术学院 珠宝首饰设计专业学生 灵感来源搜集图
Figure3-56　Source of inspiration for students majoring in jewelry design in Shanghai visual arts college

or objects, we can find fruit pods, flowers and seeds with features of shape, color, texture and tissue structure, by observing the surrounding landscape, things, people, news, we can collect old and memorable materials and articles; by searching life experiences and memories for stories, we can narrat and translat into visual forms; and by watching movies, magazines, books, we can collect patterns, patterns and color samples (Figure 3-56).

总之,有了丰富的素材,设计时才会得心应手。平时注重观察和记录,设计师是能够处处发现设计灵感的,这是自己独有的设计语言。每个设计师都是独特的个体,各自带着独特的人生经历与成长环境,使设计作品具有旺盛的生命力。

In a word, only with rich materials, may the design be handy. Pay attention to observation and record at ordinary times, stylist is able to discover design inspiration everywhere, This is his unique design language. Each designer is a unique individual, each with a unique life experience and growth environment, so that design works have exuberant vitality.

同时设计要符合需求,即适用原则。设计要和谐、美观,是既要首饰自身各部分之间构建和谐,又要使首饰与佩戴者之间显现和谐、美观,这是在为消费群体设计时首先要考虑的。

At the same time, the design must meet the requirements, that is, the applicable principles. The design should be harmonious and beautiful. It is necessary to ensure the harmony between the various parts of the jewelry itself, and to make the jewelry and the wearer appear harmonious and beautiful. This is the first consideration when designing for the consumer group.

设计的成果体现,是在设计过程中,对所选材料是否合适,是否能达到设计预期的主题,运用材料的加工技术和工艺要求是否达到设计的目标。同时,设计既要考虑形式要素,也要考虑感觉要素;前者是设计对象的内容、目的及须运用的形态和色彩基本要素;后者是从生理学和心理学的角度对这些元素组合搭配的规律,最终是为了创造形象典雅,结构巧妙,色彩协调,给人美的震撼和享受的好的设计。

The result of the design is whether the design of the selected material is appropriate, whether it can meet the design of the intended theme, whether the processing technology and process requirements of the material meet the design goals. At the same time, the design must consider both the formal elements and the sensory elements; the former is the content of the design object, the purpose and the basic elements of the form and color to be applied; the latter is the combination of these elements from the perspective of physiology and psychology. In the end, it is to create an elegant image, a clever structure, a harmonious color, and a beautiful design that is shocking and enjoyable.

设计在最终变为实物过程中的可行性,包括所选材料是否合适于表现预期的实际主题,材料的加工技术和工艺要求是否达到。设计既要考虑形式要素,也要考虑感觉要素。前者指设计对象的内容、目的及必须运用的形态和色彩基本要素;后者指从生理学和心理学的角度对这些元素组合搭配的规律,最终都是为了创造形象典雅,结构巧妙,色彩协调,给人美的震撼和享受的好的设计。

The feasibility of designing in the process of eventually becomes physical, including whether the selected material is suitable for the actual theme of the expected performance, whether the material processing technology and process requirements are met. Design must consider both formal elements and sensory elements. The former refers to the content and purpose of the design object, the basic elements of the form and color that must be used, and the latter refers to the combination of these elements from the perspective of physiology and psychology. The ultimate goal is to create elegant images, ingenious structures, and coordinated colors, giving people a good design of shock and enjoyment.

2. 首饰形式美的表现

The performance of jewelry form beauty

（1）单纯齐一：这是最简单的形式美。在色彩和款式上表现为齐一或反复，风格上的一致性给人有序感，体现节奏感。

Simple and trim: This is the simplest form of beauty. In terms of color and style, it is homogeneous or repetitive, and consistency in style gives a sense of order and rhythm.

（2）对称均衡：对称是指形式中以一条线为中轴线，线的两端或左右相等或旋转相等的一种组合，它常给人稳定和庄重之感。均衡由对称演化而来，其中轴线两侧的形状并不完全相同，但视觉重量却相等或相近。与对称相比，它更自由和富有变化。一般首饰中的戒指，手链等物件的设计都要重视对称均衡，如中间镶较大的钻石，两侧镶等量的小钻石或有色宝石就是一种均衡；冷色的铂金镶冷色的蓝宝石就是一种均衡，暖色的K黄金配红宝石也是一种均衡。首饰的形体叠加，纹饰变换，节奏变化中仍要注意保持均衡，以体现首饰的稳定感。

Symmetry Equilibrium: Symmetry refers to the form in which a line is the central axis, and the two ends of the line are equal or equal to each other in a kind of combination. It often gives people a sense of stability and dignity. The equilibrium evolv from symmetry, where the shapes on both sides of the axis are not exactly the same, but the visual weights are equal or similar. Compared to symmetry, it is more free and varied. The jewels in general jewelry, bracelets and other objects must be designed with symmetrical balance. For example, if a large diamond is set in the middle, equal amounts of small diamonds or colored gems are balanced on both sides. The cool platinum sapphire is set with a cool sapphire. A balanced, warm-colored K gold with ruby is also an equilibrium. The shape of the jewelry is superimposed, the ornamentation is transformed, and the rhythm changes still need to be balanced to reflect the stability of the jewelry.

（3）调和对比：这是反映矛盾的两种状态。调和是在差异中趋于一致，对比是在相似中突出不同。调和是把相近的两者并列。如色彩中红与橙，橙与黄，青与紫等。在同一色彩中深与浅，浓与淡的组合也属于调和。对比是把极不相同的两者并列，使人感到鲜明，振奋，活跃。如构图的虚与实，形态的方与圆，位置的远与近，颜色中的黑与白等。过分的对比显得纷杂，刺激；而过分的调和则显得不够生动。

Reconcile the Contrast: This is the two states that reflect contradictions. Reconciliation tends to be consistent in differences, and contrasts are prominently different in similarities. Reconciliation is the parallelization of the two. such as the color red and orange, orange and yellow, blue and purple and so on. Deep and shallow in the same color, the combination of thick and light also belongs to reconciliation. The contrast is to juxtapose the two very different, making people feel bright, excited and active. such as the composition of the virtual and real, the shape of the square and circle, the location of the far and near, the color of the black and white and so on. Excessive contrasts appear to be mixed and irritating, while excessive reconciliation is not vivid enough.

（4）比例适当：指事物整体与局部或局部与局部的数量关系。如戒指的戒面大而托细，项链粗而吊坠小都属于比例失调而显得不美。

The proportion of appropriate: refers to the whole thing and the local or partial and local quantitative relationship. For example, the rings with large ring face and care fine, and thick necklaces with small pendants are all disproportionate and not beautiful.

（5）多样统一：多样是指形式中各个整体之间和整体中各个部分之间因差异而具有的一种组合。统一是指各个组成部分的协调及和谐。这是形式美法则的高级形式，产生的整体效果既体现事物的千差万别又体现事物的共性，首饰既表现的丰富生动，又富有秩序和规律而不杂乱。这在系列首饰或大件组合首饰设计中必须考虑。

Diversity and Unity: Diversity refers to a combination of differences between the various parts of the form and between the various parts of the whole. Unity refers to the coordination and harmony of various components. This is a high-level form of the rules of beauty. The overall effect produced reflects both the diversity of things and the commonness of things. Jewelry is both rich and lively, rich in order and law, and not cluttered. This must be considered in the design of series jewelry or large-scale combination jewelry.

三、首饰材料
C. Jewelry materials

首饰发展到现在，所用的材料也不断地在发展与创新，出现了除贵重金属以外的各种新型材料，极大地丰富了首饰的样式与风格，也满足了各个层面的消费需求。

首饰材料的演变包含了两个内容：一是对传统材料的创新研究，如各种彩色金的使用，各种有着优良品质的合金的采用；二是新颖别致的材料的使用，如皮革、纺织材料、通过高科技技术而获得的人造材料等。

Since the development of jewelry, the materials used have continued to develop and innovate, new types of materials have emerged in addition to precious metals, and the earth has enriched the style and style of jewelry, and has also met consumer needs at all levels.

The evolution of jewelry materials includes two aspects: first, innovative research on traditional materials, such as the use of various colored gold, the adoption of various alloys with good quality, and second, the use of novel and unique materials, like leather, textiles, materials, and man-made materials obtained through high-tech technology, etc.

1. 贵金属材料
Precious metal materials

首饰所用的传统金属，一般具备以下几个条件：金属外观需美观，不易被酸碱等化学物质所腐蚀；金属需耐用，要有一定的硬度，有优良的延展性；贵金属属于稀少品种，具有一定的收藏、保值价值。

随着科学技术的不断发展，由贵金属与普通金属制成的合金材料，现已大量使用，繁荣了首饰品的材料市场。

The traditional metals used in jewelry generally have the following conditions: The appearance of the metal must be beautiful, not easily corroded by chemical substances such as acid and alkali; the metal must be durable, have a certain hardness, and have excellent ductility; the precious metal is a rare species, with A certain collection, preservation value.

With the continuous development of science and technology, alloy materials made of precious metals and common metals have been used in large quantities and have prospered the market for jewelry materials.

（1）黄金。
Gold.

黄金是化学元素金（化学元素符号Au）的单质形式，是一种软的，金黄色的，抗腐蚀的贵金属。金是最稀有、最珍贵的金属之一。国际上黄金是以盎司为单位。中国古代是以"两"作为黄金单位，是一种重要的金属。它不仅是首饰业、电子业、现代通信、航天航空业等部门的重要材料，同时又是用于储备和投资的特殊通货。黄金的颜色为金黄色，金属光泽，难分解。硬度2~3，纯金19.3，熔点1064.4℃；具良好的延展性，能压成薄箔，具极高的传热性和导电性，纯金的电阻为2.4p。纯金具有良好的抗化学腐蚀性，是最好的电镀材料。黄金作为一种贵金属，有良好的物理特性，不容易熔化。其密度大，手感沉甸，韧性和延展性好，具有良好导性。纯金具有艳丽的黄色，但掺入其他金属，颜色变化较大，如金铜合金呈暗红色，含银合金呈浅黄色或灰白色。金易被磨成粉状，这也是金在自然界中呈分散状的原因，纯金首饰也易被磨损。

Gold is a simple form of the chemical element Au (chemical element symbol Au), a soft, gold-colored, corrosion-resistant precious metal. Gold is one of the rarest and most precious metals. International gold is in ounces. In ancient China, "Liang" was used as a gold unit and was an important metal. It is not only an important material for jewellery, electronics, modern communications, aerospace and other sectors, but also a special currency for storage and investment. The color of gold is golden yellow, metallic luster, hard to break down. Hardness 2~3, pure gold 19.3, melting point 1064.4 degrees; with good ductility, can be pressed into a thin foil, with high heat conductivity and electrical conductivity, the resistance of pure gold is 2.4p. Pure gold has good chemical resistance and is the best electroplating material. As a precious metal, gold has good physical properties and is not easily melted. Its density is high, its feel is heavy, its toughness and ductility are good, and it has good conductivity. Pure gold has a bright yellow color, but it incorporates other metals and changes in color. For example, a gold-copper alloy is dark red, and a silver-containing alloy is light yellow or grayish white. Gold is easily ground into powder, which is why gold is scattered in nature. Pure gold jewelry is also susceptible to wear.

同时黄金是一种很柔软的金属，在纯金上用指甲可划出痕迹。这种柔软性使黄金非常易于加工，但很容易使装饰品蹭伤，使其失去光泽以至影响美观。所以在用黄金制作首饰时，一般都要添加铜和银，以提高其硬度。

At the same time, gold is a very soft metal. Finger marks can be used to mark the pure gold. This kind of softness makes gold very easy to process, but it is easy to make the decoration bruise, make it lose its luster as well as affect the appearance. Therefore, when making jewelry from gold, copper and silver are generally added to increase its hardness.

常见的有金条、金块、金锭和各种不同的饰品、器皿、金币以及工业用的金丝、金片、金板等。由于用途不同，所需成色不一，或因没有提纯设备，而只熔化未提纯，或提的纯度不够，形成成色高低不一的黄金。人们习惯上根据成色的高低把熟金❶分为纯金、赤金、色金3种。

经过提纯后达到相当高的纯度的金称为纯金，黄金一般指达到99.6%以上成色的纯金。

Commonly there are gold bars, gold nuggets, gold ingots and a variety of different accessories, utensils, gold coins and industrial gold, spun gold, gold plate and so on. Due to different uses, the required color is not the same, or because there is no purification equipment to pure but melting, or the purity of the proposed is not enough to form different grades of gold. People are used to dividing cooked gold❶ into pure gold, fine gold, and color gold depending on the level of fineness.

The gold that has been purified to a very high purity is called pure gold, and gold generally refers to pure gold that is more than 99.6% pure.

①赤金。

Fine gold (999 gold).

即纯金。界定赤金的标准国际与国内有所不同，国际市场出售的黄金，成色达99.6%的称为赤金。而国内的赤金一般在99.2%～99.6%（图3-57）。

Fine gold is also named pure gold. The definition of fine gold is different from the domestic and international standards. The gold sold in the international market is known as red gold with a purity of 99.6%. The domestic red gold is generally between 99.2% and 99.6% (Figure 3-57).

图3-57 赤金
Figure 3-57 Fine gold

---

❶ 熟金是生金经过冶炼，提纯成黄金，一般纯度较高，密度较细，有的可以直接用于工业生产。The cooked gold is the gold after smelting and puritying. Generally, the purity is high and the density is low. Some of them can be directly used in industrial production.

②色金。

Color gold.

也称"次金""潮金",是指成色较低的金。这是黄金由于其他金属含量不同,成色高的达99%,低的只有30%。

按含其他金属划分,熟金又可分为清色金、混色金、K金等。

Color gold is also known as "second gold" or "tidal gold," referring to gold with lower fineness. Due to the difference in the content of other metals, the color of gold is as high as 99% and that of low is only 30%.

Divided by other metals, cooked gold can be divided into clear gold, mixed gold, k gold and so on.

③清色金。

Clear gold.

指黄金中只掺有白银成分,不论成色高低统称清色金。清色金较多,常见于金条、金锭、金块及各种器皿和金饰品。

Clear gold refers to the gold, only mixed with silver components, regardless of the level of color known as clean gold. More clear gold, common in gold bars, gold ingots, gold nuggets and various utensils and gold jewelry.

④混色金。

Mixed color gold.

是指黄金内除含有白银外,还含有铜、锌、铅、铁等其他金属。根据所含金属种类和数量不同,可分为小混金、大混金、青铜大混金、含铅大混金等。

Mixed color gold refers to gold, containing silver, with other metals like copper, zinc, lead, iron etc. According to the different types and amounts of metals contained, they can be divided into small mixed gold, large mixed gold, bronze mixed gold, lead-containing large mixed gold and so on.

⑤K金。

K gold.

是指银、铜按一定的比例,按照足金为24K的公式配制成的黄金。一般来说,K金含银比例越多,色泽越青;含铜比例大,则色泽为紫红。

It refers to the proportion of silver and copper in certain proportions, and is made of gold in accordance with the 24K formula of pure gold. In general, the greater the proportion of silver in K-gold, the greener the color; the larger the proportion of copper, the purple in color.

K值所表示的百分数,只是一个大致的数,并不要求十分准确。而习惯上又多数是使用偶数K值,如24K、22K、20K、18K等。18K的意思即指24份合金中含金18份,相当于75%左右的含量。K金折合含金量的计算公式是:K值÷24×100%(即K值×4.1667%)。

The percentage represented by the K-value is only a rough number and does not require exactness. In practice, most of them use even-numbered K values, such as 24K, 22K, 20K, and 18K. The meaning of 18K refers to 18 parts of gold in 24 parts of the alloy, which is equivalent to about 75% of the content. K gold calculation method: The calculation formula of K gold equivalent gold content is: K value ÷ 24×100% (ie K value × 4.1667%).

K金首饰的分类:

K gold alloy jewelry classification:

a. 黄色系列的K金首饰。

Yellow Series K Gold Jewelry.

黄色系列的K金首饰(简称K黄),是黄金和银、铜的合金,按金的含量可以制成不同K数的K金系列首饰,主要有22K、18K、14K、10K和8K。亚洲人喜爱高K数的K金首饰,如22K和18K,欧美人喜欢14K金和10K金首饰。黄K金系列首饰颜色的深浅,与K金中金的含量和银、铜的比例有关。18K黄金的颜色比14K、10K的黄金首饰颜色黄,而同样K数的黄金首饰,如果银比铜多黄色就浅,如果铜比银多黄色就深(图3-58)。

The yellow series of K gold jewelry (abbreviated as K yellow) is an alloy of gold, silver and copper. The content of the deposit can be made into K series jewelry of different K number, mainly 22K, 18K,

14K, 10K and 8K. Asians love K-gold jewelry with a high K number, such as 22K and 18K. Westerners like 14K gold and 10K gold jewelry. The depth of the color of yellow K gold jewelry is related to the content of gold in K gold and the ratio of silver and copper. 18K gold is yellower than 14K and 10K gold jewelry, and the same K number of gold jewelry is lighter if silver is more yellow than copper, and deeper if it is yellow than silver (Figure 3-58).

铂金	platinum
钯金	palladium
铑板	rhodium plate
22ct黄金	22ct yellow gold
18ct白金	18ct white gold
18ct绿金	18ct green gold
18ct黄金	18ct yellow gold
18ct玫瑰金	18ct rose gold
9ct白金	9ct white gold
9ct黄金	9ct yellow gold

图3-58 K金
Figure 3-58 K gold

b. 白色系列的K金首饰。
White Series K gold Jewelry.

白色K金系列的黄金首饰（简称K白），呈略带青黄的白色，印记均标有WG印记，按组合可分以下两种：

The white Series K gold jewelry (K white for short) is slightly yellowish white. The stamps are marked with WG imprints and can be divided into the following two types by combination:

第一种以黄金为主和银、镍、锌组合的合金，为K白金。18K的K金是由75％的黄金和25％的银、镍、锌组合而成。14K的K白金是由58.5％的黄金和41.5％的银、镍、锌组合而成的合金。

第二种以黄金和钯为主，再加上铜、镍、锌组成的合金，在国际上采用按每种金属在K金中所占的份额作为称谓的依据，目前只有两个品种，即K白金是由3份钯金和4份铜、镍、锌组成的K白金，226K白金是2份黄金、2份钯金和6份铜、镍、锌组成的K白金。

The first type of alloy, mainly gold and silver, nickel and zinc, is K white gold. The 18K K gold is composed of 75% gold and 25% silver, nickel and zinc. The 14K K Platinum is a combination of 58.5% gold and 41.5% silver, nickel and zinc.

The second type is mainly composed of gold and palladium, together with alloys composed of copper, nickel and zinc. In the international market, the share of each metal in K gold is used as the basis for the title. At present, there are only two types of alloys. K Platinum is made from 3 parts of palladium and 4 parts of copper, nickel and zinc. White gold of 226K is 2 parts of gold, 2 parts of palladium and 6 parts of white gold consisting of copper, nickel and zinc.

c. 红色系列的K金首饰。
Red Series K gold Jewelry.

红色系列的K金首饰是以黄金和铜为主，加入少量银的合金，颜色呈淡红色。

The red Series K gold jewelry is mainly gold and copper, adding a small amount of silver alloy, the color is light red.

（2）铂金。
Platinum (Pt).

是一种天然形成的白色贵重金属。铂金早在公元前700年就出现，在人类使用铂金的2000多年历史中，它一直被认为是最高贵的金属之一。铂金首饰的纯度非常高，纯度通常都高达90%～95%，常见的铂金首饰纯度有Pt900，Pt950。根据国家规定，只有铂金含量在85%及以上的首饰才能被称为铂金首饰。铂金首饰纯度极高，因此也不会使皮肤过敏。铂金的白色光泽天然纯净，赋予了铂金首饰独特的外观。即使每天佩戴，铂金始终留有纯净如初的纯白光泽。

Platinum(Pt) is a naturally occurring white precious metal. Platinum emerged as early as 700 B.C. and has been considered one of the most noble metals in the 2000years history of human use of platinum. The purity of platinum jewelry is very high, the purity is usually as high as 90%~95%, and the purity of common platinum jewelry is Pt900, Pt950. According to national regulations, only jewelry with a platinum content of 85% or more can be called platinum jewelry. Platinum jewelry is extremely pure and therefore does not cause skin allergies. Platinum's white sheen is pure and natural, giving platinum jewelry a unique look. Even if worn every day, platinum always has a pure white luster.

铂金比黄金稀有，在全球极少数地方才得以被开采。3克仅相当于一枚较小的铂金戒指的重量，而提取这些铂金大约需要8周的时间和大量的精力。

Platinum is rarer than gold and can only be exploited in very few places around the world. The three grams is only equivalent to the weight of a smaller platinum ring, and the extraction of these platinum takes about eight weeks and a lot of energy.

铂金的性质非常稳定，不会因为日常佩戴而变质或褪色，它的光泽始终如初。即使和生活中常见的酸性物质接触，如温泉中的硫，漂白剂，泳池中的氯，或是汗水等，都不会因此受到影响。同时无论佩戴多久，铂金能够始终保持其天然纯白的光泽，并且永不褪色。

The properties of platinum are very stable and will not deteriorate or fade due to everyday wear. Its luster is always the same, even contacting with acidic substances commonly found in life, such as sulfur in hot springs, bleach, chlorine in the pool, or sweat, will not be affected. At the same time, regardless of how long it takes, platinum can always maintain its natural white luster and never fade.

铂金坚韧性和延展性意味着即使是很小的铂金爪也能安全牢固的镶嵌每一颗钻石。这可以在很多国际品牌的经典设计中得到印证——铂金能够胜任各种繁复的镶嵌设计及细节，使得各种设计灵感成为现实。铂金出众的坚韧性和延展性使其能够雕琢出精美的线条，各大顶级珠宝品牌熟知铂金的这些特质，从而创造出无数惊世工艺，尤以欧洲皇室的定制珠宝著名。从来铂金和钻石都是这些传世巨作的完美呈现，成为著名珠宝品牌与皇室的挚爱（图3-59）。

The toughness and ductility of platinum means that even a small platinum claw can safely and securely mount each diamond. This can be confirmed in the classic design of many international brands—platinum can handle a variety of complex mosaic design and details, making a variety of design inspiration become a reality. Platinum's outstanding toughness and extensibility enable it to exquisitely crafted beautiful lines. The top jewelry brands are familiar with these qualities of platinum, creating countless stunning craftsmanship, notably European custom jewelry. Never since platinum and diamonds are the perfect presentation of these handed down masterpieces, it has become the love of famous jewelry brands and royalty (Figure 3-59).

图3-59　皇室与贵族的宠儿
Figure 3-59　The darling of the royal family and nobles

①铂合金。

Platinum alloy.

指铂与其他金属混合而成的合金，如与钯、铑、钇、钌、钴、铱、铜等。尽管铂硬度比金高，但作为镶嵌之用尚且不足，必需与其他金属合金，方能用来制作首饰。

首饰业使用铂、钌合金和铂、铱合金较多。在欧洲和香港使用铂、钴合金于浇铸。在日本用铂（85%），钯合金制造链条。国际上铂金饰的戳记是Pt、Plat或Platinum的字样，并以纯度之千分数字代表之，如Pt900表示纯度是900‰。在日本铂金饰品的规格标示有Pt1000、Pt950、Pt900、Pt850（图3-60）。

Platinum alloy refers to an alloy made of a mixture of platinum and other metals, such as palladium, rhodium, ruthenium, osmium, cobalt, rhodium, copper, and the like. Although the hardness of platinum is higher than that of gold, it is not enough for the use of inlays and it must be alloyed with other metals before it can be used to make jewelry.

The jewellery industry uses platinum, tantalum alloys, and platinum and tantalum alloys. Platinum and cobalt alloys are used for casting in Europe and Hong Kong. In Japan, the chain is made of platinum (85%) and palladium alloy. The international stamp of platinum decoration is Pt, Plat or Platinum, and it is represented by the thousandth of purity. If Pt900, the purity is 900‰. The specifications of platinum jewelry in Japan are Pt1000, Pt950, Pt900 and Pt850 (Figure 3-60).

图3-60　铂合金
Figure 3-60　Platinum alloy

②钯金。

Palladium.

是铂族的一员，银白色，符号Pd，比重12，轻于铂，延展性强，比铂稍硬，不溶于有机酸、冷硫酸或盐酸。但溶于硝酸和王水。常态下不易氧化和失去光泽，温度400℃左右表面会产生氧化物，但温度上升至900℃时又恢复光泽。目前钯比铂便宜，首饰业界拿来单独使用，或作为金、银、铂合金的组成部分。有时参入一些钌，来增加其硬度。市场上常见金、钯的K金和铂、钯的合金。规格标示有Pd1000、Pd950、Pd900（图3-61）。

Palladium is a member of the platinum group, silver white, symbol Pd, specific gravity 12, lighter than platinum, strong ductility, slightly harder than palladium , insoluble in organic acids, cold sulfuric acid or hydrochloric acid, but soluble in nitric acid and aqua regia. It is not easy to oxidize and lose gloss under normal conditions. Oxide will be produced on the surface at a temperature of about 400°C, but the gloss will be restored when the temperature rises to 900°C. At present, palladium is cheaper than platinum, which is used by the jewelry industry alone, or as an integral part of gold, silver and platinum alloys. Sometimes add some helium to increase its hardness. Common gold, palladium K gold and platinum, palladium alloys are all in the market. Specifications include Pd1000, Pd950, Pd900 (Figure 3-61).

图3-61　钯金
Figure3-61　Palladium

（3）银：纯银，纹银（925）。

Silver: Fine Silver (999), silver pattern (925).

银为过渡金属的一种。化学符号Ag。银是古代就已知并加以利用的金属之一，是一种重要的贵金属。银在自然界中有单质存在，但绝大部分是以化合态的形式存在于银矿石中。银的理化性质均较为稳定，导热、导电性能很好，质软，富延展性。其反光率极高，可达99%以上。有许多重要用途。

925银一般是指含银量92.5%左右的银质品，纯度在92.5%左右即认定为纯银。因为纯度过高的银柔软并且容易被氧化，925银加入了7.5%的其他金属，使其具有理想的硬度，能更好地塑型，镶嵌各种宝石，制作出造型多样的银制品。从此银首饰以色泽光鲜、款式别致、工艺精美、价格中档的时尚品位，迅速风靡全世界（图3-62）。

Argentum is a kind of transition metal. Chemical symbol Ag. Silver is one of the metals known and used in ancient times and is an important precious metal. Silver exists in nature, but most of it exists in silver ore in a chemical state. The physicochemical properties of silver are relatively stable, with good thermal and electrical conductivity, soft and ductile properties. Its reflective rate is extremely high, reaching more than 99%. There are many important uses.

925 silver generally refers to a silver product containing about 92.5% of silver, and the purity is about 92.5% which means that it is regarded as pure silver. Because the silver with too high purity is soft and easily oxidized, 925 silver is added with 7.5% of other metals. It has the ideal hardness, can better shape, inlaid various gemstones, and create a variety of silver products. Since then, the silver jewelry has become popular in the world with its bright color, elegant style and exquisite price (Figure 3-62).

2. 非贵金属材料
Non-precious metal materials
（1）皮革，绳索，丝绢类。
Leather, ropes, silkworms.
（2）塑料，橡胶类。
Plastics, Rubbers.
（3）非金属，动物骨骼，贝壳类。
Non-metal, animal bones, shellfish.
（4）木料，植物果实类。
Wood, plant fruits.
（5）宝玉石及各种彩石类。
Gemstones and Various Color Stones.
（6）玻璃，陶瓷类（图3-63～图3-66）。
Glass, Ceramics (Figure 3-63~ Figure 3-66).

图3-62 银
Figure3-62 Silver

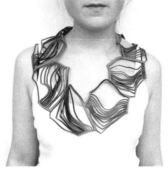

图3-63 纤维材料首饰
Figure3-63 Fiber material

图3-64 纸质材料首饰
Figure3-64 Paper material jewelry

图3-65 皮革首饰
Figure3-65 Leather jewelry

图3-66 树脂和木头综合材料首饰
Figure3-66 Resin and wood composite material jewelry

四、首饰设计常用材料、工具与设备
D. Materials、tools and equipment

1. 首饰设计基本材料

   Basic materials

   （1）设计用材料：金属材料、非金属材料、宝石材料。

   Materials: Metallic Materials, Non-Metallic Materials, Gemstone Materials.

   （2）设计用工具：设计用纸、铅笔、橡皮擦、各种规板及云形规、彩色铅笔、水彩（粉）颜料、水彩（粉）笔。

   Design Tools: Paper, Pencil, Eraser, Various Gauge Boards, Cloud Shape Gauges, Color Pencils, Watercolor (Flour) Pigments, Watercolor (Pink) Pens.

2. 首饰制作工具与设备

   Workshop furniture and equipment

   （1）工作台。

   Workbench.

   首饰制作工作台的后面一般离地面距离为90cm，在台面正中向前伸出一块长约30厘米，宽约10厘米，厚约1厘米的木板，便于手工进行锯、钻及锉等工作。工作台台面钉上厚度为1厘米的铝皮（或白铁皮），以便进行金粉的收集和防火（图3-67）。

图3-67 珠宝工作台

Figure 3-67 Jewelry bench

The back of the jewelry table is usually 90 cm away from the ground, with a board projected torward in the middle of table, which is about 30 centimeters long, about 10 centimeters wide, and about 1 centimeter thick, convenient for hand sawing, drilling, and boring. A 1 cm thick aluminum skin (or tin plate) is nailed to the table top for the collection and fire protection of gold powder (Figure 3-67).

（2）火吹套件。

Fire Blow Kit: Fire.

焊接工具是首饰制作中的重要工具，它的作用主要有熔金，退火，焊接等。火吹套件是由鼓风器，燃料容器和火枪等部件组成，各部分之间用软管连接。鼓风器可以是人力的（如风球），也可以是电力的（如空压机）。其作用是产生足够压力和流速的气流，使燃料容器中的燃料（如汽油或乙炔气，煤气，天然气，氢气等）与空气中的氧气充分混合，到达火枪后被点燃产生火焰（图3-68～图3-70）。

Soldering tools is an important tool in jewelry production. His role is mainly molten gold, annealing, welding and so on. The torch is composed of a blower, a fuel container and a musket, etc. Each part is connected with a hose. The blower can be manpower (such as wind ball), it can also be electric (such as air compressor). The role is to generate a sufficient pressure and flow rate of the gas flow, so that the fuel in the fuel container (such as gasoline or acetylene gas, gas, natural gas, hydrogen, etc.) and the oxygen in the air to fully mix, reach the musket after being ignited to produce a flame (Figure 3-68~ Figure 3-70).

（3）火枪。

Firearms.

一般有一个调节阀门，可调整火焰的粗细；有的火枪有两个调节阀门，一个调整火焰的粗细，另一个调整混合气体的混合比例。一般来说，在进行精细部件的焊接时通常使用风球＋汽油壶＋小火枪的组合，因为这种组合可以比较灵活的利用手脚的配合，调整火焰的大小，粗细；熔金和配焊药时则经常使用空压机＋煤气＋大火枪的组合，这种组合火焰猛

第三章　常用服饰配件设计

图3-68　焊接工具
Figure 3-68　Soldering tools

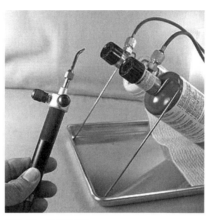

图3-69　氢氧熔焊机
Figure 3-69　Hydrogen and oxygen welding machine

图3-70　皮老虎（汽油）熔焊机
Figure 3-70　Leather tiger (gasoline) welding machine

烈，温度高，熔金速度快。此外在焊接和融化高燃点的贵金属（如铂金）时，通常采用高压氧气＋高压氢气＋专用火枪的组合，这种组合产生的温度可以达到2000℃以上。根据操作环节的具体要求选择火吹套件的适当组合是有必要的。目前教学用的实验室里用的是风球＋汽油壶＋小火枪的组合（图3-71）。

　　There is usually a regulating valve that can adjust the thickness of the flame; some guns have two regulating valves, one adjusting the thickness of the flame, and the other adjusting the mixing ratio of the mixed gas. In general, the combination of wind ball + petrol pot + squid gun is usually used in the welding of delicate parts, because this combination can be more flexible use of the cooperation of hands and feet, adjust the size and thickness of the flame; while the combination of air compressor + gas + bonfire gun is often used in molten gold and welding drugs because this combination is fierce and the temperature is high and the molten gold

speed is high. In addition, in the welding and melting of high-firing precious metals (such as platinum), a combination of high-pressure oxygen + high-pressure hydrogen + special guns is usually used, and this combination can generate temperatures above 2000℃. It is necessary to select the proper combination of fire blowing kits according to the specific requirements of the operation. At present, the teaching laboratory uses a combination of a wind ball, a petrol pot, and a small musket (Figure 3-71).

　　（4）吊机及机针。

　　Crane and needle.

　　吊机是利用电机一端连接的钢丝软轴带动机头进行工作的。吊机一般是挂在工作台的台柱上。机头为三爪夹头，用于装夹机针。机头分两种，一种为执模机头，稍微大一些，另一种为镶石机头，稍微细小一些，且有快速装卸开关。吊机的脚踏开关内有滑动变阻机构，踏下高度的不同会使吊机产生不同的转速，适合于不同的操作情况（图3-72）。

 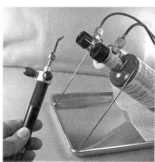 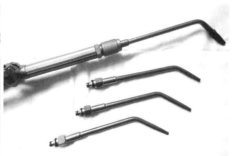

图3-71　火枪
Figure 3-71　Gas torch

095

吊机机针是首饰制作中非常重要的工具，主要用于首饰的执模，镶嵌甚至抛光等环节。根据机针针头的不同形状，主要有以下几种：粗球针，扫针，钻针，吸珠，飞碟等（图3-73）。

The crane is operated by a steel wire shaft drive head connected to one end of the motor. The crane is usually hung on the stage pillar. The head is a three-jaw chuck for clamping the needle. The machine head is divided into two types, one is a model head, a little larger, and the other is a stone inlay head, slightly smaller, and there is a quick loading and unloading switch. The foot switch of the crane has a sliding variable resistance mechanism. The difference in stepping height will cause the crane to generate different rotation speeds, which is suitable for different operating conditions (Figure 3-72).

Metal finishing tools are very important in jewelry production, mainly used for jewelry molding, inlaying and even polishing. According to the different shapes of needles, there are mainly the following: thick ball burrs, sweeping needles, drill needles, suction beads, flying saucers and so on (Figure 3-73).

（5）压片机和拉丝板。

Laminating machine and drawing board.

压片和拉丝是首饰制作中经常应用的操作环节。压片机是由两个钢质压辊来完成压片的，动力来源于手摇和电动两种。拉丝板是拉丝操作的主要工具，通常需要固定使用。其拉丝孔通常为硬质合金制造，也有采用人造金刚石的，但价格极贵。拉丝孔的形状通常为圆形，也有椭圆形，半圆形，三角形，方形等，还有专门拉制异形截面丝的拉丝板。拉丝孔的直径由粗到细（图3-74、图3-75）。

Presser and wire drawing are commonly used in jewelry making. The sheet pressing machine is made of two steel press rollers, which are powered by both hand and electric. Wire drawing board is the main tool for wire drawing operation. The drawing hole is usually made of hard alloy, and some are made of artificial diamond, but it is very expensive. The shape of the drawing hole is usually round, but there are also oval, semi-round, triangle, square, etc. The diameter of the drawing hole ranges from thick to thin (Figure 3-74、Figure 3-75).

图3-72　吊机
Figure3-72　Pendant moto

图3-74　拔线板　　　　　图标3-75　压片机
Figure 3-74　Draw plate　　Figuer3-75　Rolling mill

图3-73　机针
Figure3-73　Metal finishing tools

五、首饰设计与制作的基本工艺
E. The basic process of jewelry design and production

首饰制作基本上是人工工艺和现代机械加工工艺的结合。人工工艺有：花丝工艺、錾花工艺、烧蓝工艺、点翠工艺、打胎工艺、镶嵌工艺、平填工艺等；现代机械加工工艺有：浇铸工艺、冲压工艺、电铸工艺等。以下对部分工艺作基本讲解：

Jewelry making is basically a combination of artificial craftsmanship and modern machining processes. Artificial production processes include filigree craftsmanship, blue-burning craftsmanship, flower craftsmanship, point jade craftsmanship, abortion craftsmanship, inlaying crafts, flat filling crafts, etc. Modern craftsmanship includes casting, stamping and electroforming. The following is a basic explanation of some of the processes:

1．花丝工艺
Silk filament craft

花丝工艺是将金银加工成丝，再经盘曲、掐花、填丝、堆垒等手段制作金银首饰的细金工艺。根据装饰部位的不同可制成不同纹样的花丝、拱丝、竹节丝、麦穗丝等，制作方法可分掐、填、攒、焊、堆、垒、织、编等。

The filigree craftsmanship is a fine gold craft that processes gold and silver into silk and then produces gold and silver jewelry by means of coiling, silking, filling, and stacking. According to the different parts of the decoration different patterns of filigree, arch silk, bamboo silk, wheat silk, etc. can be made. Production methods can be divided into rake, fill, rake, welding, heap, base, weaving, knitting and so on.

（1）掐丝就是将用花丝制成的刻槽，掐制成梅花、牡丹花、飞鸟、龙凤、亭台楼阁等各种纹样。

Cocoon silk is a groove made of filaments, and made from plum blossoms, peony flowers, birds, dragons and phoenixes, pavilions, and other pavilions.

（2）填丝是将撮好扎扁的花丝填在设计轮廓内。常用的种类有填拱丝、填花瓣等。

Filling the wire is to fill the design filaments with a flattened filament. Commonly used types include arched wire and petal filling.

（3）织编是将金银丝编织边缘纹样和不同形体的底纹，在底纹上再黏以用各种工艺方法制成的不同花形纹样，通过焊接完成。

Welding is the process of splicing the finished patterns together to form a complete jewelry.

（4）堆垒是用堆炭灰的方法将码丝在炭灰形上绕匀，垒出各种形状，并用小筛将药粉筛匀、焊好的过程。

The stacking process is to use the method of stacking charcoal to homogenize the code strands on the charcoal, and to form a variety of shapes, and use a small sieve to sieve the uniform powder and weld it.

2．錾花工艺
Flower processFlower process

錾花工艺通常使用钢制的各种形状的錾子，用小锤将钢錾花纹锤在过火后的条块状金银的表面。錾花工艺用錾、抢等方法雕刻图案花纹。

Repousse usually uses steel chisels in various shapes, using a small hammer to hammer steel chisel patterns on the surface of bars of gold and silver after the fire. Chasing uses chisel, grab and other methods to carve patterns.

3．烧蓝工艺
Burning blue process

烧蓝工艺又称点蓝工艺，与点翠工艺相似，都是景泰蓝工艺。烧蓝工艺不是一种独立的工种，而是作为首饰的辅助工种以点缀、装饰，增加色彩美而出现在首饰行业的。

The blue-burning process, also known as the blue process, is similar to the point Cui process and is a cloisonne process. The blue-burning process is not an independent type of work, but it is used as an accessory work of jewellery to embellish, decorate, and add color beauty to the jewelry industry.

4．浇铸工艺
Casting process

浇铸工艺是用铸造机进行首饰的成批生产的方

法。该方法具有提高工效，降低成本的优点。流程基本如是：根据首饰设计样本制成橡胶模具；用橡胶模具通过注蜡制成蜡模具；将蜡模具种成蜡树；将放有蜡树的筒注入石膏，制成石膏模具；将石膏模具放入烘炉内烘干，并加热至石膏模具脱蜡；将呈熔融状态的金注入石膏模具中；清洗去石膏，再进行抛光、镶嵌等流程即可。

The casting is a method of mass production of jewelry using a casting machine. This method has the advantages of improving ergonomics and reducing costs. The procedure is basically as follows: a rubber mold is made according to a jewelry design sample; a wax mold is made by injection of wax using a rubber mold; a wax mold is grown into a wax tree; a tube containing a wax tree is injected into plaster to make a plaster mold; put a plaster mold in the oven to dry and heat it dewax it; Inject the gold in the molten state into the plaster mold; clean the plaster, and then polish, inlay and so on.

5. 冲压工艺
Stamping process

冲压工艺是指完全用机器完成金属的切割、锉磨等过程。

Blanking dies refers to the complete use of machinery to complete the metal cutting, honing and other processes.

6. 电铸工艺
Electroforming process

电铸工艺属现代技术，其原理与电镀相同。是利用金属在电明液中产生阴极沉积的原理获得制件的特种加工方法。

The electroforming process is a modern technology with the same principle as electroplating. It is a special processing method using the principle of cathodic deposition of metal in electrolyte to obtain parts.

7. 表面处理工艺
Surface treatment process

现代表面处理工艺除了传统的抛光和电镀之外，还增加了磨砂、定砂、喷砂工艺等。

In addition to the traditional polishing and electroplating, the modern surface finises also add frosting, noisy, sandblasting, and the like.

六、首饰设计与制作的基本步骤
F. The basic steps of jewelry design and production

设计→起版→倒模→执模→镶石→抛光→电镀。

Design → Sample make→Inverting the molding→ Exeaution of molds → In lay Store →Polishing → Eleatroplating.

（1）设计：由设计师，先设计好草图，出设计图。

Design: Firstly, the designer should finish the sketch or the drawing.

（2）起版：由起版师根据设计图，制作实物样板。

Sample make: From the version of the division according to the design, production of physical model.

（3）倒模：样板出来后由该部门制作和样板同样的蜡样，然后再进行真空浇铸，制成半成品，即首饰的粗模。

Inverting the mold: After the sample comes out, the department produces the same wax sample as the sample, and then vacuum casting to make the semi-finished product, namely the rough mold of the jewelry.

（4）执模：对分件、整件的半成品进行初步修整。

Execution of molds: Initial trimming of sub-components and whole semi-finished products.

（5）镶石：要镶嵌的首饰要经该部门加工，素金类非镶嵌的首饰可直接进行抛光部。

Inlay (stone setting)stone: jewelry to be inlaid through the Department of processing, non-inlaid gold jewelry can be directly polished Department.

（6）抛光：相对执模工序，抛光工序是更细致的加工。

Polishing: relative to the mold process, polishing

process is a more detailed processing.

（7）电镀：根据客户或设计师的要求对首饰进行表面上的光、亮、色及耐磨保护处理（图3-76）。

Electroplating: According to the requirements of customers or designers, jewelry on the surface of the light, color and wear protection (Figure 3-76).

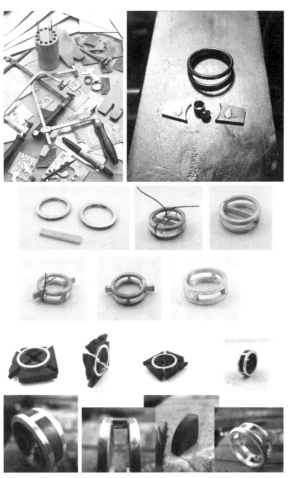

图3-76　珠宝设计制作过程
Figure3-76　Jewelry making process

首饰的出现是一种普遍的社会艺术文化现象，它是反映生活的一面镜子，同时也是对文化的表征和思想情感的体现。首饰设计是一直伴随着人类而存在的，在现今服装与首饰已经深入社会的每个层面，对于服装设计专业学生来说，本章在内容设置上由浅到深、由兴趣到专业的循序渐进的过程，使学生从身边的事物开始慢慢将服装服饰设计理论掌握并逐渐运用到自己的设计中。

The appearance of jewelry is a common social, artistic and cultural phenomenon. It is a mirror reflecting life, and also a reflection of culture and thoughts and feelings. Jewelry design is always accompanied by the existence of human beings. Today's clothing and jewelry have been deep into every level of society. For dress design major students, the content setting of this chapter is from shallow to deep, from interest to professional step by step, leading the students to start from their surroundings slowly, and practicing clothing apparel design theories into their designs gradually.

第五节　潮流与趋势——3D打印对首饰设计行业的影响

Section 5　Trends and directions—the impact of 3D printing on the jewelry design industry

一、3D打印技术与珠宝首饰行业
A. 3D printing technology and jewelry industry

3D打印技术已经有30余年的历史。1989年，美国广播公司（ABC）在"早安美国"节目中介绍了他们的"光固化"技术，并将这一技术术语科普成"3D打印"。

3D printing technology has enjoyed its name more than 30 years. In 1989, ABC introduced their Photo-curing technology on "Good Morning America" and popularized the technical term into "3D printing."

在过去的几十年里，3D打印技术对传统制造方式实现了"质"的变革。当前，3D打印技术已悄然成为各行业的技术变革的关键环节，如从医疗设备到航空旅行等，涉及了许多制作行业。珠宝行业正在将3D打印成为其主流制造技术（图3-77）。

In the past few decades, 3D printing technology has revolutionized the traditional manufacturing methods. At present, 3D printing technology has quietly turned out to be a crucial link in the technological transformation of various industries,. It involves in many production industries from medical equipment to air travel. The jewelry industry is turning 3D printing into its mainstream manufacturing technology (Figure3-77).

图3-77　3D打印技术
Figure 3-77　3D printing

3D打印技术从寻找其大规模应用之路，逐渐走向普及，适时也给首饰设计行业带来了深刻的变化，在加快产品研发进度的同时，根据需求定制，既符合珠宝行业的特点，又不失为珠宝首饰设计、制作的一种新方向。

Starting from seeking for its large-scale application, 3D printing technology has gradually become popular, and it has brought profound changes to the jewelry design industry at the appropriate time. While accelerating the progress of product development, it is customized according to the requirements, which not only meets the characteristics of the jewelry industry, but also leads a new direction for the design and production of jewelry.

传统的珠宝生产成本较高，独立设计师常为此"烦恼"特别是想要赶超大师级作品的设计师，首先得有经济实力，如今富有竞争力的大订单只能在国外完成，希冀着半年后能收回成本，而且其中还存在着许多风险。

The costs for Traditional jewelry production are high, and independent designers often feel "trouble" for this. Especially the designers who are eager to catch up with masterpieces need to have a certain financial capacity. Nowadays, large competitive orders can only be fulfilled overseas, while the cost could be wished to be collected after half a year, besides there are still many risks.

如今，独立设计师无需筹集一大笔资金，就能把作品推向全球市场。设计CAD模型，然后直接打印即可。3D打印服务机构已经开始抢占这块市场，提供了各类相关服务，从单一的生产到质量保障、包装、品牌推广、销售，甚至售后服务。设计人员可以挑选最适合自己的服务机构——可以操控整个过程，也可以专注于设计环节。这是一种全新的探索途径，设计师可以充分发挥自己的想象力。与材料成本相比，设计本身的优势会更具明显，价值更高。这也是定制商品的价格高于批量生产的原因所在（图3-78）。

图3-78　电脑制作模型与设计图
Figure3-78　Computer making model and design drawing

Nowadays independent designers do not have to raise a large sum of money to launch their works to the global market. The CAD model can be sent to be print directly right after designing. 3D printing service agencies have begun to engage this market, providing a variety of related services, from a single production to quality assurance, packaging, branding, sales, and even after-sales service. Designers can choose the service organization that best suits them—they can control the entire process or focus on the design process. This is a new way to explore, and designers can fully exert their imagination. Compared with the cost of materials, the advantages of the design itself will be more obvious and its value increases. This is also the reason why the price of customized goods is higher than mass production (Figure 3-78).

二、珠宝首饰设计相关软件简介
B. The introduction of jewelry design related softwares

1. Jewel CAD

Jewel CAD是由香港电脑珠宝科技有限公司1990年开发。是用于珠宝首饰设计，制造的专业化CAD/CAM软件（图3-79）。其有高度专业化，高工作效率，简单易学的特点，是业界认可首选的CAD/CAM软件系统。

Jewel CAD was developed by Hong Kong Computer Jewelry Technology Co., Ltd. in 1990. It is a professional CAD/CAM software for jewelry design/manufacturing (Figure 3-79). It is highly specialized, highly efficient, and easy to learn. It is the industry's preferred CAD/CAM software system.

 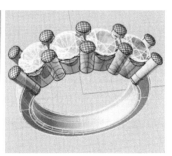

图3-79　Jewelry CAD珠宝三维建模软件
Figure 3-79　Jewelry CAD modeling software

Jewel CAD 5.1破解版已经更新到Jewel CAD 5.Pro了，内有Jewel CAD5.1（UPGATE 9）、Jewel CAD5.1（UPGATE 10）和Jewel CAD5.1（UPGATE 12）三个版本，三个版本均为免加密狗破解版，性能稳定，不死机，无任何功能限制。可以存光影图、可正常输入、输出、开启、插入、切薄片、储存档案。现在Jewel CAD Pro版本也已经稳定。

Jewel CAD 5.1 cracked version has been updated to Jewel CAD 5. Pro, there are three versions of Jewel CAD5.1 (UPGATE 9), Jewel CAD5.1 (UPGATE 10) and Jewel CAD5.1 (UPGATE 12), all three versions are dog-free version. Stable performance, no crash, no functional limitations. It can store light and shadow images, it can input, output, open, insert, slice and store files. The Jewel CAD Pro version is now stable.

（1）Jewel CAD软件特点。

Features of Jewel CAD.

①方便、易上手、能快速学习操作，只要有基本的电脑知识，就能在较短时间内学会软件。

Convenient, easy to use, easy learning operation, only basic computer skills are required to operate the software.

②有非常简单的图解用户界面和直觉功能，能做到类似像电影的播放那样去学习各种不同的技巧和方法建模。

Very simple graphical user interface and intuitive function enable users to learn different kinds of techniques and methods modeling like movie-like play.

③运用灵活的和高级的建模功能，去创造和修改曲线和曲面，并能将建模的功能，应用到更复杂的设计中去。

With flexible and advanced modeling capabilities to create and modify curved lines and surfaces, even more complex design can be made by applying modeling functions..

④用特别的功能去设计颈链，运用扭曲曲面和石头的设置。

Design the necklace with special features, using twisted surfaces and stone settings.

⑤简单而高效的功能应用于自由状态曲面去布尔运算。

Simple and efficient functions are applied to free-form surfaces to Boolean operations.

⑥固定的设计库能加快设计的速度，运用石头设备库，零件库，用户库，一旦产生新的设计灵感，即时去提取零件或组成来完成今后的设计。

The fixed design library can speed up the design, by applying the stone equipment library, parts library, user library, and once new design inspiration is generated, the parts or components can be extracted immediately to complete the future design.

⑦一个完整的设计方案，真实感的渲染，又快又容易地输出高品质的彩图，珠宝设计和高品质的图像的图书开始用Jewel CAD设计出版。

A complete design, realistic rendering, fast and easy output of high-quality color graphics, jewelry design and high-quality images of the book began to be published with Jewel CAD design.

⑧允许三维视图处理模型，在CNC加工中输出标准的GM编码和STL数据，输出标准的无缝合线的STL和SLC数据能快速地做成模型。

Allow 3D view processing model to output standard GM code and STL data in CNC machining, and output standard seamless STL and SLC data, which can be quickly modeled.

⑨在设计中能方便地计算金的重量。

It is convenient to calculate the weight of gold in the design.

（2）成型过程。

The molding process.

成型过程是指用计算机控制的喷嘴，喷成一个很薄的平面图形，每完成一幅这种平面图形后，以下降同样薄的厚度再喷（图3-80）。这样，一层层的平面图形就叠成极为复杂的部件，其实也就是相当于把首饰物件的切面一层层地喷出来。

The molding process refers to the use of a computer-controlled nozzle to spray a very thin planar pattern. After each such planar pattern is completed, it is sprayed at the same thin thickness (Figure 3-80). In this way, the plane graphics of one layer are stacked into extremely complicated parts, which is equivalent to spraying the cut surface of the jewelry object layer by layer.

Jewel CAD分为切薄片和数控加工两部分，支持的快速成型系统有Soliscape、3D Systems、EnvisionTEC、OBJET、MEIKO、Roland等公司的机器（图3-81）。

Jewel CAD is divided into two parts: slice cutting and CNC machining. The supported rapid prototyping systems are machines from Soliscape, 3D Systems, EnvisionTEC, OBJET, MEIKO, Roland (Figure 3-81).

一般钟表行业使用数控加工比较广泛，珠宝行业使用切薄片式的快速成型系统比较广泛。切薄片命令所应用的快速成型系统（图3-82）。一般来说，切薄片厚度越薄越好，这样生成的模型会更精细，效果更佳。但是，不同的机器最高的精度是不一样的。

In the general watch industry, CNC machining is widely used, and the jewellery industry uses a slit-type rapid prototyping system. The rapid prototyping system applied by the slice command (Figure 3-82). In general, the thinner the slice thickness, the better, thus the resulting model will be more refined and better. However, the highest precision of different machines differs.

2. Rhino犀牛

Jewel CAD加快了首饰款式更新的速度。目前，首饰设计软件中，Jewel CAD和Rhino的使用人数最多，Rhino的建模功能强大，而且渲染效果更佳；而JeweL CAD工作效率高、渲染速度快、针对性强、操作容易，并且能直接参与制模，适合商业用途。

Jewel CAD speeds up the update of jewelry styles. Currently, Jewel CAD and Rhino are mostly used in jewelry design software. Rhino's modeling functions are powerful with better rendering results; And JeweL CAD has high efficiency, fast rendering, high pertinence, easy operation, and can directly participate in modeling, suitable for commercial use.

Rhino是美国Robert McNeel & Assoc.开发的PC上强大的专业3D造型软件（图3-83），它可以广泛地应用于三维动画制作、工业制造、科学研究以及机械设计等领域。它能轻易整合3DS MAX与Softimage的模型功能部分，对要求精细、弹性与复杂的3D NURBS模型，有"点石成金"的效能。能输出obj、DXF、IGES、STL、3dm等不同格式，并适用于几乎所有3D软件，尤其对增加整个3D工作团队的模型生产力有明显效果，故使用3D MAX、AutoCAD、MAYA、Softimage、Houdini、Lightwave等3D设计人员要认真学习使用。

so the 3D designers need serious learning to use 3D MAX, AutoCAD, MAYA, Softimage, Houdini, Lightwave, etc.

3. 3DESIGN
3DESIGN

法国Visionnumeric公司成立于1988年，一直致力于开发建模软件，从工业设计到珠宝设计，凝聚了该企业数年来的科研成果，使其应用于珠宝的3DESIGN首饰建模软件。如今该软件俨然成为后起之秀，并且在业内好评如潮（图3-84）。

Founded in 1988, Visionnumeric France has been dedicated to developing modeling softwares ranging from industrial design to jewelry design, which has brought together the company's research results over the years for 3DESIGN jewelry modeling applications. Nowadays, this software has become a rising star and has received enormous positive reviews in the industry(Figure 3-84).

图3-80　电脑后期喷蜡输出
Figure3-80　Wax spraying output in late stage of computer

图3-81　3D打印机
Figure 3-81　3D printer

图3-82　模型到金属实物
Figure 3-82　Model to metal object

图3-83　Rino 3D建模软件
Figure3-83　Rino 3D modeling software

Rhino is a powerful professional 3D modeling software developed on the PC developed by Robert McNeel & Assoc. (Figure 3-83). It can be widely used in 3D animation, industrial manufacturing, scientific research and mechanical design. It can easily integrate the model functions of 3DS MAX and Softimage. It has the effect of "Golden touch" for the 3D NURBS models, which requires to be fine, flexible and complex It can output different formats like obj, DXF, IGES, STL, 3dm,which suitable for almost all 3D software, especially for dramatically increasing the model productivity of the entire 3D work team,

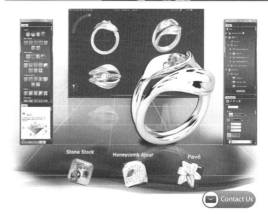

图3-84　3DESIGN珠宝专业建模软件
Figure3-84　3DESIGN jewelry professional modeling software

3DESIGN提供了一种独特的参数结构树，它记录了设计中的每一个历史参数，不需要返回草图重新开始设计，而只需要直接修改草图形状或修改3D造型参数，系统会自动全面进行更新。它可以把2D的草图转变为真实的3D图形。只要通过一般的扫描仪把画好的草图扫描成普通的图片格式，传入3DESIGN中，将自动转变为立体三视图。在3DESIGN里面可以轻松的排列宝石，它提供了自动和手动两种排列方式，可以对任意一颗宝石的属性进行修改。3DESIGN也提供了丰富的资源库，包括各类戒指、宝石、戒托、珍珠、水晶以及形态各异的2D截面形状等，并且整合了施华洛世奇所有的时尚水晶部件。也可以将自己设计的作品存入资源库中。

3DESIGN provides a unique parameter structure tree, which records every historical parameter in the design, no need to return to the sketch to restart the design. It only needs to directly modify the sketch shape or modify the 3D modeling parameters, and the system will automatically update it thoroughly. It can turn 2D sketches into real 3D graphics. As long as the drawn sketch is scanned into a normal image format by a general scanner, it will be automatically converted into a stereo three-view by passing it into 3DESIGN. The jewels can be easily arranged in 3DESIGN. It provides both automatic and manual alignments, which enables the modification for any single gem's property. 3DESIGN also offers a rich library, including rings, gems, ring settings, pearls, crystals, and 2D cross-section shapes, and incorporates all the stylish crystal parts of SWAROVSKI. You can also store your own works into the database. Work Appreciation.

结语
Conclusion

珠宝三维打印改变的不仅仅只是技术，也是整个珠宝首饰行业的变革，是新一轮的科技革命和产业革命。珠宝首饰企业的制高点是人才。因为珠宝的三维打印技术是支撑首饰智能制造的基础，在操作层面需要众多文化素质高、技术过硬的建模工程师和技术人员，是那些真正能把技术与创意相结合的应用型人才。因此，首饰行业的从业人员的结构在未来必然会因三维打印技术的普及而革新，从一种劳动密集型行业转为技术与设计占主导的生产分散型行业。让该技术更好地和传统的手工艺结合，通过智能生产、将互联网、人、机器和信息融为一体，使珠宝行业发展与时俱进。

The three-dimensional printing of jewelry has changed not only techniques, but the transformation of the entire jewelry industry. It is a new round of technological revolution and industrial revolution. The commanding parts of the jewelry companies are talents. Because the three-dimensional printing technology of jewelry is the fundamental for supporting the intelligent manufacturing of jewelry. At the operational level, Massive modeling engineers and technicians with solid techniques and high cultural quality are needed. They are the ones who can truly combine technology and creativity. Therefore, the structure of the employees in the jewelry industry will inevitably be innovated in the future due to the popularity of 3D printing technology, from a labor-intensive industry to a production-distributed industry dominated by technology and design. The technology is better integrated with traditional craftsmanship, and the jewelry industry is advancing with the times through intelligent production, integration of the Internet, people, machines and information.

第四章 非金属材料在服饰配件领域的制作与工艺

Chapter 4　Production and Process of Non-metallic Materials in the Field of Apparel Accessories

课题名称：纺织纤维材料在服饰配件上的制作与工艺

Project Name: textile fiber materials in the production of accessories and processes

1. 基础知识

学习目的：通过本章的学习，使学生初步了解纺织纤维材料、纱线、织物等服饰品材料的类别、性质以及运用范畴；初步了解服饰品设计的基础知识，并能运用设计元素与服饰品材料的功能相结合来设计与制作服饰品。

本章重点：本章重点为服饰配件设计的基础、服饰品设计语言及工艺。

课时参考：16课时。

Learning Objectives: Through the study of this chapter, students will get a preliminary understanding of the types, nature and application of textile fiber materials, yarns, fabrics and other apparel materials; preliminary understanding of the basic knowledge of apparel design, and the possibilities of designing and manufacturing the furnishings by combining the use of design elements and the function of the apparel materials.

The focus of this chapter: This chapter focuses on the basis of apparel accessories design, apparel design language and technology.

Class time reference: 16 class hours.

2. 各类材料设计案例解析

学习目的：通过本章的学习，使学生了解在设计各类服饰配件中材料的艺术语言及运用方式。能够合理地运用不同的材料，正确表达设计的艺术理念。

本章重点：本章重点为纺织纤维材料在服饰配件中的运用。

课时参考：4课时。

Analysis of various material design cases

Learning Objectives: Through the study of this chapter, students will be able to understand the artistic language and application of materials in the design of various accessories. Ability to use different materials reasonably and correctly express the artistic concept of design.

The focus of this chapter: This chapter focuses on the use of textile fiber materials in apparel accessories.

Class time reference: 4 class hours.

第一节 纺织纤维材料在服饰配件上的制作与工艺

Section 1 Fabrication and process of textile fiber materials on clothing accessories

服饰所用的材料——纺织纤维，随着时间的变迁，发展迅速。根据记载，在7000多年前的新石器时代，我国已经用葛纤维织布制衣，从出土文物中，就有蚕茧及丝绣品。在马王堆中，发现了用提花机控制万余根纱线制成的线织饰物。在宋代，棉花已用于纺织，到晚清时期，黄道婆的纺织技术已经能够生产各种用途的纺织品，包括用织物增强漆胶器和纺织铜丝以增强陶瓷的景泰蓝复合材料。从19~20世纪的100年间，随着工业技术的发展，出现了再生纤维素纤维。由于生物技术的应用，使改性羊毛、有色棉花进入实用阶段。同时，高功能、高性能的纤维问世。21世纪以来，可循环的绿色纤维、新型环保纤维、超仿真、功能化、差别化纤维不断"涌现"。特别是信息技术、微电子技术运用到纺织纤维中去，进一步拓展了服饰领域材料运用的深度和广度。

The materials used in clothing, textile fibers, have developed rapidly over time. According to the records, in the Neolithic Age more than 7,000 years ago, China has used woven flax fiber for fabrics. From the unearthed relics, there are silkworm cocoons and silk embroidery. In Mawangdui, we found a wire-woven ornament made from a jacquard machine that controlled more than 10,000 yarns. In the Song dynasty, cotton was used in textiles. By the late Qing dynasty, Huang Daopo's textile technology had been able to produce textiles for various purposes, including cloisonne composites with fabric reinforced lacquers and woven copper reinforced ceramics. From the 19th century to the 20th century, with the development of industrial technology, regenerated cellulose fibers appeared. Due to the application of biotechnology, modified wool and colored cotton entered the practical stage. At the same time, high-performance fibers became available. Since the 21th century, recyclable green fibers, new environmentally friendly fibers, super-simulation, functionalization, and differentiated fibers have been "emerged". What's move, the application of information technology and microelectronics technology to textile fibers further expands the depth and breadth of materials used in the apparel sector.

纺织纤维是指直径一般为几微米到几十微米，长度比直径大百倍、千倍以上的细长柔软的物质，是可用来进行纺织加工和制作的纤维。纺织纤维的特点是具有一定的长度、细度、弹性、强力等良好物理性能，具有可纺性和良好的染色性能。应用在服饰配件上，服饰设计者们常运用常规的纺织纤维。

The textile fiber refers to a slender and soft material having a diameter of usually several micrometers to several tens of micrometers and a length of a hundred times or more than a diameter, which can be used for textile processing and production. Textile fiber is characterized by a certain length, fineness, elasticity, strength and other good physical properties, with spinnability and good dyeing properties. Applied to apparel accessories, apparel designers often use conventional textile fibers.

一、常用服饰用纺织纤维的分类及性能
A. Classification and performance of textile fibers for common apparel

1. 生物质原生纤维的性能
Performance of biomass virgin fibers

（1）棉纤维。

Cotton fiber.

耐碱不耐酸，在一定浓度的NaOH溶液中处理后纤维横向膨胀，截面变圆，天然转曲，纤维呈现丝一样的光泽，手感好，公定回潮率为8%~13%，吸湿性强。纤维柔软、透气性好，染色性能佳；适合做各类、各季服装，也可以做帽子、围巾、鞋子等服饰配件（图4-1）。

Alkali resistance, but not acid resistance. After treatment in a certain concentration of NaOH solution, the fiber expands laterally, the cross section becomes round, the natural rotation, the fiber has the same luster, the hand feel is good, the standard regain rate is 8%~13%, and the hygroscopicity is strong. The fiber is soft, breathable, and has good dyeing performance; it is suitable for all kinds of seasons and clothing, and can also be used for hats, scarves, shoes and other clothing accessories (Figure 4-1).

（2）麻纤维（以苎麻为例）。

Hemp fiber (taking ramie as an example).

麻织物吸汗后不易沾身，具有快干、挺爽、环保、可降解等特性。但麻纤维之间抱合力差，不易捻合，纱羽多，手感粗硬，有刺激感。但弹性差，不耐磨（图4-2）。

The hemp fabric is not easy to get wet after sweat absorption, and has the characteristics of quick drying, cool, eco-friendly and degradability. However, the hemp fiber has poor cohesion, not easy to mix with many yarn feathers, and has a rough hand feel and a stimulating feeling. While the elasticity is poor and it is not wearable (Figure 4-2).

（3）毛绒纤维（羊毛为例）。

Plush fiber (for example, wool).

天然的动物毛纤维，其主要由蛋白质组成。耐酸不耐碱，吸湿性、染色性能好。纤维细软、富有弹性、耐磨并具有缩绒特性；与其他纤维混纺做服装、服饰配件都很合适（图4-3）。

Natural animal hair fiber, which is mainly composed of protein. Acid but not alkali resistant, hygroscopic and good dyeing capacity. The fiber is soft, elastic, wear-resistant and with fluffy characteristics; blended with other fibers, suitable to make clothing, clothing accessories (Figure 4-3).

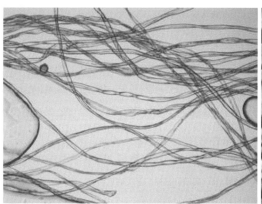

图4-1 棉纤维
Figure 4-1 Cotton fiber

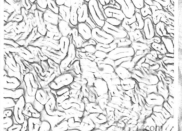

图4-2 麻纤维
Figure 4-2 Hemp fiber

图4-3　毛绒纤维
Figure 4-3　Plush fiber

图4-4　丝纤维
Figure 4-4　Silk fiber

（4）丝纤维（以家蚕丝为例）。

Silk fiber (for example, silkworm silk).

有较好的伸长度，纤维细而柔软、平滑、有弹性、光泽好、吸湿性好。耐光性较差，容易泛黄，长时间日照，强度下降50%左右。

蚕丝纤维，光泽柔和、质地细柔、穿着舒适，是高档服装的常用材料（图4-4）。

It has good elongation, and the fiber is fine and soft, smooth, elastic, good in gloss and moisture absorption. Light resistance is poor, easy to turn to yellow, the strength decreased by about 50% after long-term sunshine.

Silk fiber, soft and comfortable to wear, is a common material for high-end clothing (Figure 4-4).

2. 生物质再生纤维的性能
Performance of biomass regenerated fibers

（1）再生纤维素纤维。

Regenerated cellulose fiber.

①以黏胶纤维为例。

Taking viscose fiber as an example.

是从棉短纤维、木材、芦苇、甘蔗渣中提取纯净纤维素经过烧碱、二氧化硫处理后制备的纺织纤维。它具有良好的吸湿性，吸湿后显著膨胀，纤维直径增加50%，因此织物手感硬，耐碱不耐酸。尺寸稳定性差，容易伸长，耐磨、耐疲劳性差。但染色性能好、色谱全、色泽鲜艳、染色牢度较好。适合做女装裙料（图4-5）。

It is a textile fiber prepared by extracting pure cellulose from cotton staple fiber, wood, reed and bagasse by caustic soda and sulfur dioxide treatment. It has good hygroscopicity, significantly expands after moisture absorption, and the fiber diameter increases by 50%, so the fabric feels hard, alkali resistance, but not acid resistance. Poor dimensional stability, easy elongation, poor wear resistance and fatigue resistance. However, the dyeing performance is good, the chromatogram is complete, the color is bright, and the color fastness is good. It is suitable for women's skirts (Figure 4-5).

②以Lycell/Tencel为例。

该纤维被誉为21世纪最有前途的高科技绿色纤维。从原料、生产、工艺环保，产品可降解，具有优良的力学性能和服用性能。其强度与涤纶纤维相近，吸湿性能是棉纤维的2倍。适合做高档服装的面料（图4-6）。

The fiber is hailed as the most promising high-tech green fiber of the 21th century. From raw materials, production, process and environmental protection, the product is degradable, with excellent mechanical properties and wearability performance. Its strength is similar to that of polyester fiber, and its moisture absorption performance is twice that of cotton fiber. Suitable for high-end clothing fabrics (Figure 4-6).

③再生蛋白质纤维。

Regenerated protein fiber.

a. 以大豆蛋白质纤维为。

Take soybean protein fiber as an example.

以去油脂的大豆豆粕做原料，采取生物工程等技术手段处理而成。它具有单丝细、比重轻，耐酸、碱性强、吸湿、导湿性好。有蚕丝般的柔和光泽及蚕丝的优良性能，同时又有合成纤维的机械性能、尺寸稳定、挺括抗皱、保型性好。适合作各类服装面料、服饰配件等（图4-7）。

It is made by the degreased Soybean meal, processed by bioengineering and other technical means. It has a fine monofilament and a light specific gravity. It is strong in acid resistance, alkalinity, moisture absorption and moisture permeability. It has the soft luster of silk and the excellent performance. At the

图4-5 黏胶纤维
Figure 4-5 Taking viscose fiber

图4-6 以Lycell/Tencel为例
Figure 4-6 Lycell/Tencel as an example

大豆白纤维

图4-7 以大豆蛋白质纤维为例
Figure 4-7 Take soybean protein fiber as an example

same time, it has the mechanical properties, dimensional stability, scratch and wrinkle resistance and good shape retention of synthetic fiber. Suitable for all kinds of apparel fabrics, clothing accessories and so on (Figure 4-7).

b. 以牛奶蛋白质纤维为例。

Taking milk protein fiber as an example.

该纤维以牛乳为基本原料，经脱水、脱油、脱脂、分离、提纯，成为一种具有线型大分子结构的乳酪蛋白，再与丙烯腈共聚，经工艺处理而成。该纤维比棉、丝强度高，耐穿、耐洗、防霉、防蛀、易贮藏。对皮肤有良好的保养作用，有天然持久的抑菌功能，广谱抑菌达80%以上，适合作内衣等（图4-8）。

The fiber is made from cow's milk as a basic raw material. After being dehydrated, deoiled, degreased, separated and purified, It turns into a kind of cheese protein with a linear macromolecular structure, which is then copolymerized with acrylonitrile and processed. The fiber has higher strength than cotton and silk, and is durable, washable, mildewproof, smashproof and easy to store. It has a good maintenance effect on the skin, has a natural and lasting antibacterial function, and has a broad spectrum of antibacterial activity of more than 80%. Suitable for underwear, etc (Figure 4-8).

3. 常规合成纤维

Conventional synthetic fibers

该纤维利用煤、石油、天然气等为原料，经人工合成并经机械加工制成的纤维。

The fiber is made of coal, petroleum, natural gas, etc., which are artificially synthesized and mechanically processed.

（1）聚酯纤维：以涤纶（PET）为例。

Polyester fiber: Take polyester fiber (PET) as an example.

涤纶的学名为聚对苯二甲酸乙二酯纤维。该纤维的吸湿性差，公定回潮率为0.4%。干、湿状态下，纤维性能变化不大。对酸较稳定，挺括性、抗皱性、保型性好，耐磨性、耐光性、耐虫蛀性好，但易起毛起球，吸附灰尘。该纤维可与其他纤维混纺做服装面料、各类饰品（头饰、项链、耳环、帽子、提包、鞋等）（图4-9）。

Polyester fiber is known as polyethylene terephthalate fiber. The fiber has poor hygroscopicity and a predetermined moisture regain of 0.4%. In dry and wet conditions, the fiber properties change lit-

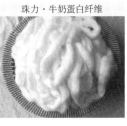

图4-8　牛奶蛋白纤维
Figure 4-8　Milk protein fiber

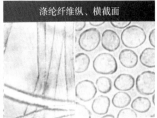

图4-9　涤纶纤维（PET）
Figure 4-9　Polyester fiber (PET)

tle. It is stable to acid, stiff, wrinkle-resistant, good shape retention, good abrasion resistance, light resistance and insect resistance, but it is easy to pilling and absorb dust. The fiber can be blended with other fibers to make fabrics and various accessories (headwear, necklaces, earrings, hats, bags, shoes, etc.) (Figure 4-9).

（2）聚酰胺纤维：以锦纶为例。

Polyamide fiber: Take nylon as an example.

耐碱不耐酸，强度高伸长能力强，弹性好，弹性回复率接近100%。吸湿性、耐磨性、耐疲劳性在合成纤维中是最好的。但耐光性差，长时间在光照下，纤维会发黄、发红、强力下降。

该纤维主要用于与棉、毛或其他纤维混纺，可用于制作袜子、围巾、衣料等（图4-10）。

The polyamide fiber is alkali resistant, but not acid resistant, high strength, high elongation, and good elasticity, and its elastic recovery rate is close to 100%. Hygroscopicity, abrasion resistance, and fatigue resistance are the best among synthetic fibers. However, the light resistance is poor. When exposed to light for a long time, the fiber will turn yellow, red and less strong.

The fiber is mainly used for blending with cotton, wool or other fibers, and can be used for making socks, scarves, clothing, and the like (Figure 4-10).

总之，合成纤维除了具有较高的强度，在防腐、保温、去污等方面的性能较强而且具有柔润光滑的表面和明艳的色彩，缺点是吸湿性和透气性差。

In summary, in addition to having higher strength, the synthetic fiber has strong performance in anticorrosion, heat preservation, decontamination, etc., and has a soft and smooth surface and bright color, and has the disadvantages of poor hygroscopicity and gas permeability.

二、纺织类纤维服饰配件的设计
B. Design of textile fiber clothing accessories

由于各种纤维的高分子化学的构成不一样，纤维的不同，服用性能、耐久性能和美学的表现力，从而带来的设计表现迥然不同。这些纤维都具备制作服饰配件的性能，并能对人们的视觉和触觉心理产生直接影响。同时，用纺织纤维纺、织不同的工艺，如针织、机织、编织、纤维集合，以及植绒、刺绣等，能构成服饰品肌理各异、特色鲜明的品种来满足各类服饰设计的需求。

Because of the different polymer chemistry of various fibers, the difference in fiber, the performance, the durability and the expressiveness of aesthetics, the design performance is quite different. These fibers have the performance of making apparel accessories and have a direct impact on people's visual and tactile psychology. At the same time, by applying different textile fiber processes, like spinning and weaving, such as knitting, weaving, weaving, fiber collection, flocking, embroidery, etc., we can constitute different varieties of clothing, distinctive

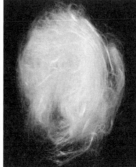

图4-10 聚酰胺纤维

Figure 4-10 Polyamide fiber

characteristics to meet the various apparel design requirements.

用设计者的"眼光"来审视，无论是天然纤维还是化学纤维，都能给人独特的"亲和感"，让人感受到亲近、舒适与温馨。人们对纤维艺术的喜爱来自远古时期纤维材料留下的历史"痕迹"并带有浓厚的感官"意识"。

With the designer's "eyes" to examine, whether it is natural fiber or chemical fiber, a unique sense "affinity" is provided, making people feel close, comfortable and warm. People's favortowards fiber art comes from the historical "marks" left by fiber materials in ancient times and has a strong sense of "consciousness."

1．设计语言
Design language

纺织纤维材料的设计"语言"，主要从四个方面来展现：光泽、色彩、肌理和可塑性。正确地运用纤维材料的设计"语言"，能使服饰配件的设计发挥出独特的艺术效果。

The design "language" of textile fiber materials is mainly expressed in four aspects: gloss, color, texture and plasticity. The correct use of the design "language" of the fiber material enables the design of the clothing accessories to exert a unique artistic effect.

①光泽。

Gloss.

是由光投射到材料表面的反射所形成的一种视觉效果。根据材料表面对光的反射能力，把材料分为强光泽、柔光泽和无光泽几个等级。强光泽材料又称光泽型材料，这类材料表面光滑闪亮，具有强烈"刺激"视觉的效果。比较典型的有漆皮效果的皮革和具有"缎结构"（织造工艺）的缎料。柔光泽材料是具有柔软光泽感的面料，一般给人柔美的感觉，丝绸柔和的光泽感就是最好的说明。无光泽材料是不具备有反射光的特质，会在视觉上给人温暖与质朴感，如纯棉织物和毛织物等。一般来说，质地粗糙的材料对光的反射少，光波吸收多，光泽感较弱。质地光滑平整的材料对光反射强，光波吸收弱，光泽感强烈。就光泽而言，天然纤维的光泽较化学纤维的光泽弱，给人亲和、自然、生态环保的视觉心理感受。而化学纤维给人的视觉感受，则相对地"工业感"和"无机感"。如不同光泽效果的项饰作品。天然丝纤维（图4-11）的丝光光泽；欧根纱面料（图4-12）的亚光光泽；尼龙类纤维（图4-13）的强光光泽，给人形成不同的视觉心理感受。

图4-11 Elsbeth joy Nielsen，项饰，丝绸故事
Figure 4-11 Elsbeth joy Nielsen, neckpiece, a silk story

图4-12 项饰，图片来至网络
Figure 4-12 Neckpiece, Picture from the internet

图4-13 Rowan Mersh，项饰
Figure 4-13 Rowan Mersh, neckpiece

第四章 非金属材料在服饰配件领域的制作与工艺

Gloss is a Avisual effect formed by reflection of light projected onto the surface of a material. Depending on the ability of the surface of the material to reflect light, the material is divided into several grades of strong gloss, soft gloss and matt. Strong gloss materials, also known as glossy materials, have a smooth, shiny surface with a strong "irritating" visual effect. More typical leather with patent leather effect and satin with "satin structure" (weaving process). The soft gloss material is a fabric with a soft luster, which generally gives a soft feeling. The soft luster of silk is the best illustration. Matte materials do not have the features of reflected light, which will give people a warm and simple feeling, such as cotton fabrics and wool fabrics. In general, a material with a rough texture has less reflection of light, more light waves, and a weaker luster. The smooth and smooth material has strong light reflection, weak light absorption and strong luster. In terms of gloss, the luster of natural fibers is weaker than that of chemical fibers, giving a friendly, natural, eco-friendly visual psychological feeling. The visual perception of chemical fiber is relatively "industrial" and "inorganic". The following items have different gloss effects. The silky luster of natural silk fiber (Figure 4-11); the matt gloss of organza fabric (Figure 4-12); the glare gloss of nylon fiber (Figure 4-13), which gives people a different visual psychological feeling.

②色彩。

Color.

广义的色彩学涵盖了美学、心理学和民俗学等多种学科知识，几乎每一色彩都会具有一定的象征意义。当视觉接触到了某种颜色，大脑神经便会接收到色彩发放的信号，即刻产生联想，例如，高明度的淡色调象征着明媚、清澈、轻柔、成熟、透明、浪漫、爽朗；高色相饱和度的鲜色调象征着艳丽、华美、活跃、外向、发展、兴奋、刺激、自由、激情；色相饱和度低的暗色调象征着稳重、刚毅、干练、质朴、坚强、沉着、充实；而浅灰调则象征着温柔、轻盈、柔弱、消极、成熟等。

The broad color science covers a wide range of subject knowledge such as aesthetics, psychology and folklore, and almost every color has a certain symbolic meaning. When visually exposed to a certain color, the brain's nerves receive the signal of color distribution, which instantly produces associations. For example, the light color of high brightness symbolizes bright, clear, soft, mature, transparent, romantic, and openness; high saturated color with fresh tones symbolize gorgeous, active, outgoing, development, excitement, stimulation, freedom, passion; dark shaded colors with low hue saturation symbolize stability, fortitude, skill, simplicity, strength, calmness, and fulfillment; while grey tuned color symbolizes gentleness, lightness, weakness, negativity, maturity, etc.

经验丰富的设计师能够借助色彩的运用，引起一般人心理上的联想，从而达到设计的目的。色彩是影响人心理感觉变化的重要因素，不同材料的色彩给人的感觉完全不同。色彩也是决定材质风格情调的重要因素，因此对色彩的应用是考验设计师设计能力的关键所在。天然纤维除本身具有的色彩之外，可以通过有机或无机着色工艺，形成丰富的色彩效果。高纯度的色彩给人温暖的感觉，高明度的色彩给人冰冷的感觉。如图4-14较图4-15在色彩的表达上更加活泼、艳丽，时尚感也更强烈。图4-15则给人以稳健、冷静、内敛的感觉。

Experienced designers are able to arouse the psychological association of the average person to achieve the purpose of design by applying colors. Color is also an important factor affecting the change of people's psychological feelings. The different materials'colors give people a completely different fee designer's ability. lings, Color is also an important factor in determining the style of the material, thus, the application of color is the key to testing Natural fibers, in addition to their own color, can carry out the rich color effects through organic or inorganic dyeing processes. High-purity colors gives a warm feeling, and high-brightness colors give people a cold feeling. Figure 4-14 is more lively, gorgeous and fashionable in color expression than Figure 4-15. Figure 4-15 gives a feeling of stability, calmness, and restraint.

图4-14 菲利普·特里西，帽饰
Figure 4-14 Philip Treacy, hat

图4-15 图片来源于网络，帽饰
Figure 4-15 Picture from the internet, hat

③肌理。

Texture.

纤维材料的视觉肌理指的是通过处理纤维材料表面而产生的视觉纹理效果。

视觉肌理的表面效果并不局限在平面构造上，还包括材料表面的光滑程度、织纹结构特点和图案（纹样）肌理的粗犷细腻效果、以及立体肌理的造型感觉和光影效果等。如当代首饰艺术家莫娜·卢辛的作品《纤维艺术雕塑》（图4-16），利用纤维材料的不同形态创作出多种肌理效果混合的项饰。

The visual texture of a fibrous material refers to the visual texture effect produced by treating the surface of the fibrous material.

The surface effect of the visual texture is not limited to the planar structure, but also includes the smoothness of the surface of the material, the texture characteristics of the texture the rough and delicate effect of the texture of the patterns and the modeling feeling, and light and shadow effect of the three-dimensional texture.

For example, the contemporary jewellery artist Mona Lusin's *Fiber Art Sculpture* (Figure 4-16) uses a variety of shapes of fiber materials to create a variety of textures and blends.

图4-16 莫娜·路易斯，软雕塑
Figure 4-16 Mona Luison, art sculpture textille

④可塑性。

Plasticity.

天然类纤维的可塑性较化学类纤维的可塑性要差。应用棉、麻、丝、毛等天然纤维材料进行服饰配件的造型设计中多采用黏胶、缝制、刺绣、扎捆等手法。而尼龙、人造棉等化学类纤维可利用高温、高压等手法进行热熔塑造、立体塑型等。如英国著名服装品牌亚历山大·麦昆2008年春夏时装秀中的帽饰（图4-17）作品，即利用尼龙类纤维的韧性和可塑型性创作出立体的帽饰效果。天然羊毛类的帽饰（图4-18）则采用一体成型黏合缝制工艺而成。

The plasticity of natural fibers is worse than that of chemical fibers. The application of cotton, hemp, silk, wool and other natural fiber materials for the design of apparel accessories is often carried out by means of gluing, sewing, embroidery and binding. Chemical fibers such as nylon and rayon can be hot-melted and molded by high-temperature and high-pressure methods. For example, the famous British clothing brand Alexander McQueen's spring/summer 2008 fashion show hat (Figure 4-17) works, which uses the toughness and plasticity of nylon fiber

第四章　非金属材料在服饰配件领域的制作与工艺

to create a three-dimensional hat effect. The natural wool cap (Figure 4-18) is made by a one-piece adhesive stitching process.

图4-17　亚历山大麦昆2008年春
Figure 4-17　Alexander McQueen spring 2008 ready-to-Wear fashion show

图4-18　钟形女帽，卡夏尔，2012秋
Figure 4-18　cloche , Cacharel, Fall 2012

2. 工艺
Techniques

纤维材料在制作产品中多采用编织、编钩、缝缀、缠绕、包裹等基础制作工艺，同时也可以运用综合表现的手法，如印染、刺绣、粘贴、折叠、并置、拼贴等。设计师通过编、织、绣、缠、捆、扎、裹、卷等多种工艺手段，赋予材料特定意义的审美价值。如丝、毛等纤维材料，质地柔软、色彩丰富、可塑性强且富有韧性，灵活运用多种编织技法，可表现出细腻、富有光泽感、色彩丰富、层次微妙等特点；麻纤维由于较粗硬、拉力强，所以采用捆绑、缠绕等技法来表现粗犷的肌理效果尤佳；而对于一些片状材料则应采用叠压、黏合、镶嵌、缝缀等表现手法。由于材料在可选择性上丰富多样，也就决定了纤维艺术的创作工艺手法灵活各异。

In the production of fiber materials, we use the basic production techniques such as weaving, knitting, sewing, wrapping, etc. At the same time, we can also use the comprehensive expression techniques such as printing and dyeing, embroidering, pasting, folding, juxtaposition and collage. The designer gives the aesthetic value of the material a specific meaning through various techniques such as knitting, weaving, embroidering, wrapping, bundling, tying, and rolling. For example silk, wool and other fiber materials, with soft texture, rich color, strong plasticity and toughness, and flexible use of a variety of weaving techniques, can show delicate, lustrous, rich colors, subtle features, etc.; Because hemp fiber is coarser and strong, it is particularly good for the use of binding, winding and other techniques to express its rough texture effect; for some sheet materials, they should be used for lamination, bonding, inlaying, stitching and other expression techniques. Due to the variety of materials available, the flexibility of creative techniques of fiber art is determined.

三、纺织纤维类服饰配件的设计原则
C. Design principles for textile fiber apparel accessories

1. 服饰美原则
The principle of clothing beauty

服饰配件作为人体的装饰品，处于与服装相互搭配、相互"映辉"的地位，在设计的过程中受到服装风格、颜色、款式等的影响与限制。因此，纺织纤维类服饰配件的设计应首先遵守服饰美的原则。

As the ornament of the human body, the accessories of the clothing are in a position of mutual matching with the clothing and mutual "reflection". In the process of design, they are influenced and restricted by the style, color and style of the clothing. Therefore, the design of textile fiber accessories should first comply with the principle of beauty.

（1）服饰配件的风格与服装的风格相统一。

The style of the clothing accessories is consistent with the style of the clothing.

服饰配件作为服装的装饰部分，起到渲染和强调服装风格的效果，只有将服饰配件与服装的风格相统一才能突出服装的整体风貌，展现穿着者的风格特征。对每一件服饰配饰而言，它们都具有自己

115

的风格和魅力，将两者很好地统一起来，使服装的风格更加突出，服饰配饰才能起到最好的效果。

As a decorative part of the clothing, the clothing accessories play the role of rendering and emphasizing the style of the clothing. Only by unifying the styles of the clothing accessories and the clothing can the overall style of the clothing be highlighted and the style characteristics of the wearer can be displayed. For each piece of clothing accessories, they all have their own styles and charms, we will unite the two well, so that the style of the clothing is more prominent, and the clothing accessories can achieve the best results.

（2）服饰配件的色彩与服装的色彩相协调。

The color of the clothing accessories is coordinated with the color of the clothing.

服装的色彩通过视觉接受系统，给人"先声夺人"的第一印象。色彩搭配得是否合理，会直接影响服饰穿戴者的整体美观度和受关注程度。因此，在进行服装与服饰配件搭配的时候，不仅需要通过配色学的原理和手法进行服饰色彩整体搭配，还要综合考虑服饰配件的色彩与服装色彩的相协调性。

The color of the garment is visually accepted by the system, giving the first impression of "first screaming". Whether the color is matched properly will directly affect the overall aesthetics and attention of the wearer. Therefore, when the clothing and clothing accessories are matched, the overall color of the clothing color can be matched Not only through the principle and method of color matching, but also the coordination of the color of the clothing accessories and the color of the clothing should be comprehensively considered.

（3）服饰配件的款式与服装的款式相呼应。

The style of the clothing accessories echoes the style of the clothing.

对服饰配件而言，在设计时不仅要考虑服装的风格，同时还要考虑服装的款式特点。因为不同服装款式搭配的配饰款式，不尽相同。服装配饰需要整体地与服装款式相结合，来考虑服饰配件的款式，使服饰配件的款式与服装的款式相呼应。如2016纳伊·姆汗春/夏时装秀（图4-19）中，纤维材料制作的耳饰与服装款式及头饰都协调的搭配在一起。2018春/夏米兰时装周丹妮拉·格雷吉斯的包包（图4-20）和项饰（图4-21）都应用了天然纤维材

图4-19 纳伊·姆汗，春夏时装秀2016
Figure 4-19 Naeem Khan, spring and summer fashion show 2016

图4-20 丹妮拉·格雷吉斯，春夏米兰时装周2018
Figure 4-20 Daniela Gregis, spring and summer milan fashion week 2018

图4-21 丹妮拉·格雷吉斯，2018春夏米兰时装周
Figure 4-21 Daniela Gregis, 2018 spring and summer milan fashion week

料进行设计与制作,其款式造型与模特的服装款式"和谐"搭配,构成完整的服饰美视觉效果。

For the accessories, it is necessary to consider not only the style of the clothing, but also the style characteristics of the clothing. Because the styles of accessories vary with the styles of the clothes they cope. Clothing accessories need to be integrated with the clothing style to consider the style of the clothing accessories, so that the style of the clothing accessories echoes the style of the clothing. As shown in the 2016 Naeem Khan Spring and Summer Fashion Show (Figure 4-19), the earrings made of fiber materials are coordinated with the styles and headwear. Daniela Gregis's bag (Figure 4-20) and accessories (Figure 4-21) in the spring and summer of 2018 are all designed and produced with natural fiber materials. The style and style of the model are harmoniously matched with the model's clothing style to form a complete "Clothing Beauty" visual effects.

2. 材料美原则
Material beauty principle

从现代纤维艺术的视角看,不难发现设计师在创作服饰配件的过程中,更注重追求材料自身的美感。纤维材料的纹理、光泽、质感等特征,被人的视觉和触觉所感知。在现代服饰配件作品中,不同的纤维材质给人以不同的艺术感受,如羊毛的粗厚、柔软,丝绒的细腻、华贵,麻线的粗犷、大方,棉线的质朴、素洁。在艺术创作时,构思设计图稿后,需先确定材料,再选择相应的表现方法。由于材料的粗细、软硬、滑涩等各种物理性都能影响并决定工艺制作技法。因此,选择合适的表现方法,才能充分发挥材料的独特魅力,从而更好地表现作品的内在魅力。

From the perspective of modern fiber art, it is not difficult to find that in the process of creating costume accessories, designers pay more attention to the aesthetics of materials. The texture, luster, texture and other characteristics of the fiber material are perceived by human vision and touch. In the modern clothing accessories works, different fiber materials give people different artistic feelings, such as the thick and soft wool, the delicate and luxurious velvet, the rough and generous twine, the simple and clean cotton thread. In the process of avtistic cveation, after designing the artwork, you need to determine the material first and then select the corresponding performance method. Due to the various physical properties such as the thickness, hardness, and slipperiness of the material, which determine the making process techniques. Therefore, by choosing the appropriate expression method, the unique charm of the material can be fully utilized to exert the inner charm of the design work.

在艺术首饰的语境中,材料的多元视角是设计师孜孜以求的目标。《那年,她来过》(图4-22)解读和把握材料的艺术语言特征,并将其运用在创作中。在艺术首饰作品中,利用化学类纤维柔软、"可塑"的材料特质,通过热加工处理工艺,改变了化学纤维材料所构成平面的视觉印象,塑造出立体的形制效果。低色彩饱和度的化学纤维传递出柔和朦胧的美感,为整件作品奠定纯洁、唯美的艺术格调。

图4-22 王书利,那年,她来过
Figure 4-22 Wangl Shuli, That year, She came

In the context of art jewellery, the multiple perspectives of materials are the goals that designers are striving for. "That Year, She Came" (Figure 4-22) interprets and grasps the artistic and linguistic features of materials and applies them to their creations. In the art jewellery works, the use of chemical fiber soft, "plastic" material characteristics, through the hot processing process, changed the visual impression of the plane formed by chemical fiber materials, shaping a three-dimensional shape effect. The low saturated color chemical fiber will convey a soft and blurred-sense, which will provide a pure and beautiful artistic style for the whole work.

3. 肌理美原则
Principle of texture beauty

追求肌理感是艺术体验的表现语言之一，它会给人们带来丰富的视觉感受。

肌理是指物体表面的组织纹理结构，即各种纵横交错、高低不平、粗糙平滑的纹理变化。它的表现形式丰富、表现风格具备特色，运用好肌理效果会使服饰配件的设计更加出色。不同的纤维材料在形状、疏密、大小、颜色等方面的差异，会形成不同的肌理外观，从而带来不同的美感效果，产生不同肌理效应，带给人不同的心理反应。如稀疏与密集的肌理，因为张力的强弱变化，产生松弛与紧张的心理感；凹凸与起伏的肌理，因张力的变化冲突，诱导退缩与扩展的心理感受；条理与节奏的肌理，因张力的规则有序给人以整齐舒展的心理感受；细腻与粗犷的肌理，则是张力在结构的隐与露中形成了含蓄明快的心理感受（图4-23）。

Pursuing a sense of texture is one of the expressions of the artistic experience, which brings a rich visual experience to people.

Texture refers to the texture structure of the surface of the object, that is, the texture change like various criss-crossing, uneven, rough and smooth. It has a rich form of expression and a distinctive style of expression. The appropriate application of texture effect will make the design of the accessories more outstanding. The difference in shape, density, size, color, etc. of different fiber materials will result in different texture appearances, which will bring different aesthetics effects, producing different texture effects, and different psychological reactions, such as the different psychological feelings of relaxation and tension caused by sparse and dense texture because of the change of tension; the different psychological feelings of inducing withdrawal and expansion caused by the texture of the bumps and undulations because of the conflict of tension; neat and stretched psychological feelings caused by the texture of rhythm due to tension; and the subtle and bright psychological feelings caused by the delicate and rough texture in the hidden and exposed structure (Figure 4-23).

图4-23-（1） 图片来源于网络，项饰
Figure 4-23-1 Picture from the internet neckpiece

图4-23-（2） 丽达配饰，项饰
Figure 4-23-2 Lida Accessories, necklace

因纺织纤维材料种类多样，质地各异，所以针对不同的材料运用不同的工艺制作手法来表现丰富的肌理外观。我们利用纤维材料这一特点，对其再加工处理，创造出纤维材料独特的表现语言。如图4-24的首饰作品，蓬松、柔软的毛纤维通过编织、扎捆缝制手法并与纤维面料并置，创造出凹凸不平的肌理效果，给人温暖的视觉效果。而图4-25的项饰则大面积地应用麻纤维本身的色彩，通过编织工艺营造出粗糙的肌理效果，给人质朴、自然的视觉心理感受。

Due to the variety of textile fiber materials and the different textures, different techniques are used

第四章　非金属材料在服饰配件领域的制作与工艺

for different materials to express and enrich the texture appearance. We use the characteristics of fiber materials to reprocess the materials cord, create a unique expression language of fiber materials. As shown in the jewellery work of Figure 4-24, the fluffy and soft wool fibers are woven and tied together and juxtaposed with the fiber fabric to create an uneven texture effect, giving a warm visual effect. The item of Figure 4-25 applies the color of hemp fiber itself to a large area, and creates a rough texture effect through the weaving process, which also gives a simple and natural visual psychological feeling.

图4-24　小泽洋子，项饰
Figure 4-24　Yoko Lzawa, necklace

图4-25　Denise Julia Reytan，项饰
Figure 4-25　Denise Julia Reytan, necklace

第二节　运用纺织纤维类材料制作服饰配件的案例解析

Section 2　Casestudy of making clothing accessories with textile fiber materials

一、首饰
A. Jewelry

日本首饰设计师岩泽洋子经常使用氨纶纱、尼龙、有机玻璃和聚丙烯等纤维材料来制作首饰。图4-25最初的设计灵感来自于古代日本人用来包裹食盒的裹布。通过这一灵感，有了将物件容纳、覆盖、包裹的想法。尼龙和氨纶纱都具有较强的弹性、质地轻、易染色、易成型等特性。通过编织机器将极细的氨纶纱和尼龙编织起来，使织物具有较强的拉伸性及透光度，这符合将物体包裹和覆盖起来的理念。再将较硬的透明有机玻璃和聚丙烯作为织物中间的填充物，既起到了固定的作用，又具有一定的透明度。将尼龙和氨纶纱染色后，包裹起透明的有机玻璃，使作品具有含蓄、朦胧、神秘的美感。

Japanese jewellery designer Yoko Lzawa often uses fiber materials such as spandex, nylon, plexiglass and polypropylene to make jewelry. The original design of Figure 15 was inspired by the wrapping of the ancient Japanese used to wrap the food box. Through this inspiration, the author has the idea of accommodating, covering, and wrapping objects. Nylon and spandex yarns have strong elasticity, light texture, easy dyeing and easy molding. The extremely fine spandex yarn and nylon are woven by a weaving machine to give the fabric a strong stretchability and transparency, which is in line with the idea of wrapping and covering the object. The hard transparent plexiglass and polypropylene are used as the filler in the middle of the fabric, which plays a fixed role and has a certain transparency. The nylon and spandex yarns are dyed and wrapped in transparent plexiglass to give the work a subtle, flawless and mysterious aesthetic.

德国首饰设计师丹尼斯-茱莉娅·雷坦的作品洋溢着青春和现代的美感。她的作品非常直观地将各种色彩鲜艳的纤维材料和硅胶、金属、塑料、半宝石等材料混搭在一起。她的首饰制作方法非常有趣，她将珠子、徽章和其他小珍宝放进硅胶中，通过硅胶翻模的方式来提取这些物件的形状，然后再与其他材料结合。她的作品中大部分是使用纤维材料，而纤维材料的易加工性，也给予设计师很多的发挥空间。作者通过此件作品将色彩鲜艳的材料搭配在一起，产生了丰富的趣味性，成功地将观赏者带入到了她的幻想世界中。

The work of Germany jewellery designer Denise Julia Reytan is full of youth and modern beauty. Her work is very intuitive to mix a variety of colorful fiber materials with silica gel, metal, plastic, semi-precious stones and other materials. Her method of making jewelry is very interesting. She puts beads, badges and other small treasures into the silica gel, and uses silicone to mold the shape of the objects and then combine them with other materials. Most of her works use fiber materials, and the ease of processing of fiber materials gives designers a lot of room to play. Through this piece of work, the authors combine colorful materials to create a rich and interesting taste that successfully brings viewers into her fantasy world.

中国台湾设计师李恒的作品（图4-26），灵感来源于服装上的刺绣图案。他将服装载体抽离后，把具象图案放大，形成失焦的马赛克图案。在金属板上利用镭射工艺，切割出马赛克边缘不规则的形状后，再将柔软的手工刺绣融入坚硬的金属框中。其作品综合运用了雕、绣、烧、编等技法，形成了"镂空风貌"，在打破作品整体沉闷感的同时增强了材料的艺术欣赏性。镂空工艺与刺绣技法相结合，使原材料焕然一新，给人留下奇妙的视觉印象。

The work of China Taiwanese designer Li Heng (Figure 4-26) is inspired by the embroidery pattern on the clothing. After he pulled the garment carrier away, he magnified the figurative pattern to form an out-of-focus mosaic pattern. After the laser process is used on the metal plate to cut the irregular shape of the mosaic edge, the soft hand embroidery is incorporated into the hard metal frame. His works use a combination of carving, embroidering, burning, and knitting techniques to form a "small-looking style", which enhances the artistic appreciation of the material while breaking the overall dullness of the work. The combination of the hollowing process and the embroidery technique gives the raw materials a new look, leaving a wonderful visual impression.

中国美术学院教师段燕俪的作品《云和山的彼端》（图4-27），表达了作者在西藏游历中被当地纯净雄浑的自然风景、质朴的人文生活、神圣的宗教文化所震撼的情感。在西藏服饰中天然面料、穿戴方式及浓艳的色彩、富含寓意的图腾纹样等都是其历代文化传承的标志象征。设计师对藏服进行解构和重组，用西藏独有的动物牦牛、藏羚羊等的塑料模型作为视觉主体，结合日常用品如自行车上的LED灯、塑料玩具等作品"集合"，看似毫不相关的材料聚集在一起，实则是通过和谐而有内涵的逻辑，来表达对西藏的赞叹、崇拜和敬仰之情。作者打破民俗和国际、传统和当代之间的界限，给少数民族文化新的诠释。

图4-26 李恒，胸针
Figure 4-26　Li Heng, brooch

图4-27 段燕俪，项饰
Figure 4-27　Duan Yanli, necklace

首饰设计师维克·梅森的创作灵感来源于澳大利亚园林中带有异国情调并极具装饰意味的植物形象。首饰作为一种便携式工具，不仅可以打开人与人之间的对话，也可以用于思想之间的交流。她的作品（图4-28、图4-29）成功地激发了佩戴者和观众之间的互动反应。作者通过该作品表达对植物的热爱、探索地方归属和生命周期的概念。

Duan Yanli's work "The Other Side of Cloud and Mountain" (Figure 4-27) of the Chinese Academy of Fine Arts expresses the emotions that the author was shocked by the natural and pure natural scenery, the pristine human life and the sacred religious culture. Natural fabrics, wearing methods, rich colors and rich totem patterns in Tibetan costumes are all symbols of their cultural heritage. The designer deconstructed and reorganized the Tibetan clothes, using the plastic models of the unique animal yak and Tibetan antelope in Tibet as the visual subject, and combining the daily collections such as LED lights and plastic toys on bicycles to "integrate". The related materials are gathered together, but in reality, through the logic of harmony and connotation, to express the admiration, worship and admiration of Tibet. The author breaks the boundaries between folklore and international, traditional and contemporary, and gives a new interpretation of minority culture.

Jewelry designer Vicki Mason's creations are inspired by the exotic and highly decorative botanical image of Australian gardens. As a portable tool, jewelry enables the dialogue between people, in addition, the communication between ideas. Her work (Figure 4-28、Figure 4-29) successfully stimulated the interaction between the wearer and the audience. Through this work, the author expresses the concept of love for plants, exploring local ownership and life cycle.

英国设计师霍娜拉，将首饰设计与编织技术结合运用在她具有科学和数学灵感的可穿戴艺术作品中。在其作品中经纬线通过穿插、交叉、重叠、缠绕等方式在长短、间距、方位之间做变化调整，形成具有不同疏密关系的、节奏感的编织纹样。不同编织针法的组合可形成千变万化的艺术效果，图4-30中的项饰作品体现出了轻盈、浪漫的柔美风格。

British designer Nora Fok combines jewellery design and weaving techniques in her science and math-inspired wearable art. In her works, the latitude and longitude lines are adjusted in length, spacing and orientation by interspersing, intersecting, overlapping, winding, etc., forming a woven pattern with different senses of density and rhythm. The combination of different knitting methods can create an ever-changing artistic effect. The item in Figure 4-30 reflects a light, romantic and feminine style.

图4-28　维克·梅森，胸针
Figure 4-28　Vicki Mason, brooch

图4-29　霍纳拉，项饰
Figure 4-29　Nora Fok, necklace

图4-30　维克·梅森，项链
Figure 4-30　Vicki Mason, necklace

二、包袋
B. Bag

包袋是服饰配件中应用最广泛的类别之一。因其实用性远大于装饰性,因此包袋设计既要有装饰的"美感"又要有实用的"功能"。针对包袋的设计,可以从材质、色彩、结构、装饰等不同设计元素方面进行考虑,其中材质的运用在包袋设计中影响尤为重要。材质是体现包袋造型的物质基础,材质不同,其表面肌理、手感、光泽不同,做出来的包袋风格也迥然不同。如包袋采用野性而又奢华的蛇皮纹理,能够让佩戴者显得高贵出众;前卫和帅酷的包袋则大量采用铆钉、链条、锁扣等金属装饰物,摇滚味十足;塑胶包袋则使女性富有新鲜而精彩的时尚感觉,清爽透亮、手感亲切,且易于去污养护;编织感觉的包袋如冬日里帽子或围巾一样给人以温暖亲切的感觉,适合秋冬季节使用等。在设计手法上多种技巧"叠加",在造型、色彩上加以变幻,使包袋饰品更加美观、引人入胜。

Bags are one of the most widely used categories in apparel accessories. Because its function is far greater than the decorative nature, the bag design must have both the aesthetic feeling of decoration and a practical "function". For the design of the bag, it can be considered from different design elements such as material, color, structure and decoration. The application of materials and design elements is especially important in the design of the bag. The material is the material basis that reflects the shape of the bag. If the material is different, the surface texture, feel and luster are accordingly different, and the style of the bag is also very different. For example, the bag uses a wild and luxurious snakeskin texture to make the wearer look noble; the avant-garde and cool bags are made of metal ornaments such as rivets, chains, locks, etc. The bag gives women a fresh and wonderful fashion feeling, refreshing and translucent, feeling intimate and easy to decontaminate; the bag of weaving feels like a winter hat or scarf gives a warm and intimate feeling, suitable for autumn and winter seasons. In the design method, a variety of techniques "overlay", in the shape, color changes, making the bag jewelry more beautiful and fascinating.

1. 棉纤维
Cotton fiber

各种织物的主要原料,棉织物是包袋设计中经常使用的材料。通过不同的加工工艺,棉纤维可以形成风格迥异的外观,可以硬挺粗犷,也可以细腻光滑。不同的风格外观,为使用棉纤维进行包袋设计提供了广阔的设计空间。如香奈儿的包包中便是利用棉毛纤维,呈现出粗犷的面料肌理效果(图4-31)。

Cotton fiber is the main raw material of various fabrics, and cotton fabric is a material often used in bag design. Through different processing techniques, cotton fibers can form a different style of appearance, which can be rough and rough, and can also be fine and smooth. Different styles of appearance provide a broad design space for bag design using cotton fibers. For example, Chanel's bag uses cotton wool fibers to present a rough fabric texture effect (Figure 4-31).

图4-31 香奈儿,包包
Figure 4-31 Chanel, bag

第四章 非金属材料在服饰配件领域的制作与工艺

2. 麻纤维
Hemp fiber

是世界上最早被人类使用的纤维，麻织物较棉织物粗硬且挺括干爽，非精纺麻织物表面有粗结纱和大肚纱，会构成麻织物独特的粗犷风格，运用在包袋的设计上，有一种特殊的自然淳朴美感。MARC BY MARC JACOBS（MMJ）2008秋/冬便结合时尚与环保，推出了以有机苎麻纤维制成的Miss Marc Goes Green系列包款（图4-32），来呼吁大家关爱拯救地球。这一系列包袋不但选用有机材质，而且袋上可爱的Miss Marc或企鹅图案，是用不含化学物质、对环境无害的染料印染，可以说整个包的制作过程都不破坏地球生态。通过此类包袋的推广，人们不再追求华丽的合成纤维，而天然、环保、前卫已成为现代包袋设计领域的新宠，在人们体现时尚的同时更体现生态化设计。

Hemp fiber is the earliest fiber used in humans in the world. Hemp fabric is thicker and crisper than cotton fabric. The surface of non-refined woven fabric has thick knots and big belly yarn, which constitute the unique rough style of hemp fabric. Used in the design of the bag, there is a special natural and simple beauty. MARC BY MARC JACOBS (MMJ) 2008 combines fashion and environmental protection with the launch of the Miss Marc Goes Green series of organic ramie fibers (Figure 4-32) to appeal to everyone to care for and save the planet. This series of bags is not only made of organic materials, but also the cute Miss Marc or penguin pattern on the bag is printed with dyes that are chemical-free and environmentally friendly. It can be said that the whole package is not destroyed by the earth's ecology. Through the promotion of such bags, people no longer pursue gorgeous synthetic fibers, and natural, environment friendly, and avant-garde have become the new fond of modern bag design, and they reflect the ecological design while reflecting fashion.

3. 丝纤维
Silk fiber

光滑柔软，富有弹性，但耐晒和耐磨性都不如棉织物和麻织物，价格也较为昂贵，因而在包袋设计中多用于工艺包，而较少用于休闲包的设计。

Silk fiber is smooth, soft and elastic, but its exposure resistance and abrasion resistance are not as good as cotton fabrics or hemp fabrics, and the price is also relatively expensive. Therefore, it is mostly used to design craft bags, and less used in leisure bag design.

4. 皮革
Leather

在包袋设计中运用较多，不同的原料皮（真皮、革、合成革等），经过不同的加工方法，可以形成不同的外观风格，适合不同风格包袋的设计。由于真皮成本较高，多用于高档皮包的设计。在大众包袋的设计，真皮较多用于肩带、提手等部位，及与其他面料进行拼接设计使用，主体包袋用合成革材料。

Leather is used more in bag design. Different raw materials (leather, synthetic leather, etc.) can be formed into different appearance styles through various processing methods, which are suitable for different styles of bag design. Due to the high cost of leather, it is mostly used in the design of high-end

图4-32 马克·雅可布，2008秋冬
Figure 4-32 MARC BY MARC JACOBS (MMJ) autumn and winter 2008

leather bags. In the design of the mass bag, the leather is used for shoulder straps, handles and other parts, and is used in combination with other fabrics. The main bag is made of synthetic leather.

西西里风情是杜嘉班纳的主题。设计师完美地将民风、民俗融入奢华感之中，如2013春/夏女装秀场的配饰使用了大量的藤编元素，无论是彩色藤编包袋（图4-33），还是高跟鞋的藤编镶边，都透露出自然气息的淳朴风情。

Sicilian amorous feelings are the theme of Dug arbanna. The designer balances the folk customs and folk customs into a sense of luxury. For example, the accessories for the spring/summer 2013 women's show displayed a large number of rattan elements, whether it is a colored rattan bag (Figure 4-33) or a rattan edging of high heels illustrate the simple and natural style.

5. 多种纤维组合
A variety of fiber combinations

天然纤维材质可以单独使用，也可以与其他材料组合后使用，这样能够发挥不同材料的优势，起到加强或柔化的作用。如图4-34香奈儿2017秋/冬款包包，棉毛纤维营造出粗犷风格，与皮革搭配柔化了棉毛纤维的粗糙感，整体增加了细腻的感觉。

Natural fiber materials can be used alone or in combination with their materials, which can take advantage of different materials to enhance or soften. As shown in Figure 4-34, Chanel's autumn and winter 2017 bag, cotton wool fiber creates a rough style, and with the leather to soften the rough feeling of cotton wool fiber, the overall increase of the delicate feeling.

6. 色彩搭配
Color matching

不同的色彩可以使人产生不同的情感，图案的色彩搭配，也是包袋风格的决定因素之一。天然材质包袋在色彩上的运用主要为自然色调，以表现返璞归真的风格。如图4-35杜嘉班纳的包袋中，天然纤维的自然色调与高纯度的红、蓝、白等颜色形成强烈的视觉冲击力，具有浓郁的民族风情。

Different colors can stir different emotions, and the color matching of patterns is also one of the decisive factors of bag style. The color applied on the natural material bags is mainly natural tones, in order to express the origin returning style. As shown in Figure 4-35, Dolce & Gabbana's bags have a strong visual impact on the natural color of natural fibers and high-purity red, blue and white colors, and have a strong ethnic flavor.

图4-33 杜嘉班纳，2013春夏
Figure 4-33 Dolce & Gabbana, spring and summer 2013

图4-34 香奈儿，2017年秋冬
Figure 4-34 Chanel, autumn and winter 2017

图4-35 杜嘉班纳，2013春夏
Figure 4-35 Dolce & Gabbana, spring and summer 2013

7. 装饰手法
Decorative techniques

装饰是表现包袋风格的基础，表现手法丰富多样，以下仅列举其中几种来说明：

Decoration is the basis for the expression of the bag style, and the expression techniques are rich and varied. The following are just a few examples:

①编结是利用条绳类形态的材料，手工或机器进行编织。在天然材质中，如皮绳、麻绳、草绳以及纸捻绳都可以直接用于包袋的编结。包袋的设计除了可以完全使用编结手法，也可以与面料相结合形成另一番美感。此外，将面料二次加工制成条状之后再进行编结，也可以作为一种思路方法运用在包袋设计中。如图4-36以针织著称的意大利品牌米索尼包包，几何抽象图案、多彩线条再加上优良的编结工艺，使米索尼的包袋设计具有一定的艺术感染力。

图4-36 米索尼，手包
Figure 4-36 Missoni, handbag

图4-37 香奈儿，2010年春夏
Figure 4-37 Chanel, spring and summer 2010

The knitting is done by hand or machine, using strip-like rope materials. In natural materials, such as leather rope, hemp rope, straw rope and paper reins can be directly used for the knitting of bags. In addition to the complete use of the knitting technique, the bag can also be combined with the fabric to create another aesthetic. In addition, the secondary processing of the fabric into strips and then knitting, can also be used as a method of thinking in the bag design. As shown in Figure 4-36, the Italian brand Missoni bag with knitting, geometric abstract patterns, colorful lines and excellent knitting process make Missoni's bag has a certain artistic appeal.

香奈儿2010年春夏时装发布会上的编织包，运用编、织、结多种工艺制作成手提包、斜挎包等如图4-37所示，让人目不暇接。使用草、藤、纸编织包袋，表现出田园、休闲的风格，可搭配休闲装，供户外使用。编织包袋除了可以使用同质的编织手柄外，还可以使用竹、木、藤制的手柄，再搭配刺绣、流苏、穗饰，表现出民族风格。

The woven bag of Chanel's spring/summer 2010 fashion show was made into a handbag, a cross-body bag, etc. by using various techniques such as knitting, weaving and knotting (Figure 4-37). It uses grass, rattan and paper woven bags to show the pastoral and casual style. It can be used with casual wear and outdoors. In addition to the same woven handle, the woven bag can also be used with bamboo, wood and rattan handles, and then with embroidery, tassels to demonstrate the national style.

②刺绣是包袋装饰中主要的运用手法。刺绣种类繁多，不同种类的刺绣手法和图案会呈现不同的艺术风格。在包袋设计中运用刺绣手法时，要考虑到包袋的造型和材料，例如，刺绣图案放置的方位，选用的图案、色彩等，都要和包袋的整体风格协调。如（图4-38）古驰刺绣包中将刺绣图案的融入到整体的包袋造型中。

Embroidery is the main application method in bag decoration. There are many kinds of embroidery, and different kinds of embroidery techniques and patterns will present different artistic styles. When using the embroidery technique in the design of bag, the shape and material of the bag should be taken into consideration, For example, the orientation of the embroidery pattern, the selected pattern, the color, etc., should be coordinated with the overall style of the bag. For example (Figure 4-38), the embroidery pattern in the Gucci embroidery bag is integrated into the overall shape of the bag.

③镶嵌手法在包袋设计中也是较为重要的方法，运用得当可以成为点睛之笔。镶嵌物选择广泛，如木材、石材（这两种主要指制成的各种珠片）、小型贝壳、兽骨等，还有将烘干定型的植物（主要是花卉）来镶嵌装饰。镶嵌方法可以是黏连或钉绣（珠片类镶嵌物）。如图4-39杜嘉班纳以西西里神秘花园为灵感的"小姐"系列手包，采用了金属、宝石、水晶、手工木雕等各种名贵物料，利用卓越的手工刺绣及镶嵌技术，打造出一系列配以锁扣的盒子手包。

Inlaying method is also a more important method in bag design, and it can be the crowning touch if used propert. Inlays are widely used, such as wood and stone (the two mainly refer to various kinds of beads), small shells, animal bones, etc., as well as plants that are to be dried and shaped (mainly flowers) to be inlaid and decorated. The damascene method can be glued or studded (bead inlays). As shown in Figure 4-39, Dolce & Gabbana's Miss series handbags inspired by the Sicilian mysterious gardens use a variety of precious materials such as metal, gemstones, crystals, hand-carved wood carvings, using a variety of hand-embroidered and inlaid techniques to create a series of locks.

④流苏手法在包袋设计中虽然不占主体地位，但会产生特别的效果，甚至在某些风格的包袋设计中成为主要装饰方法。流苏应用的部位不仅在包袋的边缘处，甚至也有整个包身使用流苏装饰的案例，给人强烈的视觉冲击力。流苏手法采用的材料形态主要为条状或珠串状，材质也可以是多种类别。条状材质主要是皮革，如图4-40普拉达羊羔皮流苏手袋和纳帕鸡蛋包，包袋通体采用流苏来装饰，营造出了浓郁的时尚感。

The tassel technique does not occupy the main position in the bag design, but it will have a special effect, and even become the main decoration method in some styles of bag design. The tassel application is not only at the edge of the bag, but also has a case where the entire body is decorated with tassels, giving a strong visual impact. The material shape of the tassel technique is mainly strip or bead string, and the material can also be of various types. The strip material is mainly leather, as shown in Figure 4-40. PRADA lamb skin fringed handbag Nappa Frange Bag, the bag is decorated with tassels to create a rich sense of fashion.

图4-38 古驰，刺绣包
Figure 4-38 Gucci, Embroidery package

图4-39 杜嘉班纳Miss系列手包
Figure 4-39 Dolce & Gabbana, miss series handbag

图4-40 普拉达，纳帕扇形包
Figure 4-40 Prada, nappa frange bag

第四章 非金属材料在服饰配件领域的制作与工艺

8. 运用移借法
The use of the moving and transferring method

是将同类或不同类的已有造型或组成部分，进行有选择的转移借用，形成新的造型或新的装饰。移借的对象也可以是其他设计领域的主要造型。采用移借法进行设计能表现别出心裁、富有创意的效果。由于移借对象多为人们熟悉的内容，最后的作品面貌往往给人一种"熟悉的陌生感"，因此较其他常规设计作品的认可速度更快。如图4-41为2016年巴黎世家和图4-42 2007款路易威登2包袋。

The existing form or component of the same type or different types is selectively transferred and borrowed to form a new shape or a new decoration. The object being moved can also be the main shape of other design fields. Designing with the use of moving and transforming method can deliver a creative effect. Since the borrowed objects are mostly familiar, the final works often give people a "familiar strangeness" and therefore will be recognized more fastly than other conventional designs. Figure 4-41 shows the Balenciaga of 2016 and the Louis Vuitton 2007 bag of Figure 4-42.

（1）运用替换法指替换原来物件的某一部分状态，使之产生新的形态。

Use the replacement method: The method refers to replacing a certain part of the original object to a new form.

（2）装饰手法变换。

The transformation of decorative method.

例如，在刺绣手法上可以改变传统刺绣的丝线，用其他材料如麻线、棉线或毛线等代替，可以呈现出特别的效果。古驰白色刺绣手包中（图4-43-1）的刺绣改变了传统刺绣方法，将丝线变换为棉线或毛线替代，打破以往平面装饰的图案，微立体凹凸的刺绣方式给人更强的视觉冲击力。

For example, in the embroidery technique, the thread used in the traditional embroidery can be changed, and other materials such as twine, cotton thread or wool can be used instead, and a special effect can be exhibited. As shown Figure 4-43-1, the embroidery in the Gucci white embroidered handbag changes the traditional embroidery method, transforming the silk thread into a cotton thread or a wool thread to replace the pattern of the previous flat decoration, and the microscopic three-dimensional embroidery method gives people a stronger visual impact.

9. 色彩变换
Color conversion

天然材质休闲包袋的色彩多为自然色调，在设计中，可以打破这一常规，应用其他色调，给予天然材质休闲包袋新的色彩感觉，在（图4-43-2）爱马仕盒形拼皮手拎藤编包的设计中就削弱了原本滕材料的质朴感，增添了时尚简洁感。

Natural materials casual bags are mostly natural in color. In the design, you can break this routine and apply other colors to give the natural material a casual bag a new color feel, as shown in Hermes box-shaped hand-woven rattan bag Figure 4-43-2. The design has weakened the simplicity of the original material, adding a sense of style and simplicity.

10. 工艺变换
Process change

不同的制作工艺也会使得包袋具有不同的风格。如英国品牌装饰品的索菲亚大号绣花拉链包，将粗犷风格休闲包的外贴袋结构变换到民族风格的包袋上，呈现出混合的特殊风格（图4-43）。

图4-41　巴黎世家，2016
Figure 4-41　Balenciaga, 2016

图4-42　路易威登，2007
Figure 4-42　Louis Vuitton, 2007

Different production processes will also render the bags with different styles. As shown in the British brand Accessorize Sophia large embroidered zipper bag, the outer patch bag structure of the rough style casual bag is transformed to the ethnic style bag, presenting a mixed special style (Figure 4-43).

由于构成包袋的材料选用相对于服装来说范围较大，因此在包袋的设计中，合理地运用材质，运用不同材料的质感对比，是完成设计的必要手段。包袋与服装的搭配可根据不同需求心理、审美情趣作相应的变化。例如，当服装的面料较为细腻时，可选择质感粗犷而奔放的包袋如图4-44所示，也可选择质感硬挺的包袋如图4-45所示；当服装面料较为厚重而凹凸不平时，则可选择一些肌理光润柔滑的手袋，与服装面料形成鲜明对比如图4-46所示。总之，从服饰的整体材质效果来看，两者之间既可相互对比也可相互补充，既可互相衬托又可相互协调，在搭配变化中产生出特有的视觉美感。此外，包袋设计除了要考虑人、自然等因素，包袋的款式、色彩、材料要与着装人的年龄、职业、性格等状况相协调，同时也要考虑到民族习俗、季节场合，并要有所区别。再则，包袋设计不论造型、色彩和材料，都是在表现服饰的个性和特征。包袋从属于大的服装范畴，在包袋设计中，要做到与服装搭配相互补充，构成整体的装饰美感，这样才能真正地传达出服饰的内涵。对于包袋，应既要与着装的整体风格相配，又要具有独特的个性。

Since the material constituting the bag is relatively large in comparison with the clothing, in the design of the bag, the rational use of the material and the comparison of the texture of the different materials are necessary means for completing the design. The combination of bags and clothing can be changed according to different needs psychology and aesthetic taste. For example, when the fabric of the garment is more delicate, the bag with rough texture and unrestrained can be selected as shown in Figure 4-44, and the bag with firm texture can be selected as shown in Figure 4-45; when the fabric is thick and uneven, the texture can be selected. The silky handbag is in sharp contrast to the fabric (Figure 4-46). In summary, from the perspective of the overall material effect of the clothing, the two can be compared with each other or complement each other, which can set off each other and coordinate with each other, and convey a unique visual beauty in the matching change. Besides, It can be seen that in addition to factors such as human and nature, the design, color and materials of the bag should be coordinated with the age, occupation and personality of the dresser. At the same time, national customs and seasonal occasions should be considered. And there must be a difference. In addition, the bag

图4-43-（1） 古驰，白色刺绣手包
Figure 4-43-1 Gucci, white embroidery bag

图4-43-（2） 爱马仕盒形拼皮手拎藤编包
Figure 4-43-2 Hermes box spear handler wines

图4-43-（3） 配饰-索菲亚大号绣花手包
Figure 4-43-3 Accessorize-sophia large embroidered handbag

第四章 非金属材料在服饰配件领域的制作与工艺

图4-44 苏菲·安德森
Figure 4-44 Sophie Anderson

图4-45 杜嘉班纳
Figure 4-45 Dolce & Gabbana

图4-46 图片来源于网络
Figure 4-46 Picture from the internet

design, regardless of shape, color and material, is the expression of the personality and characteristics of the clothing. Bags belong to the category of large clothing. In the bag design, it is necessary to complement the clothing to form the overall decorative beauty, so as to truly convey the connotation of the clothing. For the bag, it should be matched with the overall style of the dress, but also have a unique personality.

三、帽饰
C. Hat decoration

在帽饰的设计中，材料的选择同样也是重要的，它决定着帽饰的整体造型、风格以及帽子的功能。不同材料的帽饰有着不同的风格，如运用柔软的纱、绸、蕾丝等材料，会给人浪漫唯美心理感受；皮革、树脂、金属等材质给人严肃冷峻感。夏季的帽饰一般选用轻快、飘逸的面料来凸显帽饰的功能性，而冬季的帽饰一般选用保暖、厚实的面料来实现帽饰的保暖功能。如英国凯特王妃在不同季节佩戴不同纤维材质的帽饰（图4-47、图4-48）。从设计效果的角度来看，不同的材料组合可形成更丰富的肌理，并能为设计提供更广阔的设计空间。随着人们对个性化装扮的需求，依靠帽饰来表达个性和搭配风格的手法受到越来越多人追捧。无论从T台还是在日常生活中，帽饰设计都呈现个性化、创新化的趋势。

In the design of the hat decoration, the selection of materials is also important, which determines the overall shape, style and function of the hat. Different materials would create different styles of hats, such as the use of soft yarn, silk, lace and other materials, carrying a romantic and beautiful psychological feeling; leather, resin, metal and other materials give people a serious sense of cold. Summer hats generally use light, flowing fabrics to highlight the functionality of the hat, while winter hats generally use warm, thick fabrics to achieve the warmth of the hat. As shown in the United Kingdom, Princess Kate wears different fiber caps in different seasons(Figure 4-47、Figure4-48). From

图4-47 图片来源于网络
Figure 4-47 Picture from the internet

图4-48 图片来源于网络
Figure 4-48 Picture from the internet

129

the perspective of design effects, different material combinations can create a richer texture and provide a broader design space for the design. With the demands for personalized dressing, the method of relying on hat ornaments to express individuality and matching style has been sought after by more and more people. Whether it is from the T-stage or in daily life, the hat design is personalized and innovative.

佩戴不同材质的帽饰会带给人完全不同的服装风格效果。如华丽的晚宴服装适合佩戴光泽型纤维材料帽饰，若配毛纤维编织帽饰会给人一种搭配混乱的印象；而休闲运动装不适合佩戴丝毛光泽性纤维帽饰。所以帽饰材质的选择在服饰搭配整体造型中是关键要素。

Wearing different material hats will give you a completely different style of clothing. For example, the gorgeous evening dress is suitable for wearing a glossy fiber material hat, and if it is equipped with a woolen fiber woven hat, it will gives a mismatched impression; while the casual sportswear is not suitable for wearing a silky glossy fiber cap. Therefore, the choice of the material of the hat is a key element in the overall shape of the clothing.

1. 棉麻、丝绒、真丝类纤维
Cotton, velvet, silk fiber

由于它们自身所带的光泽可以给帽饰增添许多亮点，这类纤维质感和手感都比较好。运用这些材料本身的光泽度就可以在帽饰造型上制作丰富的变化效果，并且能与时装形成反差或者呼应等。这类纤维材质也常用于时尚夸张的帽饰设计风格中，同时它的舞台表现力也非常强，可以根据其垂坠度进行线条的造型，能表现出纤维线条的流动感。如图4-49和图4-50中轻薄、垂坠感好，线条光滑，轮廓自然舒展是这类帽饰的重要优点。质地轻薄、透明度高，最适合于表达柔和感觉或若隐若现效果，在现代时装帽饰中常属于优雅唯美型的设计风格。但其纤维质地太柔软，则不太适合单独作为帽饰材料，需进行特别处理或面料改造后，才能呈现所需要的外观效果。

Because of their own luster, they can add a lot of bright spots to the hat decoration. These fibers have better texture and feel. By using the gloss of these materials, you can make rich changes in the shape of the hat, and can be contrasted with fashion or echoes. This kind of fiber material is also often used in the fashion exaggerated hat design style, and its stage performance is also very strong, capable of being lined according to its drape, and showing the flow of fiber lines. Such as Figure 4-49 and Figure 4-50, the thin, drape, smooth lines and natural stretch are the important advantages of this type of hat. The texture is light and the transparency is high. It is most suitable for expressing the soft feeling or the effect that is looming. It is often an elegant and aesthetic design style in modern fashion hats. However, the fiber texture is too soft, so it is not suitable for use as a cap material alone, and special treatment or fabric modification is required to present the desired appearance.

图4-49　图片来源于网络
Figure 4-49　Picture from the internet

图4-50　卡米拉，帽子
Figure 4-50　Comilla, hat

第四章　非金属材料在服饰配件领域的制作与工艺

2. 毛呢、皮类纤维
Wool, leather fiber

这类纤维线条清晰硬挺，比较有质感，厚重挺括，具有较强的定型和造型效果，能为帽饰设计提供坚挺的线条轮廓，特别是造型夸张类的帽饰，给人视觉上的饱满膨胀感纤维材料，易达到设计的预期效果。化学类纤维同天然丝纤维一样具有良好的柔软性，但其具有一定的可塑性，较之天然真丝纤维更适合制作帽饰。这类纤维在现代时装帽饰设计风格中都频繁地运用。如图4-51和图4-52中，英王室成员在秋冬季节佩戴的帽饰多采用毛呢纤维类材料制作。

These fiber lines are clear and stiff, relatively textured, thick and sturdy, with strong shaping and styling effects. They can provide a strong line outline for the hat design, especially the exaggerated type of hat decoration, giving the full expansion of the fibrous material on the surface, visually easy to achieve the desired effect of the design. Chemical fibers have the same softness as natural silk fibers, but they have a certain degree of plasticity and are more suitable for making hats than natural silk fibers. These fibers are frequently used in modern fashion hat design styles. As shown in Figure 4-51 and Figure 4-52, the caps worn by members of the British royal family in the autumn and winter seasons are mostly made of woolen fiber materials.

菲利普·特里西是英国最具人气和商业价值的"帽饰王子"。他的设计突出帽子强烈的冲击力和装饰风格，如图4-53。在材料选择上有纤维纱网、羽毛、胶片、藤蔓及木材等；造型上大量撷取未来主义、超现实主义等意像；海底生物、动植物的外观造型，再将它们拼装组合，构成帽饰的基本轮廓，然后将那些突破传统的材质填充进去，形成一种独特的视觉语言。反传统的材料和抽象性造型是其最主要特点。

Philip Treacy is the UK's most popular and commercial "hat prince." His design highlights the strong impact and decorative style of the hat as shown in Figure 4-53. The material selection includes fiber gauze, feathers, film, vines and wood; a lot of imaginations such as futurism and surrealism are used in the styling; the appearance of the sea creatures, animals and plants, and then they are assembled and assembled to form the basic outline of the hat decoration, and then the materials that break through the tradition are filled in to form a unique Visual language. Anti-traditional materials and abstract shapes are their main features.

史蒂芬·琼斯有非常多独特的设计创意，他曾与克里斯汀·迪奥、约翰·加里亚诺、让·保罗·高尔蒂埃等设计大师进行过合作，为他们的高级女装设计相应的女帽。史蒂芬·琼斯的帽子将一种复古的情绪融入现代时装帽饰的设计精神之中，而且能够完全将这种风潮表达出来（图4-54和图4-55）。从戴安娜王妃到摇滚明星，都成了他最为忠实的拥护

图4-51　英女王，帽子
Figure 4-51　Queen of England, hat

图4-52　梅根，帽子
Figure 4-52　Meghan Markle, hat

图4-53　菲利普·特里西，帽饰
Figure 4-53　Philip Treacy, hat

者。他也多次为迪奥的高定时装秀设计帽饰。

Stephen Jones has a lot of unique design ideas. He has worked with design masters such as Christian Dior, John Galliano and Jean Paul Gaultier to design the corresponding women's hats for their haute couture. Stephen Jones's hat incorporates a retro mood into the design spirit of modern fashion hats, and it can fully express this trend (Figure 4-54、Figure 4-55). From Princess Diana to rock star, many celebrities have become his most loyal supporters. He has also designed hats for the Haute couture show at Dior.

皮尔斯·艾特金森，2011秋冬头饰系列"巴黎"，很难被称作真正意义上的帽子，但却深受各大时尚杂志以及明星名流们的喜爱。樱桃造型的头饰（图4-56）也成了皮卡斯·艾特金森的招牌款式，不仅可爱俏皮而且刷新了帽饰廓型的记录。风格既雅致又神秘，同时还充满幽默。头饰以午夜蓝和魅黑紫为主要色调，材料有光滑的绒面革、温暖蓬松的羽毛、质感厚实的毡呢、硬朗的金属尖钉、柔软的薄纱等（图4-57）。与所有追求极致的设计师一样，皮系斯·艾特金森完全将自己的作品当作艺术品来打造，甚至是雕塑作品，这让他的设计虽然乍一看颇为怪诞，却有着强大的气场，深受明星们的喜爱。他认为"一顶美丽的帽子是人们每天的必需品，帽子能给人带来自信我关注更多的是一种风格的缔造，它不一定适合你，但能形成一股风潮，进而影响你的时尚品位。"

Piers Atkinson, 2011 Fall Winter Headwear Collection "Paris", is hard to be called a real hat, but is very popular among fashion magazines and celebrities. The cherry-shaped headpiece (Figure 4-56) also becomes the signature style of Piers Atkinson, not only displays cute and playful style, but also has refreshed the record of the hat silhouette. The style is both elegant and mysterious, full of humor. The headwear is dominated by midnight blue and charm black purple(Figure 4-57). The material is smooth suede, warm fluffy feathers, thick felt, tough metal spikes, soft tulle and more. Like all the most sought-after designers, Piers Atkinson completely uses his work as a work of art to create, even a sculpture, which makes his design quite grotesque at first glance, but it does have a strong atmosphere and quiteliked by stars. He believes that a beautiful hat is a daily necessity for people, and a hat can bring confidence —"I pay more attention to the creation of a style, it may not be suitable for you, but it can form a wave, and then affect Your fashion taste."

图4-54 史蒂芬·琼斯，帽饰-1　　图4-55 史蒂芬·琼斯，帽饰-2
Figure 4-54 Stephen Jones, hat-1　　Figure 4-55 Stephen Jones, hat-2

图4-56 皮尔斯·艾特金森，2011秋冬巴黎系列头饰　　图4-57 皮尔斯·艾特金森，2011秋冬巴黎系列头饰
Figure 4-56 Piers Atkinson, autumn and winter 2011"Paris"hat　　Figure 4-57 Piers Atkinson, autumn and winter 2011"Paris"hat

Kreuzzz是中国设计师胡钰珑，2012年创立的帽子品牌，其帽子作品中多使用兔毛、麻布、真丝、马尾、花朵、金属链条及蜡等材料。运用定型、手工缝制、上胶和防水处理、中式大漆、轻微灼烧等工艺，如图Kreuzzz 2015春/夏系列深色青苔。帽子不论从材质上、造型上还是色彩上都与深色青苔衣服的调性形成了和谐的搭配（图4-58）。

第四章 非金属材料在服饰配件领域的制作与工艺

大多数纺织纤维材料都可以制作或者装点帽饰，如毛绒、丝绒、棉麻等，同时结合其他的羽毛、绸带、金属等材料。工艺成形通过缝制的方法将布片缝合而成，也有的利用棉线或者麻线编织而成的帽子，针织因其工艺成品的弹性比较大，也是帽子常见的制作工艺。此外，利用模型和工具改变材料组织结构也可以制作出立体感的帽子（图4-59、图4-60）。

Kreuzzz is a hat brand created by Chinese designer Hu Wei in 2012. His hat works mostly use rabbit hair, linen, silk, horsetail, flowers, metal chains and wax. Use stereotypes, hand-stitching, gluing and waterproofing, Chinese lacquer, light burning, etc., as shown in the deepmoss × Kreuzzz 2015 spring and summer series. The hat forms a harmonious match with the tonality of deepmoss clothes in terms of material, shape and color.

Most textile fiber materials can be made or decorated with hats, such as plush, velvet, cotton and linen, combined with other feathers, ribbons, metals and other materials. The process is formed by sewing the cloth piece by sewing method, and some of the hat is woven by cotton thread or twine. The knitting is also a common manufacturing process of the hat because of the relatively high elasticity of the finished product. In addition, the model and the tool are used to change the material. The organizational structure can also create a three-dimensional hat (Figure 4-59, Figure 4-60).

帽饰的种类比较多，帽饰作为配饰需根据服装的变化而变化。从服装风格上来说，帽饰应与服装相符，因为它们是一个整体，受到相互的制约。服装休闲宽松时，帽饰的造型可以适当夸张；服装修身紧致时，帽饰的造型也应该紧凑精致；衣服造型干练、简洁时，帽饰的式样也应该是干练、简洁的风格。设计师设计的帽饰无论是面向市场还是展示个性都应该从服装整体的角度考虑。通过帽饰的功能性或装饰性，体现设计师的创作理念、服饰作品的风格并能够表达出穿戴者的气质和风度。

There are many types of hats, and the hats as accessories need to be changed with the clothing. In terms of clothing style, the hats should conform to the clothing because they are a whole and are subject to each other. When the clothing is loose and relaxed, the shape of the hat decoration can be

图4-58　Kreuzzz深色青苔，2015春夏系列
Figure 4-58　Kreuzzz for deepmoss, 2015 A/W

图4-59　菲利普·特里西，帽饰
Figure 4-59　Philip Treacy, hat

图4-60　马里奥斯·施瓦布，无关工作室新娘帽
Figure 4-60　Marios Schwab, unconnected-studio brida

exaggerated properly; when the clothing is slim and firm, the shape of the hat decoration should also be compact and exquisite; when the clothes are sleek and simple, the style of the hat decoration should also be a simple and concise style. The designer's hats should be considered from the perspective of the overall clothing, whether it is facing the market or displaying personality. Through the functional or decorative nature of the hat decoration, the designer's creative concept, the style of the costume work and the temperament and grace of the wearer can be expressed.

优秀的帽饰作品在与服装进行风格协调时应具有独特性和一致性。此外帽饰与服装的配色也要讲究整体性和协调性。从帽饰的材料和服装相配的角度看，帽饰的材料应该尽量与服装面料一致。如毛呢的服装与毛呢的帽子搭配，毛线衣裙与毛线帽饰搭配（图4-61），特殊情况时可以根据需要适当变化。如牛仔服可以搭配牛仔帽，亦可以配草帽或者麻帽，但风格应保持一致（图4-62）。夏天穿着的针织衫、丝绸衣可以适当搭配质地细腻、编织精致的丝制帽饰（图4-63）。

Excellent hat ornaments should be unique and consistent in their style of coordination with the clothing. In addition, the color matching of the hat and the clothing should also pay attention to the integrity and coordination. From the point of view of the material of the cap and the matching of the garment, the material of the cap should be as close as possible to the fabric. For example, the woolen dress is matched with the woolen hat, and the sweater dress is matched with the woolen cap (Figure 4-61). In special cases, it can be changed as needed. For example, denim can be worn with a cowboy hat, or with a straw hat or a hat, but the style should be consistent (Figure 4-62). Knitwear and silk blouses that are worn in the summer can be paired with delicate silk woven hats (Figure 4-63).

图4-61 针织帽，图片来源于网络
Figure 4-61 Knitted hats, pictures from the Internet

图4-62 藤制牛仔帽
Figure 4-62 Rattan cowboy hat

图4-63 帽饰，图片来源于网络
Figure 4-63 Hats, picture from the internet

结语
D. Chapter summary

作为一门实用生活美学，服装配饰越来越受到人们的重视，其设计也标志着服饰设计观念的进步和开放，目前已成为设计师必须具备的基本艺术修养之一，这种素养不但具有实用意义且能培养追求美的精神境界，提高审美能力。

As a practical life aesthetic, clothing accessories get more attentions, and its design also marks the progress and opening of the concept of fashion design. At present, it has become one of the basic artistic accomplishments that designers must possess. This kind of quality is practically meaningful, in addition, it can cultivate the spiritual realm of pursuing beauty and improve the aesthetic ability.

纺织类纤维服饰配件在现代服装设计中使用的范围极其广泛，正是由于服饰配件的存在，服装的风格面貌才呈现出千变万化的形式。纺织类纤维因其丰富的材料面貌为服饰配件的设计提供了丰富的可能性，他们特有的材料艺术语言为设计师展开个性化的设计提供了广阔的空间。

服饰配件在服装整体设计中可以增强艺术效果，使款式变化丰富、促进实用功能，提高经济效益。服装的造型总在单纯的原型基础上进行变化，配件的合理搭配，才能形成有主次、有节奏、突出主题，深入浅出的对比效果，引起人们的联系回味，在视觉影响和心理感受上有突出的作用，令人悦耳赏心，富有魅力、美感。

Textile fiber accessories are widely used in modern clothing design. Due to the existence of clothing accessories, the style of clothing can present massively various forms. Textile fibers offer a wealth of possibilities for the design of apparel accessories due to their rich material appearances. Their unique material art language provides a broad space for designers to personalize their designs.

Apparel accessories can enhance the artistic effect in the overall design of the garment, make the style change rich, promote practical functions, and improve economic benefits. The shape of the clothing is always changed on the basis of the simple original line, and the reasonable matching of the accessories can form the primary, the rhythm, the prominent theme, and the contrast effect in the simple and simple way, causing people to contact and recollect, in terms of visual influence and psychological feeling. The outstanding effect is pleasing to the eye, full of charm and beauty.

第三节　玻璃材料和陶瓷材料
Section 3　Glass and ceramic materials

1. 玻璃的分类及应用特点

学习目的：通过本节的学习，使同学初步了解玻璃材料的各种分类、基本性能，以及应用特点。

本节重点：玻璃材料的应用特点。

课时参考：4课时。

Classification and application characteristics of glass

Learning Objectives: Through the study of this section, students will get a preliminary understanding of the various classifications, basic properties, and application characteristics of glass materials.

Focus of this section: the application characteristics of glass materials.

Class time reference: 4 class hours.

2. 玻璃材料制作服饰配件工艺

学习目的：通过本节的学习，使同学初步了解玻璃材料来制作服饰配件的制作工艺，主要包括热熔工艺、灯工工艺等基本知识。

本节重点：玻璃饰品的成形工艺。

课时参考：8课时。

Glass material manufacturing apparel accessories process

Learning Objectives: Through the study of this section, students will get a preliminary understanding of the manufacturing process of making apparel accessories from glass materials, including basic knowledge such as hot melt process and lamp work process etc.

The focus of this section: This chapter focuses on the forming process of glass jewelry.

Class time reference: 8 class hours.

3. 陶瓷材料制作服饰配件分类工艺

（1）陶瓷材料的分类及应用特点。

学习目的：通过本节的学习，使同学初步了解陶瓷材料的基本分类、常用材料及应用特点。

本节重点：陶瓷材料的基本分类。

课时参考：4课时。

Ceramic material manufacturing apparel accessories classification process

Classification and application characteristics of

ceramic materials.

Learning Objectives: Through the study of this section, students will get a preliminary understanding of the basic classification, common materials, and application characteristics of ceramic materials.

The focus of this section: This chapter focuses on the basic classification of ceramic materials.

Class time reference: 4 class hours.

（2）陶瓷饰品的制作工艺。

学习目的：通过本节的学习，使同学初步了解陶瓷饰品的制作工艺，包括工具材料、成形工艺、装饰手法以及烧成方式等基本知识。

本节重点：陶瓷饰品的成形工艺、装饰手法。

课时参考：8课时。

Ceramic accessories, manuracturing process.

Learning Objectives: Through the study of this section, students will get a basic understanding of the manufacturing process of ceramic jewelry, including tool materials, forming techniques, decorative techniques, and firing methods.

The focus of this section: This chapter focuses on the forming process and decorative techniques of ceramic jewelry.

Class time reference: 8 class hours.

（3）运用陶瓷材料制作服饰配件案例解析。

学习目的：通过本节的学习，使同学初步了解陶瓷饰品的创作构思、制作过程、工艺要求以及最终呈现等实际的操作流程。

本节重点：陶瓷饰品的案例分析、制作流程。

课时参考：4课时。

Case Analysis of Making Apparel Accessories Using Ceramic Materials.

Learning Objectives: Through the study of this section, students will get a preliminary understanding of the actual operation process of ceramic jewelry creation concept, production process, process requirements, and final presentation.

The focus of this section: the analysis and production process of ceramic jewelry.

Class time reference: 4 class hours.

玻璃是非晶无机非金属材料，它的主要成分为二氧化硅和其他氧化物。一般是以多种无机矿物（如石英砂、硼砂、硼酸、重晶石、碳酸钡、石灰石、长石、纯碱等）为主要原料，另外加入少量辅助原料制成的。原料在熔融时形成连续网络结构，在冷却过程中黏度逐渐增大并硬化，常温下为固态。

Glass is an amorphous inorganic non-metallic material whose main components are silica and other oxides. Generally, it is made of a variety of inorganic minerals (such as quartz sand, borax, boric acid, barite, barium carbonate, limestone, feldspar, soda ash, etc.), and a small amount of auxiliary materials are added. The raw material forms a continuous network structure upon melting, and the viscosity gradually increases and hardens during the cooling process, and is solid at normal temperature.

玻璃是一种用途非常广泛的材料，它晶莹剔透、色彩丰富，会给人以变幻莫测、出人意料的艺术效果，从而也给人们带来生活的喜悦和创作的灵感。玻璃具有的特性，使它迅速走入人们的生活中，特别是它成为当代首饰常用的材料时，它表现的已经不再是仅仅追求佩戴的美观，而是表现力的丰富。

Glass is a very versatile material. It is crystal clear and colorful, and it will give people an unpredictable and unexpected artistic effect, which will also bring people joy and creative inspiration. The characteristics of glass make it quickly enter people's lives. Especially when it becomes a common material for contemporary jewelry, it is no longer just the pursuit of wearing beauty, but the expressiveness.

当前，艺术品能应用多种材料，在人们追求精神和心理上的对社会、对自然更为精致的表现时，玻璃的材质特性，如光与影、折射与反射、内部色彩的流动等就是设计师新的着眼点，他们用光，用影像去反映微妙的内心活动。玻璃首饰中的光与影的相互映射，让首饰有着无穷魅力，成为当下人类生活情境中丰富的语言表达方式。

At present, artworks can use a variety of materials. When people pursue spiritual and psychological performance towards the society and the nature, the material properties of glass materials, such as light and shadow, refraction and reflection, internal color flow, etc are the designer's new focus. They use light and images to reflect subtle inner activities. The mutual mapping of light and shadow in glass jewellery makes jewelry have infinite charm and become a manifestation of rich language expression and life texture in the current human life situation.

一、玻璃材料的分类及应用特点
A. Classification and application characteristics of glass materials

1. 玻璃材料的分类
Classification of glass

根据玻璃自身的成分，分为非氧化物玻璃和氧化物玻璃。非氧化物玻璃：品种和数量较少，主要有硫系玻璃和卤化物玻璃。氧化物玻璃：分为硅酸盐玻璃、硼酸盐玻璃、磷酸盐玻璃等。玻璃的主要成分是二氧化硅，而根据碱金属、碱土金属氧化物的不同，又分为：

According to the composition of the glass itself, it is divided into non-oxide glass and oxide glass. Non-oxide glass: less variety and quantity, mainly sulfur glass and halide glass. Oxide glass: divided into silicate glass, borate glass, phosphate glass, and the like. The main component of glass is silica, and according to the difference of alkali metal and alkaline earth metal oxide, it is further divided into:

（1）石英玻璃。

Quartz glass.

石英玻璃的二氧化硅含量大于99.5%，其热膨胀系数低、耐高温、化学稳定性好、透紫外光和红外光、熔制温度高、黏度大、较难成型。石英玻璃多用于半导体、电光源、光导通信、激光等技术和光学仪器等。

Quartz glass has a silica content of more than 99.5%, featuring low thermal expansion coefficient, high temperature resistance, good chemical stability, high UV and infrared light, high melting temperature, high viscosity and difficult to form. Quartz glass is mostly used in semiconductors, electric light sources, optical communication, laser technology, optical instruments, etc.

（2）高硅氧玻璃。

High silica glass.

高硅氧玻璃的二氧化硅含量约96%，其性质与石英玻璃相似。烧结温度一般在1000℃。高硅氧玻璃具有类似石英玻璃的性质，耐高温、热膨胀系数低、化学稳定性好、透紫外线，在很多方面可代替石英玻璃。

High silica glass has a silica content of about 96% and is similar in nature to quartz glass. The sintering temperature is generally 1000°C. High silica glass has the properties of quartz glass, high temperature resistance, low thermal expansion coefficient, good chemical stability, and ultraviolet light transmission, which can replace quartz glass in many aspects.

（3）钠钙玻璃。

Soda lime glass.

钠钙玻璃以二氧化硅含量为主，同时还含有15%的氧化钠和16%的氧化钙，其成本低廉、易成型、适宜大规模生产，因此在建筑、日用玻璃制品方面有广泛的应用。通常所用的门窗玻璃也多属于钠玻璃及其深加工产品。它的产量占实用玻璃的大多数。钠钙玻璃可用于生产玻璃瓶罐、平板玻璃、器皿、灯泡等。

Soda-lime glass is mainly composed of silica, and contains 15% of sodium oxide and 16% of calcium oxide. It is cheap in cost and easy to form, suitable for large-scale production, so it has wide application in building and daily glass products. The window glass that we usually use is also mostly sodium glass and its deep processing products. Its output accounts for the majority of practical glass. Soda-lime glass can be made into produce glass bottles, flat glass, utensils, light bulbs, etc.

（4）铅硅酸盐玻璃。

Lead silicate glass.

铅硅酸盐玻璃主要成分有二氧化硅和氧化铅，具有独特的高折射率和高体积电阻，与金属有良好的浸润性，可用于制造灯泡、真空管芯柱、晶质玻璃器皿、火石光学玻璃等。

Lead silicate glass is mainly composed of silica and lead oxide. It has a unique high refractive index and high volume resistance. It has good wettability with metals and can be used to manufacture bulbs, vacuum tube columns, and crystalline glassware, Flint optical glass.etc.

（5）铝硅酸盐玻璃。

Aluminosilicate glass.

铝硅酸盐玻璃以二氧化硅和氧化铝为主要成分，它软化变形温度高，用于制作电灯泡、高温玻璃温度计、化学燃烧管和玻璃纤维等。

Aluminosilicate glass is mainly composed of silica and alumina. It has a high softening temperature and is used for making electric bulbs, high temperature glass thermometers, chemical combustion tubes and glass fibers.

（6）硼硅酸盐玻璃。

Borosilicate glass.

又称耐热玻璃或硬质玻璃，以氧化硅和氧化钡为主要成分，具有良好的耐热性和化学稳定性，硼硅酸盐玻璃发明于1912年，其商标名为Pyrex。

它是第一种耐高温，有较好抗热冲击能力的玻璃材料。硼硅酸盐玻璃材料制作的玻璃珠宝。硼硅酸盐玻璃可以用来制作咖啡壶、炉子、实验室用的玻璃器皿、吊灯及其他在高温环境中工作的设备。它抗酸和抗化学介质腐蚀的能力很强、热膨胀率很低，因此被用来制作天文望远镜的镜片和其他精密仪器。另外，硼硅酸盐玻璃还可以用作树脂的强化纤维。

Also known as heat-resistant glass or hard glass, with silicon oxide and yttrium oxide as main components, has good heat resistance and chemical stability. Borosilicate glass was invented in 1912 and its trade name is Pyrex.

It is the first glass material with high temperature resistance and good thermal shock resistance (as shown in the glass jewellery made of borosilicate glass). Borosilicate glass can be used to make coffee makers, stoves, laboratory glassware, chandeliers (pictured) and other equipment that operate in high temperature environments. It is highly resistant to acid and chemical media and has a low coefficient of thermal expansion, so it is used to make lenses and other precision instruments for astronomical telescopes. In addition, borosilicate glass can also be used as a reinforcing fiber of a resin.

（7）磷酸盐玻璃。

Phosphate glass.

磷酸盐玻璃以五氧化二磷为主要成分，折射率低、色散低，多用于光学仪器中。

Phosphate glass is mainly composed of phosphorus pentoxide, which has low refractive index and low dispersion, and is used in optical instruments.

（8）钾玻璃。

Potassium glass.

钾玻璃是以氧化钾代替钠玻璃中的部分氧化钠，并适当提高玻璃中氧化硅含量制成。它硬度较大，光泽好，又称作硬玻璃。钾玻璃多用于制造化学仪器和高级玻璃制品。

Potassium glass is made up of potassium oxide instead of partial sodium oxide in soda glass, and is made by appropriately increasing the content of silica in glass. It has higher hardness and good gloss, also known as hard glass. Potassium glass is mostly used in the manufacture of chemical instruments and High-grade glass products.

2. 玻璃的基本性能

Basic properties of glass

（1）强度。

Strength.

玻璃的一般强度为50~200mPa，玻璃抗拉强度较弱，抗压强度较强，它与陶瓷都是属于脆性材料。

The general strength of glass is 50~200mpa. The glass has weak tensile strength and strong compressive strength. It is a brittle material, like ceramics.

（2）硬度。

Hardness.

玻璃的硬度较高，玻璃的硬度在5～7摩氏，比一般金属硬，不能用普通刀具进行切割。根据玻璃的不同硬度可以选择磨料、磨具和其他的加工方法。

The hardness of the glass is relatively high. The hardness of the glass is between 5 and 7 Mohs. It is harder than ordinary metals and cannot be cut with ordinary tools. Abrasives, abrasives and other processing methods can be selected depending on the different hardness of the glass.

（3）光学特性。

Optical characteristics.

玻璃是一种高度透明的物质。如普通平板玻璃，能透过可见光线的80%～90%，紫外线大部分不能透过，但红外线较易通过。

Glass is a highly transparent substance. For example, ordinary flat glass can transmit 80%~90% of visible light, and most of the ultraviolet light cannot pass through, but infrared rays are easy to pass.

（4）电学性能。

Electrical performance.

常温下玻璃是电的不良导体。而温度升高时，玻璃的导电性迅速提高，熔融状态时变为较好的导电体。

Glass is a poor conductor of electricity at room temperature. When the temperature is raised, the conductivity of the glass is rapidly increased, and it becomes a good conductor in the molten state.

（5）热性质。

Thermal properties.

玻璃是热的不良传递导体，在一般情况下，不能承受温度的急剧变化。

Glass is a poor transfer conductor for heat and, under normal conditions, cannot withstand sudden changes in temperature.

（6）化学稳定性。

Chemical stability.

玻璃有较稳定的化学性质：有较高的耐酸腐蚀性，耐碱腐蚀性较差（即耐酸不耐碱）。玻璃一般不溶于酸（氢氟酸例外——与玻璃反应生成四氟化硅（SiF4），从而导致玻璃的腐蚀）；但溶于强碱，如氢氧化铯。如玻璃长期受大气和雨水的侵蚀，会在表面产生磨损，失去表面的光泽。尤其是一些光学玻璃仪器，易受到周围介质的作用，从而破坏玻璃的透光性。

在设计中常见的玻璃材料是一种硬而且脆的透明非晶体材料，具有良好的抗风化、抗化学介质腐蚀性。

Glass has a more stable chemical properties: It has high acid corrosion resistance and poor alkali corrosion resistance (ie acid resistance and alkali resistance). Glass is generally insoluble in acid (except hydrofluoric acid-reacts with glass to form silicon tetrafluoride (SiF4), which causes corrosion of the glass); but is soluble in strong bases such as barium hydroxide. If the glass is eroded by the atmosphere and rain for a long time, it will wear on the surface and lose the luster of the surface. In particular, some optical glass instruments are susceptible to the surrounding medium, thereby damaging the light transmission of the glass.

The glass material commonly used in our design is a hard and brittle transparent amorphous material with good weathering resistance and chemical resistance.

二、陶瓷材料的分类及应用特点
B. Classification and application characteristics of ceramic materials

1. 陶瓷材料的分类
Classification of ceramic materials

陶瓷，中国人早在公元前8000年～公元前2000年

（新石器时代）就发明了陶器。陶瓷则是陶器，炻器和瓷器的总称，凡是用陶土和瓷土这两种不同性质的黏土为原料，经过配料、成型、干燥、焙烧等工艺流程制成的器物都可以称为陶瓷。

Ceramics, Chinese people invented pottery as early as 8000B.C. ~2000B.C. (the Neolithic Age). Ceramics are the general term for china pottery, and porcelain. Any clay made of clay and china clay is used as raw material. The utensils made by the processes of compounding, forming, drying and roasting can be called ceramics.

（1）陶瓷材料一般分为陶和瓷两大类。

Ceramic materials are generally divided into two major categories: ceramics and porcelain.

①陶：陶一般是用陶土作胎，胎体质地疏松，有不少孔隙，有一定的吸水性，它的表面一般比较粗糙，有一种原始的质朴感觉，适合制作一些比较天然形态感觉的配饰。

Pottery: Pottery is the craft or activity of making objects out of clay The texture of the carcass is loose. It has a lot of pores and a certain water absorption. Its surface is generally rough, and it has a primitive and rustic feeling. It is suitable for making some accessories with a more natural shape

②瓷：它的坯体质地紧密，基本上不吸水，半透明，断面呈石状或贝壳状；瓷的表面细腻，玻化程度高。因此在适用于制作表面较为细腻的服饰配件。

Porcelain: Its body is compact in texture, basically non-absorbent, translucent, and has a stone-like or shell-like cross section; the surface of porcelain is fine and the degree of vitrification is high. Therefore, it is suitable for the production of more delicate clothing accessories.

（2）陶与瓷的区别。

The difference between pottery and porcelain.

二者区别，可以从原料、烧成温度、硬度、透明度、釉料五个方面进行比较：

The difference: we can compare the difference between the two from five aspects: raw materials, firing temperature, hardness, transparency, glaze:

①原料。

Raw materials.

陶使用一般黏土即可制坯烧成，瓷则需要选择特定的材料，即以高岭土作坯。

Pottery can be made from the ordinary clay, and porcelain needs to be selected from a specific material, that is, kaolin is used as a mold.

②烧成温度。

Burning temperature.

陶器烧成温度一般都低于瓷器，最低甚至达到800℃以下，最高可达1100℃左右。

The firing temperature of pottery is generally lower than that of porcelain, and the lowest is even below 800 °C, up to about 1100°C.

瓷器的烧成温度则比较高，大都在1200℃以上，甚至达到1400℃左右。

The firing temperature of porcelain is relatively high, mostly above 1200 °C, and even reaching about 1400°C.

③硬度。

Hardness.

陶器烧成温度低，当坯体并未完全烧结，敲击时声音发闷，胎体硬度较差，有的甚至可以用刀划出沟痕。

The burning temperature of the pottery is low. When the mold is not completely sintered, the sound is stuffed When knocking, the hardness of the carcass is poor, and some can even use a steel knife to draw the groove.

瓷器的烧成温度高，胎体基本烧结，敲击时声音清脆，胎体表面用一般钢刀很难划出沟痕。

The burning temperature of the porcelain is high, the carcass is basically sintered, the sound is crisp when struck, and it is difficult to draw a groove mark on the surface of the carcass with a general steel knife.

④透明度。

Transparency.

陶器的坯体即使比较薄，也不具备半透明的特点。例如，龙山文化的黑陶，薄如蛋壳，却并不透明。

瓷器的胎体无论薄厚，都具有半透明的特点。

Even if the body of the pottery is relatively thin, it does not have the characteristics of translucency. For example, the black pottery of Longshan culture is as thin as an eggshell, but it is not transparent.

The carcass of porcelain is translucent regardless of its thickness.

⑤釉料。

Glaze.

陶器有不挂釉和挂釉的两种，挂釉的陶器釉料在较低的烧成温度时即可熔融。

瓷器的釉料有两种，既可在高温下与胎体一次烧成，也可在高温素烧后的胎体上再挂低温釉，第二次再低温烧成。

Pottery has two kinds of glaze and hanging glaze. The glazed pottery glaze can be melted at a lower firing temperature.

There are two kinds of porcelain glazes, which can be fired once at a high temperature with the carcass, or low temperature glaze on the carcass after high temperature burning, and then fired at a low temperature for the second time.

2. 陶与瓷的常用材料

Common materials for pottery and porcelain

（1）陶的常用材料：黏土。

Common materials for pottery: Clay.

黏土，是颗粒非常小的（<2μm）可塑的硅铝酸盐。除了铝外，黏土还包含少量镁、铁、钠、钾和钙，是一种重要的矿物原料，一般由硅酸盐矿物在地球表面风化后形成。

Clay, is a very small (<2 μm) plastic aluminosilicate. In addition to aluminum, clay also contains small amounts of magnesium, iron, sodium, potassium and calcium. It is an important mineral raw material, usually formed by the weathering of silicate minerals on the surface of the earth.

（2）瓷的常用材料。

Common materials for porcelain.

①长石质瓷。

Feldspar porcelain.

以长石作助熔剂的"长石—石英—高岭土"三元系统瓷。瓷质洁白，薄层呈半透明，断面呈贝壳状，不透气，吸水率很低，瓷质坚硬，机械强度高，化学稳定性好。适用于餐具、茶具、陈设艺术瓷。

The feldspar-quartz-kaolin ternary system porcelain with feldspar as a flux. The porcelain is white, the thin layer is translucent, the cross section is shell-shaped, airtight, the water absorption rate is very low, the porcelain is hard, the mechanical strength is high, and the chemical stability is good, suitable for tableware, tea sets, and furnishings art porcelain.

②绢云母质瓷。

Sericite porcelain.

以绢云母为助熔剂的"绢云母—石英—高岭土"系统瓷。具有长石质瓷特点，且透明度更高，有"白里泛青"的特点。

Porcelain-quartz-kaolin system porcelain with sericite as flux. It has the characteristics of feldspathic porcelain, and its transparency is higher. It has the characteristics of "green in the white".

③骨质瓷。

Bone china.

以钙为助熔剂的"盐—高岭土—石英—长石"系统瓷。白度高，透明度好，瓷质软，光泽柔和，但瓷质较脆，热稳定性差。

"Salt-kaolin-quartz-feldspar" system porcelain with calcium as a flux. High whiteness, good transparency, soft porcelain and soft luster, but porcelain is brittle and has poor thermal stability.

④锂质瓷、镁质瓷。

Lithium porcelain, magnesia porcelain.

用锂辉石或其他含锂原料代替坯料中长石所制的瓷，具有高热稳定性，常用于耐热器皿及耐热厨房用具。镁质瓷的晶相以"氧化镁—氧化铝—二氧化硅"三元系统瓷有良好的电学性能，具有高的机械强度及热稳定性，白度好、色调柔和。

The use of spodumene or other lithium-containing materials to replace the porcelain made of feldspar in the blank has high thermal stability and is commonly used in heat-resistant utensils and heat-resistant kitchen utensils. The crystal phase of magnesia porcelain has good electrical properties with "magnesium oxide—alumina—silica" ternary system porcelain, high mechanical strength and thermal stability, good whiteness and soft color tone.

三、玻璃材料制作服饰配件工艺
C. Glass material manufacturing and apparel accessories techniques

1. 玻璃制品成型原理
Glass forming principle

根据玻璃的相关性质，进行加工成型。玻璃的相关性质：

Processing is carried out according to the relevant properties of the glass. Related properties of glass:

（1）无固定熔点，其黏度随温度升高连续变小，冷却时黏度变大而固化。

There is no fixed melting point, and the viscosity continuously decreases with increasing temperature, and the viscosity increases and solidifies upon cooling.

（2）有内聚力和表面张力，使玻璃熔化时团缩增厚，软化时吹成球状或拉延成圆柱形。

Cohesive force and surface tension make the glass shrink and thicken when it melts. When softened, it is blown into a spherical shape or drawn into a cylindrical shape.

（3）热导性差，使玻璃部件局部加热软化直至熔融，而其余部位仍处于低温，不变形且可以手持。

Poor thermal conductivity, the glass parts are locally heated and softened until molten, while the rest is still at low temperature, no deformation and can be hand-held.

（4）玻璃灯工加工的玻璃部件一般是薄壁，其热膨胀系数与热稳定性成反比关系。硬质玻璃软化点高，热膨胀系数小，可用煤气—氧或氢—氧等高温火焰加工而不致破裂。软质玻璃热膨胀系数大，热稳定性差，用高热值煤气加热即可达到灯工要求的温度。

Glass-worked glass components are generally thin-walled, and their thermal expansion coefficient is inversely proportional to thermal stability. The hard glass has a high softening point and a small coefficient of thermal expansion, and can be processed by a high-temperature flame such as gas-oxygen or hydrogen-oxygen without cracking. Soft glass has a large thermal expansion coefficient and poor thermal stability. It can be heated by high calorific value gas to reach the temperature required by the lamp.

（5）能浸润金属，浸润角愈小则黏着力越大。纯金属状态浸润角一般比其氧化物状态浸润角大，因而玻璃与表面氧化的金属更能气密地封接。

It can infiltrate the metal, the smaller the infiltration angle, the greater the adhesion. The pure metal state wetting angle is generally larger than the oxide state wetting angle, so that the glass is more hermetically sealed with the surface oxidized metal.

2. 玻璃制品成型工艺（图4-64、图4-65）
Glass product molding process (Figure 4-64、Figure 4-65)

（1）玻璃热熔工艺。

Fusing Glass.

玻璃热熔是在一定温度范围内（一般在750℃~800℃）玻璃表面呈现熔化状态，使两部分玻璃相互黏合或融为一体。

热熔的形式和用于装饰的元素丰富多样，如线状和块状玻璃的搭配、彩色釉料与金属氧化物的运用等，都可以通过多种尝试，发现一些特殊的效果。

Fusing Glass is a molten state in a certain temperature range (generally around 750℃ ~ 800℃), so that the two parts of the glass are bonded or integrated into each other.

The form of hot melt and the variety of elements

第四章　非金属材料在服饰配件领域的制作与工艺

图4-64　热熔成型玻璃首饰
Figure 4-64　Fusing glass jewellery

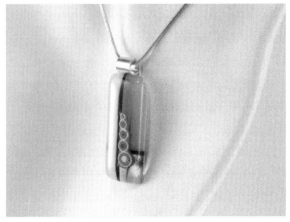

图4-65　热熔成型玻璃首饰
Figure 4-65　Fusing glass jewellery

used for decoration, such as the combination of linear and block glass, the use of colored glazes and metal oxides, can be found through a variety of attempts to find some special effects.

具体的操作过程为：

The specific operation process includes:

①玻璃切割。

Glass cutting.

将所选用的彩色平板玻璃按照图案进行切割。

The selected colored flat glass is cut in a pattern.

②拼贴玻璃图案。

Collage glass pattern.

将不同色彩的玻璃在石棉纸上依据设想拼成图案。

Different colors of glass are patterned on Asbestos paper according to the assumption.

③入窑烧制。

Into the kiln for burning.

放入电窑中，设定温度（一般在700℃~800℃）进行烧制。

Place it in an electric kiln and set the temperature (usually around 700℃~800℃) for firing.

143

④研磨边角。

Grinding corners.

作品退火后取出，研磨边角。

After the work is annealed, it is taken out and the corners are ground.

（2）玻璃灯工工艺。

The Flamework glass.

玻璃灯工工艺用的主要设备是带调节阀的喷灯。

将具有压力的煤气—空气、煤气—氧气或氢气—氧气等混合气体送入喷灯，喷射不同温度的火焰。常用操作工具有铁钳、夹具、碳棒、打孔钢针、活塞钳、吹气工具、量具以及使火焰变成扁窄状的耐火用具等。玻璃灯工大多采取手工方式进行，大型制品采用灯工车床夹持加工，成批生产的制品采用专用机床进行机械化操作。

The main equipment used in The flame glass technology is a burner with a regulating valve.

A mixed gas such as gas-air, gas-oxygen or hydrogen-oxygen with pressure is sent to the burner to spray flames of different temperatures. Commonly used operating tools include iron tongs, clamps, carbon rods, perforated steel needles, piston tongs, air blowing tools, measuring tools, and refractory appliances that make the flames flat and narrow. Most of the glass lamp workers are carried out by hand. The large-scale products are clamped and processed by the lamp lathe, and the batch-produced products are mechanized by special machine tools.

根据加工对象，灯工分以下3种基本类型：

According to the processing object, the process is divided into the following three basic types:

①玻璃变形加工将玻璃加热软化，借外力和内腔吹气进行操作，如弯曲、吹泡、拉延、翻口边、封管底等。冷凝管球芯、蛇形芯、保温瓶胆圆底、试管翻口圆底等制品的加工属于此例。

Glass deformation processing softens the glass, and operates by external force and inner air blowing, such as bending, blowing, drawing, tumbling, sealing the bottom of the tube, and the like. The processing of products such as the condensing tube core, the serpentine core, the vitreous bottom of the thermos flask, and the round bottom of the test tube belong to this example.

②玻璃部件互相焊接玻璃烧熔后焊合整形，熔成一体，如玻璃管件对口接、丁字接、内外管套口部环接等，同种玻璃焊接最为安全。不同种的玻璃焊成的制品，如果热膨胀系数存在过大差异，焊合后冷却时将出现收缩差异，造成体积效应引起结构应力，当应力超过玻璃强度时导致炸裂。石英玻璃与普通玻璃焊接属于此例。解决的办法是采用几种热膨胀系数递变的玻璃作为中间过渡玻璃。一般而言，两种玻璃的热膨胀系数差异应小于6×10^{-7}℃。

The glass parts are welded to each other and then welded and shaped, and melted into one body, such as glass tube fittings, butt joints, inner and outer tube sleeves, etc., the same kind of glass welding is the safest. For products made of different kinds of glass, if there is too much difference in thermal expansion coefficient, there will be a difference in shrinkage after cooling after welding, causing structural stress caused by volume effect, and bursting when stress exceeds glass strength. Quartz glass and ordinary glass welding belong to this example. The solution is to use several glass with a coefficient of thermal expansion as the intermediate transition glass. In general, the difference in thermal expansion coefficient between the two glasses should be less than $6\times10^{-7}℃$.

③玻璃部件与金属焊封。尽量采用热膨胀系数接近的玻璃与金属进行匹配封接。当两者热膨胀系数差异太大时，可选用中间过渡玻璃、性质较软的金属细丝或薄箔与玻璃进行非匹配封接，封接处产生的结构应力可通过金属变形得到补偿。

Glass parts and metal welding seals. Try to use a glass with a thermal expansion coefficient close to the metal to match the seal. When the difference between the thermal expansion coefficients of the two is too large, the intermediate transition glass, the soft metal filament or the thin foil may be used for non-matching sealing with the glass, and the structural stress generated at the sealing portion may be compensated by metal deformation.

四、陶瓷材料制作服饰配件工艺
D. Ceramic materials to make clothing accessories process

1. 陶艺服饰配件的制作工艺
Craftsmanship of ceramic art accessories

制作陶艺服饰配件的工具、材料与一般制陶工具基本相同。但是由于陶艺服饰配件的尺寸应适合服饰配件的佩戴需要，一般不能太大、过重，所以除了一般的工具以外，还需要增加一些适合小件器物制作的陶艺工具。有些工具并不是专业的陶艺工具，但可根据需要作为专业工具；也可以借助一些雕塑工或雕刻工具，甚至可以自己动手制作适合使用的工具。

The tools and materials for making ceramic art accessories are basically the same as those for general ceramic tools. However, since the size of ceramic art accessories should be suitable for the wearing of clothing accessories, it is generally not too big or too heavy, so in addition to the general tools, it is necessary to add some ceramic tools suitable for small parts. Some tools are not professional ceramic tools, but can be used as professional tools as needed; you can also use some sculptors or engraving tools, or even make your own tools.

（1）主要陶艺工具及应用。

Main ceramic tools and applications.

擀泥棍：用于把泥土擀成均匀的泥片，但作陶艺服饰配件需要的尺寸相对较小，故选择小型的擀泥棍，或是借用一小段圆柱木棍擀出泥片即可。

Clay stick: used to make the soil into a uniform piece of mud, but the size required for ceramic accessories is relatively small, so a small mud stick, or a small piece of cylindrical wooden stick could be chosen and used to pull out the mud.

木拍：用于拍打泥料，混合不同的泥料。

Wood pat: used to beat mud and mix different mud.

海绵：吸水、干燥、擦拭、清理坯体。

Sponge: absorb, dry, wipe, and clean the body.

不锈钢刮片、双面刀片、美工刀片等：清理坯体

Stainless steel scraper, double-sided blade, artist blade, etc.: cleaning the blank

钢丝：切割泥土、肌理制作。

Steel wire: cutting clay and texture.

木刀：切割泥土、坯体表面处理。

Wood knife: cutting soil, body surface treatment.

砂纸：打磨坯体表面和底部。

Sandpaper: sanding the surface and bottom of the blank.

锯条：切割泥土、肌理制作（图4-66）。

Saw blade: cutting clay and texture (Figure 4-66).

（2）其他工具。

Other tools.

木制雕塑刀：可做坯体表面装饰图案的雕塑。

Wooden sculpture knife: a sculpture that can be used as a decorative pattern on the surface of a blank.

木刻刀：坯体表面装饰图案的刻画。

Woodcutting knife: the depiction of the decorative pattern of the surface of the blank.

紫砂钻孔工具：坯体湿润时给泥珠、泥片钻孔。

Purple sand drilling tool: Drill holes for mud and mud when the body is wet.

小钻头：坯体干燥时大小泥珠的钻孔。

Small drill bit: Drilling of large and small mud beads when the blank is dry.

美工刀：泥土切割。

Utility knife: dirt cutting.

尖嘴钳：金属配件的安装与调配（图4-67）。

Needle-nose pliers: installation and deployment of metal fittings (Figure 4-67).

2. 成型工艺
Molding Technique

（1）泥条成型。

Mud strip molding.

以泥条为基础的成型方法，可以单独使用，也可以结合其他技法进行成型。以手工搓出的泥条与挤泥器挤出泥条来成型的陶艺服饰配件的区别是：手工搓制的泥条手工韵味十足，但是规整性较差；挤泥器制作的泥条有均匀的形态，较少变化，适合做规则的设计。泥条的大小粗细可以从几毫米到几厘米，根据需要变换即可。

The mud-based molding method can be used alone or in combination with other techniques. The difference between the handcrafted mud strips and the clay pottery extruded clay strips is that the handmade mud sticks are full of charm, but the regularity is poor; the mud strips made by the muder are evenly distributed. The shape, less change, is suitable for the design of the rules. The size of the strip can vary from a few millimeters to a few centimeters, as needed.

泥条的长短随意性较强，但是考虑到干燥和烧成的收缩，应该略微放大。泥条的卷曲也要注意干湿程度，幅度较大的卷曲需要在含水量高的泥条上进行；卷曲的幅度如果较小，则相对容易，只要把握泥条不变形、不开裂就可以了。

The length of the strip is more random, but it should be slightly enlarged considering the shrinkage of drying and firing. The curling of the mud strip should also pay attention to the degree of dryness and wetness. The curl with a large amplitude needs to be carried out on the mud strip with high water content; if the curl is small, it is relatively easy, as long as the mud stick is not deformed or cracked.

泥条与泥条之间的相互连接应在泥土湿润的情

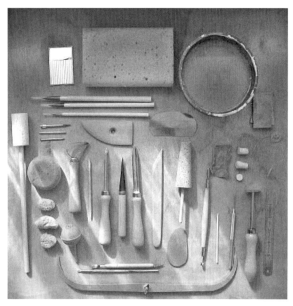

图4-66　陶艺工具
Figure 4-66　Pottery tools

图4-67　其他工具
Figure 4-67　Other tools

况下进行，这样便于各种形状、角度的选择与调配；泥条之间应用工具或是手指黏连、压紧，以防干燥和烧成过程中的开裂。干燥后的泥条连接相对比较复杂，应在两个相接的部分加水以湿润泥土，再以泥浆连接。

The interconnection between the mud strip should be carried out under the condition that the soil is wet, so that various shapes and angles can be selected and blended; tools or fingers can be adhered and pressed between the mud strips to prevent drying and burning, and cracking in the process. The connection of the dried mud strips is relatively complicated. Water should be added to the two adjacent parts to wet the soil and then connected by mud.

泥条表面基本上是没有肌理的，泥条的面积较小，肌理的制作在整个过程中不易被保留，尤其是细小的肌理比较容易在制作中被磨光和减弱。

The surface of the mud strip is basically free of texture, and the area of the mud strip is small. The production of the texture is not easy to be preserved throughout the process, especially the fine texture is easier to be polished and weakened in the production.

卷曲的泥条应在最柔软、湿润的情况下进行，卷曲过程中要保持水分，一旦失去水分，泥条就容易开裂。泥土湿润时，泥土的柔韧性很强，把泥土弯曲成任意角度都相对容易，缺点是容易变形，不易保持所需要的造型。较为干燥时弯曲泥条就要注意湿度的把握。

Curled mud strips should be curled in the softest and moist condition. Keep the moisture during the curling process. Once the moisture is lost, the strips will crack easily. When the soil is wet, the soil is very flexible, and it is relatively easy to bend the soil to any angle. The disadvantage is that it is easily deformed and it is not easy to maintain the desired shape. When drying the mud, it is necessary to pay attention to the humidity.

泥条成型配件的孔，可以在"半干"时直接穿制，或是泥条卷曲成的孔可以直接利用，不需要钻孔。

圆形泥条是最简单的泥条形态，是方形和不规整形泥条的制作基础。原则是只要搓泥条时均匀规整，尽可能避免中间存在气孔。方形泥条是在圆形泥条上用平面工具压成水平的形状，也可以使用挤泥器来制作。

The holes in the mud-formed fittings can be worn directly in the "semi-dry" or the holes in which the mud strips are crimped can be used directly without drilling.

Round mud strips are the simplest form of mud strips and are the basis for the production of square and irregular shaped mud strips. The principle is that as long as the mud is evenly regularized, it is avoided as much as possible in the middle. Square mud strips are pressed horizontally on a round mud strip with a flat tool, or they can be made using a clay mixer.

泥条卷曲可以形成变化丰富的形状，自由而多变。泥条根据需要也可以单独使用，这可以作为条形配件的坯体。泥条的制作可以是规整的，也可以是很随意的。泥条的分段切割形态最适宜环绕或黏合成为珠子，这种泥条可以是方的，也可以是圆的，甚至可以是不规则的。泥条卷曲也可以成为球状，这也为泥条成型和泥珠成型之间的变换提供了丰富的可能。

The mud strip curls can form a rich and varied shape, free and changeable. The mud strip can also be used separately as needed, which can be used as a blank for the strip fitting. The production of the mud strips can be regular or very random. The segmented cut form of the mud strip is most suitable for wrapping or bonding into beads. The strips can be square, round, or even irregular. The mud strip curl can also be spherical, which also provides a rich possibility for the transformation between the mud forming and the mud forming.

挤泥器可以制作出各种各样美丽的泥条，出泥口的形状各异，使得挤出的泥条也就有了各种的形态。根据使用的泥土的干湿程度不同，挤出的泥条也有一定的变化。

The squeezing device can produce a variety of beautiful mud strips, and the shapes of the mud outlets are different, so that the extruded mud strips have a colorful form. Depending on the degree of dryness and wetness of the soil used, the extruded mud strips also have some changes.

湿润的泥土挤出的泥条柔软平滑，较干的泥被挤出后容易有细小的开裂肌理。不同的出泥口挤出的泥条形状各异，给配饰的设计带来多种可能（图4-68）。

The mud strips extruded from the moist soil are soft and smooth, and the dry mud tends to have a fine cracking texture after being extruded. The different types of mud strips extruded from the mud outlets have different shapes, which brings many possibilities to the design of the accessories (Figure 4-68).

piece of mud, cut into a certain size, generally ranging from 1 cm to 2 cm, and finally cut and surface texture on the mud sheet (Figure 4-69).

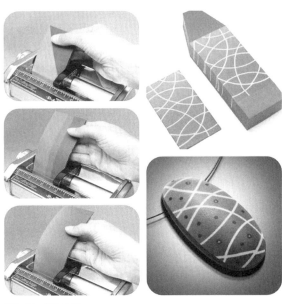

图4-69　泥板成型
Figure 4-69　Mud Forming

图4-68　泥条成型
Figure 4-68　Mud strip molding

（2）泥板成型。
Molten plate forming.

在陶艺服饰配件的制作过程中，因为尺寸一般都偏小，一般称为泥片成型。它指的是根据自己的设计构思，用工具或手工把泥拍打成薄泥片，切割成一定的尺寸，一般从1~2cm，最后在泥片上进行切割与表面的肌理制作（图4-69）。

In the production process of ceramic art accessories, because the size is generally small, it is generally called mud molding. It refers to the design of the idea, using tools or hand-made mud to make a thin

(3) 肌理的制作。
Fabrication of texture.

泥片表面的肌理制作要注意如果是在泥片成型后再刻画或进行浅浮雕而出现的肌理，应注意保证泥片的原始厚度和体积关系，防止雕刻过程中破坏泥片的基本形态并带来大的变形；如果是利用泥土本身的肌理，那么就要防止在切割和制作过程中手工操作时破坏泥土的自然肌理。

因为泥土柔软的特性，泥片的成型在制作过程中还可以借助一些外在的辅助材料来支撑形态的塑造，或利用工具刮泥形成自然堆积的肌理；也可以运用一些有肌理的材料，如纸、树叶、布料进行支撑和压印。虽然这些材料都属于有机物，入窑烧成后就会消失但也不可滥用，因为有些材料的拉伸性能较差，若是全部借助外部条件进行定型，泥土在干燥过程中不断收缩，而支撑材料并不能同步收缩，那么泥土就容易产生干裂（图4-70）。

The texture of the surface of the mud sheet should be noted that if the texture is formed after the mud sheet is formed or embossed, the original thickness and volume relationship of the mud sheet should be ensured to prevent the basic shape of the mud sheet from being damaged during the carving process. It brings great deformation; If it is to use the texture of the soil itself, it is necessary to prevent the natural texture of the soil from being destroyed by hand during the cutting and manufacturing process.

Due to the soft nature of the clay, the formation of the mud sheet can also be supported by some external auxiliary materials during the production process, or the tool can be used to form a naturally deposited texture; some texture materials can also be used, such as Paper, leaves, and fabric are supported and embossed. Although these materials are organic, they disappear after being fired in the kiln, but we can't abuse them because some materials have poor tensile properties. If all are shaped by external conditions, the soil shrinks during the drying process, and since the support material can't shrink synchronously, the soil is prone to chapped (Figure 4-70).

泥土柔软湿润时进行弯曲和定型，以达到所需的形态。在泥土半干燥时，把辅助材料脱离原始泥坯，这样既保留了泥土肌理，又不至于破坏泥片的造型。这个过程的工艺要求是要很好地把握泥土的收缩特性，在泥土达到不黏手的时候就予以拆除，以保留需要的肌理，同时也避免了开裂的危险。

We can use external auxiliary materials to curl and support the soil, bending and shaping when the soil is soft and moist to achieve the desired shape. When the soil is semi-dry, the auxiliary material is removed from the original mud, which retains the texture of the soil without destroying the shape of the mud. The process requirement for this process is to have a good grasp of the shrinkage characteristics of the soil and to remove the soil when it is not sticky, to preserve the desired texture and to avoid the risk of cracking.

把有肌理的辅助材料，如有网格的塑料布覆盖在泥土上，把泥土连同塑料布卷曲成一样的形状，在泥片具有了可支撑的硬度时就把塑料布拆除，这样泥土上既留下了网格的细致肌理，又不会在干燥过程中出现开裂。借助其他一些本身具有肌理的材料，可以把肌理转移、应用到泥片上，这种方法直接、有效。

如制作发夹等有一定弯曲程度的泥坯时，如果是平放在桌面上，干燥和烧成后是不符合发夹的使用功能的。因此，在制作时就应该把泥坯做成弯曲的形态，并放置在弯曲的支撑泥块上，缓慢地干燥，并将二者同入窑烧制，烧成后才能够最大限度地符合发夹的弯曲弧度。制作过程中还必须考虑收缩的比例，适当地放大制作的尺寸，在烧制后能够足以覆盖钢夹。

Cover the soil with a protective auxiliary material, such as a meshed plastic sheet, and curl the soil into a shape with a plastic sheet. When the sheet has a supportable hardness, remove the plastic sheet, so that the soil remains. The fine texture of the mesh is printed on the soil without cracking it during the drying process. With the help of other materials with

图4-70　白色繁星陶瓷首饰　吉姆·杰斯提斯
Figure 4-70　Ceramic Jewelry pewter starbursts by kim justice

3. 辅助材料
Auxiliary materials

可借助外部辅助材料进行卷曲和支撑泥土，在

their own texture, the texture can be transferred to the mud, directly and effectively.

If the mud mold requires a certain degree of bending, such as a hair clip, it will not conform to the function of the hairpin after drying and burning if it is placed flat on the table. Therefore, the mud should be made into a curved shape during the production, and placed on the curved supporting mud, slowly dried, and the two are fired in the kiln, and the maximum curved curvature can be matched with the clamp after firing. The proportion of shrinkage must also be taken into account during the production process, and the dimensions produced are appropriately enlarged to cover the steel clip after burning.

4. 组合构成
Combination composition

在泥片和泥片之间的组合并列构成图案化的设计，可以把泥片切割成为需要的形态，进行叠加或组合。

如发夹的泥坯就是以泥片为基本成型元素，把泥片切割、挤压并进行组合构成。把泥片切割成为相同相似的形态进行叠加、组合时要注意泥片单元之间的黏合，要在泥片湿润时结合紧密，或是在半干燥时用泥浆进行连接，并保证两个相接部位之间干湿程度的一致（图4-71）。

The combination of the mud pieces is juxtaposed to form a patterned design, and the piece can be cut into a desired form, superimposed or combined.

For example, the mud of the hairpin is made up of mud sheets as the basic forming elements, and the pieces are cut, extruded and combined. Cut the mud pieces into the same similar form for superposition and combination. Pay attention to the adhesion between the mud pieces. It should be tightly combined when the mud pieces are wet, or connected with mud when semi-drying, and ensure two phases. The degree of dryness and wetness between the joints is the same(Figure 4-71).

图4-71　泥片成型的陶瓷首饰　卢卡·特帕尔蒂
Figure 4-71　Ceramic jewelry with clay slices　luca tripaldi

（1）加法和减法。
Addition and subtraction.

泥片成型的外形，可以顺应泥片的自然边缘与外形，也可以根据需要进行相接、分割等设计，在泥片上做合适的加法和减法。

The shape of the mud piece can conform to the natural edge and shape of the mud, and can also be connected and divided according to the needs, and appropriate addition and subtraction can be done on the mud piece.

加法（主要指泥之间的黏连与结合），适合在泥土湿润的时候进行，可在泥片上增加泥团、泥片、泥条等。如果在干燥时做加法，就要注意两部分的干湿程度，如果干湿一致，那么调同种泥土的泥浆，把两个相接的位置打湿，用泥浆连接就行；如果两部分的干湿度不均匀，不能相接，务必等到两部分的干燥程度完全一致时再连接，因为两部分干湿不均容易引起干燥过程中的开裂，烧成时可能脱落和开裂。

Addition (mainly refers to the adhesion and bonding between the mud), is suitable for the soil when the soil is wet. We can add mud, mud, mud and so on on the mud. If adding during drying, pay attention to the degree of dryness and wetness of the two parts. If the dryness and wetness are the same, adjust the mud of the same kind of soil, wet the two connected positions, and connect with the mud; if the two parts are dry, the humidity is not uniform and cannot be connected. It is necessary to wait until the dryness of the two parts is completely the same, because the two parts may fall off and crack when burned if they are unevenly wet and dry, causing cracking during the drying process.

减法（主要指泥板的刻画等装饰），一般适合在泥土干燥的时候进行，因为在泥片湿的时候刻画或做装饰，都很容易把泥土连带起来，黏连在刀上，增加工作的难度。而在干燥或半干时雕刻或塑造，那就相对容易得多，可以比较容易地刻画出所需要的线条和形态。干燥时雕刻、塑造要求非常准确，如果刻画错误，那就不易补救，因为泥土干后再相接不容易。

Subtraction (mainly refers to the decoration of the mud board, etc.), is generally suitable for the drying of the soil. When the mud is wet, it is quite likely to connect the soil, and stick it to the knife, increasing the difficulty. It is relatively easy to sculpt or shape when dry or semi-dry, and it is easier to describe the lines and shapes needed. The requirements for engraving and shaping during drying are very accurate. If the sculpting is wrong, it is not easy to remedy, for it is not easy to connect after the soil is dry.

（2）泥片的穿孔。

Perforation of the mud piece.

泥片的穿孔要选择合适的时机，一般来说是在干燥初期，泥土略微失去水分，不粘连工具且具有了一定硬度的时候，用尖刀或钻孔的工具在需要穿孔的地方进行。一般不主张在干燥后期穿孔，因为那样很容易使制作完的泥片发生破裂。

The perforation of the mud piece should be selected at the right time. Generally speaking, in the initial stage of drying, when the soil loses moisture slightly, when the tool is not stuck and has a certain hardness, it is carried out in a place where it is needed with a sharp knife or a drilling tool. Perforation in the late drying stage is generally not recommended because it is easy to break the finished mud sheet.

（3）卷曲。

Curling.

泥片的卷曲特别需要注意泥土的干湿，因为一般的瓷泥可塑性较差，卷曲时容易开裂。泥片较薄，容易卷曲成型，造型较为准确，缺点是泥片太脆弱，易碎，在烧制之前不易保存。若泥片较厚，则相对安全得多，不易破损；但是还要注意泥片的重量，太厚就显得笨重，不符合服饰配件的装饰功能。泥片的开裂在干燥的初期如果得到控制就不易扩大。泥土表面的浅层裂纹是可以接受的，如果全部裂透了整个泥片就可能在烧成过程中断裂（图4-72）。

The curling of the mud sheet requires special attention to the wetness and dryness of the soil, because the general porcelain clay is poor in plasticity and is easily cracked when curled. The mud sheet is thin, easy to curl and form, and the shape is more accurate. The disadvantage is that the mud sheet is too fragile and brittle, and is difficult to store before burning. If the mud is thicker, it is relatively safer and less prone to breakage; but it should also pay attention to the weight of the mud piece. If it is too thick, it will be bulky and not suitable for the decorative function of the clothing accessories. Cracking of the mud sheet is not easily expanded if it is controlled at the beginning of drying. Shallow cracks on the surface of the soil are acceptable. If all of the mud is cracked, it may break during the firing process (Figure 4-72).

图4-72 泥片卷曲成型的陶瓷首饰

Figure 4-72 Clay-formed ceramic jewelry

（4）泥珠成型。

Mud beads molding.

泥珠成型是各成型方法中最为自由而富有表现力的一种。取一小团泥在手心里搓制，就成为一个规则或不规则的泥球，可以保持圆形，也可以压扁成为扁形珠，更可以在泥珠的表面做各种装饰，刻、画、印、雕等均可。甚至是简单地随手搓出的泥团，都可能成为泥珠造型的原坯（图4-73）。

Mud beads molding is the most free and expressive of all molding methods. Take a small handful mud in the palm and rub, it becomes a regular or irregular mud ball, you can keep it round, you can also flatten it into a flat bead, you can make various decorations on the surface of the clay, engraved, Painting, printing, carving, etc. Even the mud that is simply pulled out by hand can become a mold of mud bead shape (Figure 4-73).

图4-73　泥珠成型
Figure 4-73　Mud beads molding

泥珠的形态，可以手工搓出，也可以由泥条分段切割、揉扭，或是由泥片切割、卷曲变形来制成。因此，泥珠的形状不局限于圆形、方形、椭圆形、三角形、锥形、不规则形等，自由而富有变化。当然有时泥珠的范围更宽泛，只要是视觉上可以成为一个团形的，都可以称为泥珠，尺寸基本不受限制。

The shape of the mud can be cut out by hand, or it can be cut or twisted by the strips, or cut and curled by the mud. Therefore, the shape of the mud beads is not limited to a circular shape, and a square shape, an elliptical shape, a triangular shape, a tapered shape, an irregular shape, and the like can be freely and varied. Of course, sometimes the range of mud beads is wider, as long as it can be visually a cluster. It can be called mud beads, and the size is basically unlimited.

手工制作的泥珠一般来说都是实心的，泥珠较小的尺寸决定了制作空心泥珠是比较困难的。泥珠的制作过程相对是简单和迅速的，因此在手工制作的过程中也不需要过多地考虑形态的一致，手工制作即可；但是在产业化的过程中，就必须使用模具成型以保证形态的规整一致。

Handmade clay beads are generally solid, and the small size of the clay beads makes it difficult to make hollow clay beads. The process of making clay beads is relatively simple and rapid. Therefore, it is not necessary to consider the uniformity of the shape in the process of hand-made, and it can be hand-made. However, in the process of industrialization, mold molding must be used to ensure The shape is consistent.

把已干燥的泥珠坯体进行表面的小块切割，给小小的泥珠以空间的变化，在泥珠表面形成形状各异、大小不等的平面，产生微妙的体积感。或者可以在泥珠的表面作出或深或浅的各种肌理，都是改变泥珠表面形态的有效方式。

The dried clay bead body is cut into small pieces on the surface to change the space of the small mud beads, and planes of different shapes and sizes are formed on the surface of the mud beads, resulting in a subtle volume feeling. Or it can be made on the surface of the mud or deep or shallow texture, which is an effective way to change the surface morphology of the mud.

（5）模具成型。

Mold forming.

模具成型是借助石膏或其他材质的模具，进行注浆或印坯的成型方法。模具的使用在陶艺服饰配件的工业化生产中，对量化制作程度有着直接的影响，适应制作的是规整、统一的造型和数量上的快速积累。在进行小批量制作时，也可以根据需要使用模具成型的工艺。

用模具制作陶瓷服饰配件与陶瓷生产的一般工艺一样，首先要用泥土或石膏做出模种，然后翻石膏模。待模具干燥后，采用注或印坯的方法来制作坯体。印坯出来的坯配件可以单独使用，还可以把压印出来泥坯重新排列和组合，形成新的造型。

Molding is a method of grouting or printing by means of plaster or other material. The use of molds has a direct impact on the degree of quantitative production in the industrial production of ceramics accessories. The production is a regular, unified shape and rapid accumulation of quantity. In the case of small batch production, the mold forming process can also be used as needed.

The use of molds to make ceramic clothing accessories is the same as the ceramic production process. First, the mold is made of clay or plaster, and then the plaster mold is turned. After the mold is dried, the blank is formed by a method of injecting or printing. The blank parts from the blank can be used separately, and the embossed mud can be rearranged and combined to form a new shape.

五、陶艺服饰配件的装饰与烧成工艺
E. Decoration and burning process of ceramic art accessories

1. 绞胎
The Studded porcelain mold

泥土之间的混合使用由来已久，早在17世纪的英国就已经出现了，那时候的陶工就开始把深色泥土与浅色泥土进行混合，并在揉泥过程中有意识地保留泥土混合的痕迹（图4-74）。

The mixing between the soils has been around for a long time. It appeared in the UK in the 17th century. At that time, the potters began to mix the dark soil with the light soil and intentionally retain the soil mixture mark during the mud process (Figure 4-74).

在陶艺服饰配件的制作过程中，改变泥土表面色彩的方法有很多种。可以把泥土与泥土进行混合，也可以把泥土与色剂进行混合而达到某种所需要的色彩，还可以把透明釉与色剂进行混合而改变釉面的色彩。

There are many ways to change the color of the earth's surface during the production of ceramic accessories. It is possible to mix the soil with the soil, or to mix the soil with the toner to achieve a desired color, and to mix the transparent glaze with the toner to change the color of the glaze.

图4-74　绞胎瓷瓶
Figure 4-74　Studded porcelain bottle

因为陶艺服饰配件的尺寸一般都比较小，如果以一般的上釉方法（吹釉、浸釉）来进行，不仅浪费釉料，有时还会使泥土的细节被过分地覆盖。比如，吹釉的方法适用于面积较大的配件，一般用于大件的青花、色釉作品的上釉；但如果面对的是体积、面积都很小的陶珠、瓷珠等，这种方法就很不容易把釉上均匀。如果是流动性大的釉，又面临着烧成过程中粘连棚板的危险。

The size of ceramic art accessories is generally small, If the general glazing method (blowing glaze, glaze) is carried out, not only the glaze is wasted, but sometimes the details of the soil are over-covered. For example, the method of blowing glaze is suitable for the accessories with larger area, and is generally used for glazing of large blue and white glaze works. However, if it is faced with ceramic beads and porcelain beads with small volume and area, this method is not easy to evenly glaze. If it is a glazed glaze, it is in danger of sticking the slab during the firing process.

在这种情况下，采用色剂或装饰色泥，就比较容易避免上釉过程中出现的问题。常用的色剂有：三氧化二铁、氧化铜、氧化钴等；肉眼可见的直观色彩有红、黄、蓝、绿等较为鲜艳的色彩。

色剂的使用：色彩鲜明的色剂在和瓷泥混合后仍然具有鲜艳的色彩效果。

In this case, it is easier to avoid problems in the glazing process by using a toner or a decorative color mud. Commonly used toners are: ferric oxide, copper oxide, cobalt oxide, etc.; the visual colors visible to the naked eye are red, yellow, blue, green and other relatively bright colors.

Use of coloring agent: The bright color toner still has a vivid color effect after being mixed with porcelain clay.

色剂的使用较为简单，如果想要鲜艳的色彩，可以直接调水或与透明釉混合作为装饰色彩运用，用毛笔直接绘制即可。使用色剂直接作为装饰的色彩效果鲜艳夺目，很容易呈现出时尚的色彩，但缺点是少了中间色彩的缓和。

The use of the toner is relatively simple. If you want bright colors, you can directly transfer the water or mix it with the transparent glaze as a decorative color, and draw it directly with a brush. The color effect of using the toner directly as a decoration is vivid and dazzling, and it is easy to present a fashionable color. But the disadvantage is that there is less mitigation of the middle color.

可以把色剂直接与泥土混合，成为色泥（化妆土）。把色剂加入泥中呈现的色彩会因色剂的用量大小而有所区别。比例加大，发色逐渐加深；比例减小，色彩较为淡化。

色剂的使用，可以是单独使用而不罩面也可以和釉配合使用，在色剂上罩一层纯透明釉，那么色剂本身的色彩就可以清楚地显现，如果是用半透明的釉，成色就相对较弱。

We can also mix the toner directly with the soil to become a color mud (cosmetic soil). The color that is added to the color of the toner will vary depending on the amount of the toner used. The proportion is increased, and the color is gradually deepened; the proportion is reduced, and the color is lighter.

The use of the toner can be used alone without covering the surface or in combination with the glaze. A pure transparent glaze is applied over the toner, so that the color of the toner itself can be clearly displayed. If a translucent glaze is used, the color is relatively weak.

2. 彩绘

Painting

（1）釉下彩绘。

Underglaze painting.

以青花钴料及釉下彩料为主的釉下装饰方法，遵循陶瓷生产历来的对于釉下彩料、釉的使用方式，采用勾线、平涂、分水等各种装饰方法，绘制后罩一层透明釉，具有釉下彩装饰的一切特点。如以青花为装饰的陶艺服饰配件，色彩幽静典雅（图4-75）。

图4-75　陶瓷彩绘手镯　艾比　西摩尔和凯瑟琳　惠勒　澳大利亚　墨尔本

Figure 4-75　Porcelain bangle-These creations come from the hands of a duo Abby Seymour and Katherine Wheeler in Melbourne, Australia.

第四章　非金属材料在服饰配件领域的制作与工艺

The underglaze decoration method based on blue and white diamonds and underglaze coloring materials follows the traditional method of using the underglaze coloring materials and glazes in ceramic production, and adopts various decorative methods such as hook line, flat coating and water distribution to draw the back cover. A layer of transparent glaze incorporates all the features of the underglaze decoration. For example, the ceramic art accessories decorated with blue and white are quiet and elegant (Figure 4-75).

（2）釉上彩绘。

Glazed painting.

陶艺服饰配件的釉上彩绘可以在已烧成的透明釉上进行，也可以在颜色釉上进行。根据装饰设计的色彩和图案的设计，可以进行两次以上的多次烧成，以达到丰富、多样、细腻的色彩效果。以粉古彩或新彩为主的装饰方法，对于服饰配件的制作工艺要求是进行两次烧成，第一次以高温把坯体和釉面烧成，第二次再以低温把彩绘烧成。

The glaze painting of ceramic art accessories can be carried out on a fired transparent glaze or on a color glaze. According to the design of the color and pattern of the decorative design, it can be fired twice or more to achieve rich, diverse and delicate color effects. The decoration method based on powder ancient color or new color, the production process requirement for the clothing accessories is to perform two firings, the first time to burn the green body and the glazed surface at high temperature, and the second time to burn the color with low temperature to make painting.

第四节　颜色釉
Section 4　Color Glaze

以高温或低温的颜色釉来装饰配件，烧成后还可以再次或多次进行釉上的低温装饰，或者也可保留釉的原始色彩与肌理。以颜色釉进行装饰的服饰配件，制作时要考虑到各种色釉之间色彩的对比、呼应和协调，还要考虑到颜色釉的肌理与其他配件之间可能产生的协调与对比关系（图4-76~图4-78）。

The accessories are decorated with high-temperature or low-temperature color glaze. After firing, the low-temperature decoration on the glaze can be performed again or repeatedly, or the original color and texture of the glaze can be preserved. Apparel accessories decorated with color glaze should be considered in the comparison, echo and coordination of the colors between the various glazes, and also consider the possible coordination and contrast between the texture of the color glaze and other accessories (Figure 4-76~Figure 4-78).

图4-76　陶瓷手镯
Figure 4-76　Ceramic bracelet

图4-77　新月形项链　塞拉·高美兹（22K黄金的光泽映衬出醒目的月牙型，如同太阳光略过月球的表面）
Figure 4-77　Crescent necklace Syra Gomez (Like sunlight passing across the moon's surface, 22K gold luster illuminates the eye-catching crescent form)

155

图4-78 陶瓷上釉挂饰
Figure 4-78 Ceramic glazed ornaments

一、上釉工艺
A. Glazing Process

陶艺服饰配件的尺寸较小，如果用一般的喷施方法来上釉，非常浪费釉料。因为喷施的方法适合面积大的物件，机器喷出的釉喷洒的范围较大，人工吹釉同样也存在这样的问题。对于小量的设计制作，较为适合的方法就是手工施釉和浸釉。

The size of ceramic art accessories is small, and if the glaze is applied by a general spraying method, the glaze is very wasteful. Since the spraying method is suitable for a large-area object, the sprayed glaze sprays a large range, and the artificial blown glaze also has such a problem. For a small amount of design, the most suitable method is manual glazing and immersion glaze.

手工施釉的优点在于，简便易行，一支普通的毛笔就可以进行了。用毛笔蘸釉后均匀地涂在坯体的表面。缺点是手工的操作很难做到釉面厚度一致且平整，对于流动性大的釉，在烧成过程中，釉面能够烧得平滑；但是流动性小的釉，就难以达到平整光洁的釉面效果。

The advantage of manual glazing is that it is simple and easy, and an ordinary brush can be carried out, that is, to apply the glaze with a brush and apply it evenly on the surface of the blank. The disadvantage is that it is difficult to make the glaze thickness uniform and flat by manual operation. For the glaze with high fluidity, the glaze can be burned smoothly during the firing process; but the glaze with small fluidity is difficult to achieve a smooth glaze surface effect.

另一种较为有效而便捷的方法就是浸釉。针对陶艺服饰配件的尺寸，可以借助一定的工具，用小镊子夹着泥坯浸釉，在釉水中停留几秒即可；或者用钢丝或线把配件穿起来浸釉，但是要注意的是不可以同时几个一起浸，因为在把泥坯提起的时候，釉水会把几个配件同时粘连起来，分开时容易损伤釉面。配件的孔在浸釉过程中可能有釉流入，之后要把它清除干净，否则将影响烧成。

Another more effective and convenient method is to immerse the glaze. For the size of ceramic art accessories, you can use a certain tool to hold the glaze with small tweezers and stay in the glaze for a few seconds; or use the wire or thread to wear the glaze, but pay attention not to dip together at the same time, because when the mud is lifted, the glaze will stick several accessories at the same time, and it is easy to damage the glaze when separated. The hole of the fitting may have a glaze in the immersion process, and then it should be cleaned, otherwise it will affect the finish burning.

二、烧成工艺
B. Burning Process

1. 支烧
Branch burning

以一头尖一头粗的小泥条来支撑泥珠类的烧成，较为简便。缺点是在烧成过程中可能会出现釉的流动使泥珠和支钉粘连在一起的问题，而且出现的概率很大。手工制作小量的泥珠，多数采用这种方法，但在烧成之前应该要注意清理孔周围的釉层。

It is relatively simple to support the burning of clay beads with a small thick strip of mud. The disadvantage is that during the firing process, there may be a problem that the flow of the glaze causes the mud and the nail

第四章　非金属材料在服饰配件领域的制作与工艺

to stick together, and the probability of occurrence is large. This method is mostly applied in handmade small amounts of clay Specific care should be taken to clean the glaze around the holes before burning.

2. 吊烧
Hanging burning

这是烧泥珠最安全而高效的方法，但是需要有高温电炉丝才能进行。吊烧可以保证各种陶艺配件的正反面均有釉层，手感光滑。

This is the safest and most efficient way to burn mud, but it needs to be done with a high temperature electric wire. Hanging can ensure that all kinds of ceramic accessories have glaze on the front and back, and the hand feels smooth.

用高温电炉丝从中间的孔穿过，两端以支柱支撑，但每个配件之间要有一定的间隔。如果间隔太小或无间隔，烧成时可能引起黏连，在烧成之前也要注意清理孔周围的釉（图4-79）。

Use a high-temperature electric wire to pass through the hole in the middle, and make sure both ends supported by the pillars, but there must be a certain interval between each fitting. If the interval is too small or there is no gap, it may cause adhesion when burning, and also pay attention to clean the glaze around the hole before firing (Figure 4-79).

3. 平烧
Flat burning

这种烧成方式对于圆形泥珠以外的其他平面类陶艺服饰配件都是非常简便而适用的，把配件直接放置于棚板上就可以烧制，非常方便。缺点就是烧成的配件都只是单面施釉，即一面有釉，而另一面无釉，给人的感觉较为粗糙。

This type of firing is very simple and suitable for other flat ceramics accessories other than round clay beads. It is very convenient to place the accessories directly on the shelf. The disadvantage is that the finished accessories are only glazed on one side, that is, one side has glaze, and the other side has no glaze, giving a rough feeling.

玻璃、陶瓷材料作为人造材料的历史伴随着人类文明的进程不断地推进发展着，早期人类就开始使用这两种材料制成可佩带的饰品来装点自己，在和这两种材料不断的对话过程中人们越来越多地发现了其独特的美，它们具有其他材质不可替代的特性，把服饰配件的材料从固有的贵金属以及珠宝中解脱出来，带来了源源不断的创造力，拓展着人们的审美视野（图4-80）。

The history of glass and ceramic materials as man-made materials has been continually advancing along with the progress of human civilization. Early humans began to use these two materials to make wearable jewelry to make themselves, and to talk constantly with these two materials. In the process, more and more people have discovered their unique beauty. They have the irreplaceable characteristics of other materials. They free the materials of the accessories from the precious metals and jewelry, which brings a constant stream of creativity and expands people's aesthetic vision (Figure 4-80).

图4-79　吊烧
Figure 4-79　Hanging burning

图4-80　平烧
Figure 4-80　Flat burning

第五节 运用陶瓷材料制作服饰配件案例解析

Section 5 Case study of using ceramic materials to make clothing accessories

前面分别了解了陶瓷饰品的各种制作工艺，下面就以法国陶艺家罗米的陶瓷首饰制作过程来介绍一下陶瓷饰品的制作过程（她的制作工艺方式是以泥片成型工艺和彩绘工艺的结合来制作饰品），来进行案例解析（图4-81～图4-86）。

In the previous section, we learned about the various manufacturing techniques of ceramic jewelry. Below we introduce the process of making ceramic jewelry with the ceramic jewellery making process of French ceramic artist Loumi (Her production process is based on the mud forming process and the painting process, combining to make jewelry) for case analysis (Figure 4-81~Figure 4-86).

图4-81 杰奎琳（毕加索的妻子戴着他的陶瓷首饰）
Figure 4-81 Jacqueline (Picasso's wife wearing his ceramic jewelry)

图4-82 陶瓷项链 马尔塔 阿曼达
Figure 4-82 Marta Armada silver porcelain Más

图4-83 灰色陶瓷耳环 亚沙哈布特尔
Figure 4-83 Gray Earrings Porcelain Sterling Silver Earwires by yashabutler

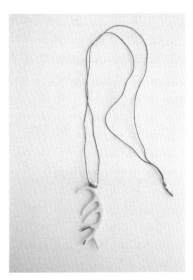

图4-84 鹿角 凡若撒工作室
Figure 4-84 Deer Collier by Atelier Vanrosa

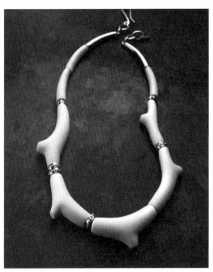

图4-85 德尔菲·纳丁 瓷项链 2009
Figure 4-85 Delphine Nardin biscuit de porcelaine, et laiton plaqué or, 2009

图4-86 概念陶瓷项链
Figure 4-86 Concept ceramic necklace

一、绘制图纸
A. Drawing

首先，设计制作陶瓷饰品需要预先设计好图纸，并根据图纸效果来选择相应的成型工艺和装饰工艺（图4-87）。

First of all, the design and manufacture of ceramic jewelry requires pre-designed drawings, and select the corresponding molding process and decoration process according to the effect of the drawings (Figure 4-87).

图4-87 绘制图纸
Figure 4-87 Drawing sketches

二、制作泥片
B. Making mud pieces

根据设计图纸，罗米选择使用泥片成型工艺来制作她的陶瓷饰品（图4-88）。

According to the design drawings, Loumi chose to use the clay forming process to make her ceramic jewelry (Figure 4-88).

图4-88 制作泥片
Figure 4-88 Making mud pieces

三、泥片烧制
C. Mud film burning

将切好的泥片饰品上好釉料，入窑烧制（图4-89）。

Put the cut clay pieces on the glaze and burn them into the kiln(Figure 4-89).

图4-89 泥片烧制
Figure 4-89 Mud burning

四、陶瓷彩绘
D. Ceramic painting

罗米在烧好的饰片上彩绘了金水（一种纯金制成的陶瓷釉），进行装饰（图4-90）。

Loumi painted gold water (a ceramic glaze made of pure gold) on the burnt plaque for decoration (Figure 4-90).

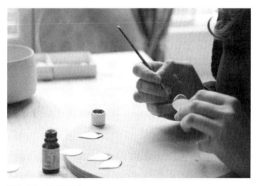

图4-90 陶瓷彩绘
Figure 4-90 Ceramic painting

五、彩绘烧制
E. Painted firing

将装饰好的瓷片再次入窑，经过约780℃的烧制，瓷片表面的金色釉料很好地附着在陶瓷表面（图4-91、图4-92）。

The decorated porcelain pieces are again put into the kiln and fired at about 780 degrees Celsius. The golden glaze on the surface of the ceramic sheets adheres well to the ceramic surface (Figure 4-91、Figure 4-92).

六、最终完成
F. Final completion

给烧成的瓷片穿上金属链，最终的饰品便完成了（图4-93）。

Put the metal chain on the burnt tiles and the final jewelry is finished(Figure 4-93).

图4-91 彩绘烧制
Figure 4-91 Painted firing

图4-92 彩绘烧制
Figure 4-92 Paint the burned products

图4-93 最终作品
Figure 4-93 Finished products

第五章　服饰配件搭配
Chapter 5　The collocation of accessory and attire

服饰配件，顾名思义即除服装之外的，所有附加在人体上的装饰品的总称。服饰配件在现代服装设计中使用的范围极其广泛，以其品类繁多、形式丰富、功能多样而越来越受到设计师的重视和青睐，也正是由于服饰配件的存在，使服装风格千变万化。

第一节　配件及其特点
Section 1　Accessories and their features

从整体配套服饰的角度来看，服饰配件是服装的一个重要组成部分，这是服饰配件最为重要的特性之一。因为服饰配件本身能成为一件单独的艺术品，同时又能包容于服饰这样一个整体之中。就服饰品而言，无论是首饰还是包袋，或是鞋帽，任何一件饰品的搭配不和谐，都可能影响整体服装效果的体现。服饰搭配艺术是人体整体性装饰的艺术，是体现服装与配件之间和谐而统一的艺术形象。如果在服饰搭配时，将服装与饰品分割思考，就会削弱服饰整体形象。

From the point of view of the overall matching clothing, clothing accessories are an important part of clothing, which is one of the most important characteristics of clothing accessories. Because the accessories themselves can become a single piece of art, and at the same time can be accommodated in the whole of clothing and decoration. As for apparel products, no matter whether it is jewelry or bags, or shoes and hats, if the matching of a piece of jewelry is not harmonious, the overall effect of clothing may greatly be affected. The art of dressing is the art of the overall decoration of the human body. It is a harmonious and unified artistic image between clothing and accessories. If you divide the idea of clothing and accessories when you try to wear and watch them, you will weaken the overall image of the clothing.

在服装的整体搭配中，服饰配件处于一个从属的位置，是服装艺术的一部分。个人形象的构建，是要通过个人外在形象以及内在修养表现出来，而外在的形象就需要包括服装、服饰品、发型、化妆等因素进行完美结合。一般而言，个人的装扮应该注重服装与个人形体、气质条件的吻合，其服饰配件、发型、化妆等都要围绕服装的总体效果来进行设计，以体现着装者的审美水平和艺术修养。但在一些特殊的场合下，如珠宝款式发布会上，也会将服装与饰品的配置关系"颠倒"，以款式简洁、色彩素雅的服装搭配华丽的首饰，以达到凸显珠宝主体的目的。又如在我国苗族民族服饰的搭配，非常注重服饰品的装扮，大量使用银质饰品来进行装饰，在服饰的整体配套中，银饰品所起到的作用是极为突出的。

In the overall collocation of clothing, the clothing accessories are in a subordinate position, which are part of the art of clothing. The construction of personal image is to be expressed through the external image and internal cultivation of the individual, and the external image needs to include the perfect combination of clothing, hair, makeup and other factors. In general, personal dressing should focus on the matching of clothing and personal form and temperament conditions. The clothing accessories, hair style, make-up, etc. should be designed around the overall effect of the clothing to reflect the aesthetic level and artistic accomplishment of the wearer. However, in some special occasions, such as the jewellery style conference, the relationship between the clothing and the jewelry will be "reversed". The styles are simple and the colors are elegant and the gorgeous jewellery is used to achieve the purpose of highlighting the main body of the jewelry. Another example is the matching of Miao national costumes in China, which pays great attention to the dressing of costumes, and uses a lot of silver ornaments to decorate. In the overall matching of costumes, the role played by silver jewelry is extremely prominent.

服饰配件的分类方法有多种，可以按照不同的要求进行分类。

There are many ways to classify apparel accessories, which can be classified according to different requirements.

（1）按装饰部位：可以分成发饰、面饰、耳饰、腰饰、腕饰、足饰、帽饰、衣饰等。

By the decorative parts: clothing accessories can be divided into hair accessories, face decorations, earrings, waist ornaments, wrist ornaments, foot ornaments, hat ornaments, clothing, etc.

（2）按工艺：可以分成缝制型、编结型、锻造型、雕刻型、镶嵌型等。

By craftmanship: clothing accessories can be divided into weaving, knitting, forging, sculpturing, setting, etc.

（3）按材料：可以分成纺织品类、毛皮类、贝壳类、金属类等。

By materials: clothing accessories can be divided into fabric, fur, shells, metals, etc.

（4）按装饰功能：有首饰品、包袋饰品、花饰品、腰带、帽子、手套、伞、领带、手帕饰品等。

By functions: clothing accessories can be divided into jewelries, bagging, floral accessories, belts, hat, gloves, umbrellas, ties, handkerchiefs, etc.

在服饰搭配艺术中，对于服饰配件的划分是按照不同的装饰效果以及装饰部位进行分类的（表5-1）：

In the art of clothing matching, the division of clothing accessories is classified according to different decorative effects and decorative parts (Table 5-1):

表5-1 不同种类的服饰配件装饰部位与装饰效果、功能

Table 5-1 In the art of dress and accessories combination, it can be classified by the effects and body parts they decorate

服饰配件种类 Accessories categories	装饰部位 Decorative parts	常见材料 Common material	装饰效果与功能 Decorative effects and functions
首饰类 Jewelries	人体的各个部位 All body parts	金属、玉石、珠饰、皮革、塑料等 Metal, gems, beads, fur, plastic, etc.	兼具装饰以及实用的性能，恰到好处点缀首饰，可以起到画龙点睛的作用，能使原本平淡无奇的服饰配套显得熠熠生辉 Both with decorative and practical performance, just the right embellishment of jewelry capable of playing a finishing touch, and making the original plain and unpleasant clothing accessories look brilliant

续表

服饰配件种类 Accessories categories	装饰部位 Decorative parts	常见材料 Common material	装饰效果与功能 Decorative effects and functions
帽饰类 Hat	头部 Hat	纺织品、皮革、绳草等 Fabric, fur, cord-grass, etc	兼具遮阳、防寒护体的实用目的及美观的装饰作用。由于处在人体极为醒目的位置——头部，在服饰搭配艺术中，帽饰类的搭配对服装的整体效果起到了重要的作用 Functions: sun shielding; cold prevention and the decoration. As the head position dominates, the effect of hats is crucial in the art of fashion combination.
鞋袜、手套类 Shoes, hosieries; gloves	手、足部位 Hands, feet	纺织品、皮革等 Fabric, fur, etc	兼具防寒、保暖之护体功能以及装饰作用。随着人体的活动，鞋袜、手套类处于一个不断变化的视觉位置，是不容忽视的重要饰品与配件之一 Cold and injury prevention together with deco purpose. With the daily activities of human bodies, the visual positions of shoes, hosieries, gloves are constantly moving; this is one of the important accessories we should care
包袋类 Bagging	因使用手法不同而不同 Varies with users	纺织品、皮革、绳草等 Fabric, furs, cord-grass, etc.	兼具放置物品的实用性能以及美观的装饰性，是服饰搭配艺术中重要的饰品之一，因材料的不同、制作方法的不同，呈现出不同的风格风貌。在服饰搭配时，其装饰性功能应符合服装的总体风格特征 Containing and deco function. It is one of the important accessories. The style and appearances vary with materials; and manufacturing methods. It should cope with the garment theme style
腰饰类 Waist accessories	腰部 Waist	纺织品、皮革、绳草、金属、珠饰等 fabric, furs, cord-grass, metal, beads	兼具绑束衣服的实用功能及装饰的美学功能 It combines the practical functions of the tie-up clothes with the aesthetic features of the decoration
领带、领结、围巾类 Tie, bowtie, scarf, etc.	颈部 Neck	纺织品、皮革等 Fabric, furs, etc.	兼具固定衬衫或防寒保暖之实用目的以及装饰的重要作用 Fasten the shirt or cold prevention purpose. The important deco function
其他类 Others	人体的各个部位 All the parts	各种材料 Various materials	包括了伞、扇子、眼镜、打火机等，有的原为实用品，有的则逐渐过渡为实用与装饰相结合 Including the umbrella, fan, glasses, ligher, etc. some used to be appliances; some turn to combine the practical purpose and deco function

一、帽
A. Hat

按照古代装饰部位的划分，帽类装饰于人体醒目的位置——头部，应该是属于首饰类装饰物。但按现代服饰搭配艺术的划分，帽类饰品则作为一个单独的单元来进行装饰。现代人常使用帽、巾，它的作用基本为装饰与御寒，其制作材料主要为布帛、各种动物毛皮等一些软质材料（表5-2）。

According to the division of ancient decorative parts, caps are decorated in the eye-catching position of the human body——the head should belong to the jewelry decoration. However, according to the division of modern clothing and art, the hat jewelry is decorated as a separate unit. Modern people often use caps and towels, and their functions are basically for decoration and cold protection. The materials are mainly soft materials such as cloth and various animal furs(Figure 5-2).

帽子的选择与个人的形体条件紧密相关，不同的脸型、不同的肤色、不同的气质所适合的帽子也有差别。人的脸型有蛋形、胖形、方形和国字形之分。一般情况下胖脸型的人戴圆顶帽会显脸部偏大，若选用宽大的鸭舌帽就比较合适；蛋型脸的人戴鸭舌帽会显得脸部更加上大下小，更见消瘦，如果选用圆顶帽就比较适宜；方脸型和国字脸型的人选择帽子的范围比较宽泛，相对容易一些。从肤色上说，帽子与人的脸部比较贴近，因此选择帽子的色彩时要尽量选择能够衬托自己肤色的，尤其是肤色不够白净的人，帽子颜色选择不当会使人的脸部色彩显得晦暗。相比帽子，头巾的搭配受到人形体条件的限制就要少得多，毕竟头巾的装饰面积较小，与人体脸部的贴合度也没有帽子"紧密"。

Selecting hat depends largely on the individuals' physical features. The appropriate hat styles differ from face shapes, complexions, and personal styles. For example, the face shapes range from egg shaped, square, round, shaped, etc. Generally, the round faced person's face would appear even bigger sized if round top hat chosen, while the wide casquette would fit.

表5-2 现代帽饰的常见分类

Table 5-2 The classification of comteparoray hats

划分标准 The standard	常见种类 Common species
按用途分 Purpose	风雪帽、雨帽、太阳帽、安全帽、防尘帽、睡帽、工作帽、旅游帽、礼帽等 Snow hat、rain hat、sun hat、safety helmet、dust cap、night cap、work cap、tourism cap、bowler hat, etc
按使用对象和式样分 Users and types	男帽、女帽、童帽、情侣帽、牛仔帽、水手帽、军帽、警帽、职业帽等 Men hat、women hat、children hat、lovers hat、cowboy hat、sailors hat、military hat、police hat、occupational hat, etc
按制作材料分 Materials	皮帽、毡帽、毛呢帽、长毛绒帽、绒绒帽、草帽、竹斗笠等 Fur hat、felt hat、woolen hat、long furred hat、cashmere wool hat、straw hat、bamboo hat, etc
按款式特点分 Syles	贝雷帽、鸭舌帽、钟形帽、三角尖帽、前进帽、青年帽、披巾帽、无边女帽、龙江帽、京式帽、山西帽、棉耳帽、八角帽、瓜皮帽、虎头帽等 beret、casquette、bell cap、Triangle tip cap、QianJin hat、youth hat、shawl hat、frameless women hat、Longjiang hat、Beijing style hat、Shanxi style hat、cotton eared hat、octagonal hat、melon skin styled hat、tiger head styled hat

Persons with egg shaped faces would appear triangle thinner faced with casquette, so round top hats are the appropriate chocice for them. Square face basically fits all hats, so it is relatively easy for square faced persons to pick hats. Regarding the complexion, the color of hats should support the face tones due to its proximity. For those with dark tones, they would appear even darker if having unappropriate hats color. Comparing with hats, the combination of head scarves have fewer limits due to its proximity to human faces and smaller decoration area.

需要注意的是，因为帽子或头巾所佩戴位置的特殊，容易给人留下比较醒目的印象，因此帽子以及头巾的使用从风格、色彩、纹样直至装饰手法，其搭配应服从服饰搭配的整体性。如图5-1所示，帽子的色彩、材质、风格与服装相映衬，起到了很好的衬托作用。

It should be noted that because the position of the hat or headscarf is special, it is easy to leave a more eye-catching impression. Therefore, the use of hats and headscarves should be in accordance with the style, color, pattern and decorative techniques. As shown in Figure 5-1, the color, material, style and style of the hat are against each other, which plays a good role in setting off.

二、包
B. Bag

包袋出现的初期，是出于实用的目的，用以放置物品。我国古代人们常常使用包袱、褡裢作为出行物品的收纳用途。由于时尚产业的推动，包袋已经成为现代人服饰搭配的重要组成部分，兼具了实用和美观的双重功能。同时，包袋的设计也向着不断求新求异的方向发展，每一季的流行服饰时尚发布都离不开包袋的点缀。

With the development of fashion industry, currently bagging turns to be an importnatn part of the fashion. The initial purpose of bagging is for daily practical containing. our ancient ancestors used to put stuff into cloth wrappers to travel. It has both practical and deco purpose. Meanwhile, the design of the bagging moves towards the innovative trend. Every fashion season is accompanied with bagging.

图5-1　帽子与服装相协调
Figure 5-1　Hats and costumes coordinate

1. 包袋常见材质与风格
Common materials and styles of bags

包袋的制作材料较为广泛，包括棉、麻、布、帛，各种人造以及天然的面料，可用作服装使用的面料都可以用作包袋的设计、制作，而很多服饰不能使用的材质，包袋也可以使用，如草、木、竹、PV等，可以通过编织、雕刻等不同的工艺，设计并制作出风格迥异的包袋。制作材料的差异，装饰手法各异，使得包袋呈现多样的款式风格。在不同的流行阶段，包袋设计师往往根据不同时期的不同流行风情来设计，从而形成独特的风格。如以抽象图案表达，花样百出的度假风情；如用洗水、漂白、喷砂洗、染洗或制造如被猫抓过的白线效果，或是以起皱的效果加点扎染技巧，造出浓浓的怀旧效果；或是从工装元素提取感觉，功能化、实用性的口袋设计以及装饰性明线、补丁，拉链的金属质感与包袋外廓明朗的线条，营造出粗犷热烈的气质。

The bagging manufacturing materials are vast: cotton, linen; cloth; silk; many sythetic and natural materals are taken into bagging design. Some materials can be made into bagging but not in attires, like :grass; wood; bamboo, PV, etc. widely ranged styles can be designed through different craftswork like knitting, sculpturing, etc. different manufactring materials, decoration methods would produce quite diversed styled finished products. In different fashion periods, fashion designers would take its specific fashion trends into the unique style in that period. For example: the vacation style is conveyed by abstracted pattern; The strong tastes of retro emotions are expressed through washing, bleaching, sand blasting, dying, the artificial white threaded effects , cat scratching or the wrinking effects with tie dying technique; Or the rough and passionate atmosphere is created from the elements from working attire, like putting the metal zippers, the practical pocket design, decorational surface lining, clear shaped lining into functional design.

2. 包袋与服饰搭配注意事项
Bags and clothing matching precautions

包与服饰的搭配，关系到年龄、职业、季节、使用者性格、着装、使用场合等诸多因素。不同年龄段的人对时尚的观点也不同，包的款式搭配和颜色搭配应该和自己的年龄段相吻合，才不会使人产生搭配不协调的感受。不同的职业对包的选择也有区别，职业女性可以选择较为简洁的款式，这样可以突出自己的品位；不同场合使用的服饰配饰是不一样的。如经常外出，可以选择休闲一些的包，会显得比较有活力。如需携带一些资料或面见客户，可以选择实用型包。包的选择还需与季节相协调，夏季的包一般应以浅色或淡纯色为好，这样不易产生"扎眼"的感觉；冬季可以选择略深颜色的包。

着装是一门艺术，运用包和衣服，作整体的搭配，款式、颜色都能和着装产生不同的"互动"效果。考虑到色彩的协调性，有三种搭配方式比较简便且可以取得不错的效果：同色类搭配法，包和衣服为同色深浅的搭配方式，可以产生非常典雅的感觉，如咖啡色着装搭配驼色包袋；对比色搭配法，包和服装可以使用明显的对比色，从而产生一种另类抢眼的搭配方式，如黑色裙子加白色皮鞋搭配黑白色间色包袋；衣服色彩呼应搭配法，可以让包的颜色和服装的色彩、花纹、配饰协调搭配，如粉白色上衣加淡紫色裙子搭配淡紫色或米色包袋。如图5-2所示，包的色彩与服装形成鲜明的对比，使服装色彩视觉效果更为突出。

The combination of bags and clothing is related to age, occupation, season, user's personality, dress, use occasions and many other factors. People of different ages have different views on fashion. The style of the package should first match the age of the package, so that it will not cause dissonance; in addition, the color and age of the package should be considered. Different occupations also have different choices for the package. Professional women can choose a more concise style, which can highlight their own taste; if you go out often, you can choose some leisure bags, it will be more energetic. If you need to bring some information or meet customers, you can choose a practical package. The choice of package also needs to be coordinated with the season. The package in summer should generally be light, so it is not easy to produce the feeling of "eye-catching"; in winter, you can choose a package with a slightly darker color.

Dressing is an art, using bags and clothes for the overall match; styles, colors and dresses can produce different "interactive" effects. Considering the harmony of colors, there are three ways to match it, which is relatively simple and can achieve good results: the same color matching method, bags and clothes are the same color shades, which can produce a very elegant feeling, such as brown dress with camel bag. Using contrast color matching, bags and clothing can use obvious contrast colors, resulting in an alternative eye-catching style, such as black skirts plus white shoes with black and white color bags. Matching with the color of the clothes, you can match the color of the bag with the color, pattern and accessories of the clothing, such as a pink-white top with a lavender skirt with a lavender or beige bag. Figure 5-2, the color of the package is in sharp contrast with the clothing, making the color visual effect of the clothing more prominent.

服饰搭配时，色彩与服饰搭配的协调固然是重要的，但更为重要的是服饰风格与包袋风格之间的融合。休闲的T恤打扮，可以搭配质地柔软的包袋，以配合悠闲的生活态度；端庄稳重的职业服饰，可以搭配廓型鲜明，外形小巧一些的包袋，以体现职业妇女干练又不失精巧的性格特点；高贵的晚礼服，包袋的配饰同样要具有优雅的气质。如图5-3所示，包袋的色彩、材质均与服装相协调，且风格与服装相契合，为服装增色不少。需要注意的是，日常生活所需的包袋与舞台表现的包袋有所差异，日常使用包袋最好控制在身上只背或提着一个包，看起来才会干净利落（图5-4）。如果要带的东西太多必须分装成两包时，最好能够一背一提，即一个用背包、另一个用手提包的方式来处理，背包和手提除了要有大小之分，也要避免使用同一系列的产品，整体造型看起来才会轻盈活泼。

When dressing, the coordination of color and clothing is important, but more important is the fusion of clothing style and bag style. Casual T-shirts can be worn with soft-bags to match the leisurely attitude of the clothes; dignified and steady professional clothing can be matched with a bag with a distinctive shape

图5-2　包的色彩与服装形成鲜明对比

Figure 5-2　The color of the bag contrasts sharply with the clothing

图5-3　包袋的款式风格与服装相呼应

Figure 5-3　The clean shaped bag optimizes the dress' effect

图5-4 包袋数量的控制使服装显得更为干练
Figure 5-4 The control of the number of bags makes the clothing look more attractive

and a small shape to reflect the professional women's skill and not Inferior personality characteristics; noble evening dresses, bag accessories also have an elegant temperament. Figure 5-3, the color and material of the bag are coordinated with the clothing, and the style is compatible with the clothing, adding a lot of color to the clothing. It should be noted that the bags required for daily life are different from the bags for stage performance. The daily use of bags is generally best controlled in one body——only the back or carrying a bag, it looks like clean and neat (Figure 5-4). If you want to bring too many things, you must pack them into two packs. It is best to carry them one by one, that is, one with a backpack and the other with a handbag. In addition to the size of the backpack and handbag. Also, if we avoid using the same series of products, the overall shape will look light and lively.

包袋与服饰如果搭配得当将会起到画龙点睛的作用，使服饰形象显得更为整体；而如果包袋与服饰的搭配不够协调，则会起到适得其反的效果，使服饰形象显得凌乱，着装者的个人魅力也会减弱。

The combination of bags and clothing, if properly matched, will play a finishing touch, making the clothing image more holistic, and if the combination of bags and clothing is not harmonious, it will have the opposite effect, making the clothing image seem messy. The personal charm of the wearer will also be weakened.

三、鞋、袜
C. Shoes and hosieries

古代的鞋子称履，历代鞋履的形制非常丰富，其区别主要反映在质料、款式以及装饰等几个方面。如质地有丝、帛、葛布、麻等多种；款式有履、靴等多种；装饰手法则有绣、镶等多种手段。我国古代对于鞋子的穿着，在色彩、材料上都有一些限制，如丧葬着白鞋、女子缠足着三寸金莲之陋习等；不过随着社会的发展，人们对于服饰的宽容度也越来越大。古代鞋制品与现代鞋制品款式、用法差别较大。现代鞋类制品的材料除了传统的布帛之外，常见的还有各种动物的皮革、各种人造皮类制品，装饰手法、款式也不尽相同。

In ancient China, shoes are called "lv". The forms of shoes in every dynasty are so diverse and rich. The major differences lie in the manufacturing materials, styles and ornaments. Materials are silk, linen, etc; styles are slippers, boots, etc. decoration methods are embroiding, setting, etc. Certain restrictions were applied for the shoes colors and materials in ancient China. For example, the white shoes were for funerals, the wrapped small feet for women, etc. Along with the social development, much more tolerance is rendered for shoes and hoiseries as well.

鞋子的面积不大，又处于人体形象的最底端，处理不得当，容易显得突兀，鞋子的选择，其色彩、材质或风格要注意与服饰相协调，见图5-5。此外，鞋与帽子相配套使用时应注意"视错现象"对个人形象产生的影响：如果着装者身高不高的话，应当避免同时使用浅色的帽子和鞋子，因为浅色的色彩具有一定的扩张感以及膨胀感，会造成视觉上的上下压缩，使人显得更加矮小。

The size of the shoes is not large, and it is at the bottom of the human body image. It is not suitable for handling. It is easy to look awkward. The choice of shoes, color, material or style should be coordinated with the costumes, as shown in Figure 5-5. In addition, the use of shoes and hats should pay attention to the impact of "visual error phenomenon" on the personal image: if the wearer is not tall, you should avoid using light-colored hats and shoes at the same time, because the light color has a certain sense of expansion and the sense of expansion can cause visual up-and-down compression, making people appear shorter.

袜在古代被称作内衣解，我国现存的早期的袜子大多为两汉时期的遗物，质地不仅有罗，还有绢、麻、织锦等。由实物观察可见，当时的袜子制作都比较宽松不具有弹性，这种袜子穿着在胫部，很容易滑落，因此人们在穿着时往往在袜子的上端另系袜带。现代具有弹性的袜子是从西方传过来的，1928年，杜邦公司展示了第一双尼龙袜，同时拜尔公司推出丙纶袜。1940年，高筒尼龙袜在美国创造历史最高销售纪录，并开始成为普通日用品。按照长短划分，现代袜品可分为长袜、中长袜、短袜等；按照使用的对象划分，有男袜、女袜、童袜等。古代的袜子属于不外显的内衣的一部分，但在现代时尚设计师的手中，它也成了服饰装扮的一部分。

Hoiseries are categorized into underwear in ancient ages. The 'antique hoiseries' are mainly from the Two Han dynasties. The materials range from silk to linen, etc. By observing the real objects, the socks in that time were relaxing but not elastic. They were worn on the shin, easy to slip down. Thus, the fasten ribbon was put on the top. The modern elastic stockings are from western. In 1928, the first pair of nylon socks was displayed by DuPont. Meanwhile, Bayer displayed polypropylene fiber socks. In 1940, high stockings achieved the historically highest sales volum in U.S., and turned out to be the daily consuming products. The modern socking are classifed by the length,to be long, mid-long, or short socks. If classified by wearers, they are men, women, children's socks. The hoiseries are now the part of the fashion in the modern designers' hands while they belonged to the underwear category before.

图5-5 鞋的款式与服装相称
Figure 5-5 The style of the shoes is commensurate with the clothing.

1. 男性服饰搭配时袜子使用注意事项
The notes for men wearing socks

袜子在服饰搭配中起到的作用常常被人们所忽视，尤其是男士的服饰装扮，有时一个细小环节的疏忽造成的失礼是难以补救的。常用的男袜多为短袜，可分成两大类：深色的西装袜和浅色的休闲袜。

In fashion, socks are usually ignored, especially in men wearing. Sometimes a neglection in details is unretrievable. Men's socks are normally short. There are two categories in men's socks: dark colored suit socks and light colored leisure socks.

男士穿袜子最重要的原则是讲求整体搭配，多数情况下，裤边会直盖鞋面，只有在不经意中才能见到袜子的存在。此时，袜子的色彩、质地、清洁度就会为穿着者的品位提供打分依据。男袜的颜色应该是基本的中性色，并且比长裤的颜色深。男袜的颜色与西装相配是最简单，也是最时髦的穿法，如果西装是黑色的，可以选择黑色的袜子；深蓝色的西装就应该配深蓝色的袜子；米色西装配棕色或深茶色袜子。当一位男士坐着的时候，从他的西裤裤腿和西装皮鞋之间露出来一截雪白的棉袜或一截光腿，这种将正装和休闲袜混搭的现象是十分不雅的。一般情况下，白棉袜只用来配休闲服和便鞋，标准西装袜的颜色是黑、褐、灰、藏蓝的，以单色和简单的提花为主。材质多为棉和弹性纤维，冬季增加羊毛来保暖。标准的西装袜袜筒应该到小腿处，以保证袜边不会从裤管里露出来，而且尽量挑不醒目的浅颜色；若做纯白的运动打扮，袜子一定要是纤尘不染的白色运动袜。

The top rule for men to wear shocks is the intergration of dress. In most cases, the shoes' surface would be covered by pants ends. Socks are only peered in few cases. Therefore, the texture, cleaness and color of socks could add credits to the wearer's tastes. The color of men's socks should be neutral, darker than the pants. The easiest and most up-to-date matching style is the color of socks coordinating suit. If the suit is black, the socks would better be black; the dark navy suit should go with the same colored socks; cream suit should go with brown or dark tea colored socks. When a man sits, a part of white cotton socks or part of naked legs would be revealed between the pants and shoes, that would be ungraceful if the suit is combined with leisure styled socks. Generally, white cotton socks are only for leisure clothes and shoes. The standard suit socks' colors are black, brown, grey or navy blue in pure single color or simple pattern. The materials are mostly cotton and elastic fiber. Wool is used to keep warm in winter. The standard length of suit socks should go upto the shin in case the edges won't be revealed from the pants. The light unspectacle colors are preferred. If the pure white sports style is carried, the cleaness of white sports socks must be Guaranteed.

2. 女性服饰搭配时袜子使用注意事项
The notes for female socks

相对于男士而言，袜子在女性服饰形象塑造上所起到的效果更为重要。现代女袜形式多样，不仅有长袜、连裤袜、中长袜、丝袜、棉袜等不同的长度与材质，色彩也非常丰富，包括彩印、彩织的多色袜、单色袜，且质地亦各不相同，不但有透明的或不透明的袜子，还有袜边、袜背或整双袜子饰有花纹的，令人眼花缭乱。如图5-6所示，袜子的外廓或松散或与肌体贴合；色彩或是鲜明跳跃或是与服装相协调，均是服饰形象塑造不可或缺的一个部分。

女性袜子的选择要根据个人的腿部生理条件、服饰的整体造型要求来进行。袜子选用得当，可以使人的腿部显得修长。腿部短粗，就不适合穿具有外扩感的浅色袜、花式厚袜、不透明袜、羊毛袜以及金银色袜、花边袜、大朵提花袜等，而以深色的袜子为宜。如果双腿又细又瘦，则穿浅色或肉色的袜子为好。如果腿部肤质较差或是汗毛较重，过薄的袜子会令汗毛更加明显（图5-6）。

Comparing with men, socks are more important in women's dressing. The types of curent female socks are various: long stockings, leggings, medium long socks, silk socks, cotton socks, etc. Materials and colors are so diversed, including the multi-colored or single colored socks by printing, color weaving; transparent or opaque socks; with clear socks' decorated edges, and patterns on socks top or both sides. The individual's legs physical properties should be considered for females' socks selecting as well as integrating with the dress. The legs would appear slim if the appropriate socks are chosen. The light colored visually expanding socks, patterned thick socks, opaque socks, chiffon bordered socks, silver gold colored socks, large jacquard socks should be avoided if the legs are short and thick, dark colored socks are suitable. If legs are slim, then the light or nude colored stockings are good. If the skin quality of legs is weak or heavily haired, too thin socks would exaggerate the visibility of hair (Figure 5-6).

女性的袜子款式和色彩多种多样，但在裙装里占主导地位的还是丝袜。正式场合套裙应当配长筒袜或连裤袜，不能光腿或穿彩色丝袜、短袜，袜子不能有网眼、花纹、图案，穿上去后要平整。穿裙子时，应配长筒丝袜或连裤袜，袜子的袜口一定要高于裙子下部边缘，不仅在站立时袜口外露显得不雅，即便在行走或就座时袜口外露也不合适，袜子和裙子中间露一段腿肚子，这种穿法术语称为"恶性分割"，容易使腿显得又粗又短，在国外往往会被视为是没有教养的妇女的基本特征。事实上，鞋、袜、裤或裙颜色出现三种或正好被袜子分成三部分时就会产生"三截腿"的效果，是不雅的。此外，袜子不得有丝毫破洞、抽丝、染色现象。白色袜子在正式社交场合中不多见，一般只适合在运动场中出现，小姑娘穿白色袜子时，显得活泼可爱。

The women socks' styles and colors are various, but the silk socks dominate in skirt wearing. In formal occasion, the suit dress should be with the long stockings, naked legs, colorful socks, short socks, meshed, patterned socks are not allowed. The stockings should be worn smoothly. The long stockings should be with dress. The border of socks should be higher than the skirt. It is inappropriate if the edges of the socks revealed when standing or walking. The terminology 'malicious cutting' is used for part of legs revealed between socks and skirt that would make the legs appear-

图5-6 袜与服装的协调
Figure 5-6 Socks and clothing coordination

ing short and thick. It is regarded as the sign of uneducated women in western countries. In fact the colors of shoes, skirts, pants are divided into 3 sectioned legs by the socks. Besides, there should not have any broken holes, drawn out, dying in the stockings. In formal occasions, white socks are not common; they only appear in the sports occasion. Young girls appear lovely and active wearing white socks.

职业装不适宜搭配厚质的棉袜，职业女性的丝袜以单色或带隐约细小的花纹为好，颜色以肉色、黑色为宜。作为职业着装，袜子不宜过于突出，而只要起到配合整个仪表的作用即可。袜子和鞋子或裙、裤的颜色一致，会使整个腿部看起来修长和统一，也同时显现出主人是有一定审美能力的人。如果袜子能够有效地与服饰相搭配不仅可以使人的腿部显得修长，而且能够成为装扮美丽的"载体"。一个基本通则是：身上穿得越复杂，腿上穿的丝袜就越应该简单、清爽。

袜子与服装相配，裙子和裤子的织物质地是袜子选择的先决条件。一些厚质的呢以及羊毛织物类可搭配不透明袜或花式袜子，半透明的薄袜可搭配薄型的呢及羊类织物，而丝织衣料、雪纺、棉织物之类的质地以搭配透明薄袜为佳。另外还应注意袜子和鞋子的配合协调，通常鞋跟越高，袜子应越薄，厚袜不要配细高跟鞋，薄袜不要配球鞋（图5-7）。

Thick cotton socks are not suitable for professional dress. Pure colored silk stockings or tiny patterned ones are good. The colors are better to be nude or black. For the professional dress, stockings should not be conspicuous, but have to be supportive for the dress instead. The color of stockings should coordinate with shoes, skirts, or pants. Then the legs would appear slim and integrated, reflecting the good tastes of the wearers.

Thick cotton socks are not suitable for professional dress. Pure colored silk stockings or tiny patterned ones are good. The colors are better to be nude or black. For the professional dress, stockings should not be conspicuous, but have to be supportive for the dress instead. The color of stockings should coordinate with shoes, skirts, or pants. Then the legs would appear slim and integrated, reflecting the good tastes of the wearers (Figure 5-7).

一些外形宽松、色彩夸张的袜子或袜套，只适合在休闲的时间穿着，可搭配一些短裙、短裤穿用，根据服装的色彩、款式风格进行搭配，营造出可爱、活泼的服饰特征，穿着者以少女为宜。

Some relaxing fancy stockings or socks are only for the leisure time. They can match short skirts or shorts. They can create the lovely and actively appearance when combining with shorts and short skirts. They are especially suitable for young girls.

图5-7　衣服和袜子的颜色协调
Fig 5-7　The colors of dress and stockings coordinate

四、围巾
D. Scarf

围巾常见于女性的服饰搭配领域。围巾的面料品种很多，棉、毛、丝、麻等都是围巾常见的面料品种，根据面料的厚、薄程度不同，适用于不同的季节；围巾的款式也非常丰富，如大方巾、小丝巾、三角巾、长巾、异形巾等，根据其风格的不同，可以搭配不同的服装款式。如图5-8所示，围巾花型取自裙装，且与服装同属一类材质，两者风格协调相得益彰。围巾面料的花型丰富多彩，几乎服饰适用的一切花型都可适用于围巾。围巾与服饰的搭配，色彩的调和是首要的条件，如可以与服饰相融合，也可以有意识地与服饰拉开色彩对比。

Scarf is used more in dress combination. The fabrics vary from cotton, wool, silk and linen. The thickness of scarf varies with seasons. The styles are very diversed. For instance: big squared scarf, small sized silk scarf, triangle shaped scarf, long scarf, irregular shaped scarf, etc. They can match different styled dress. Figure 5-8, the pattern of scarf is from the dress, so their styles perfectly match. The patterns of the scarf fabric are vivd because almost all the patterns on dress can be applied on scarf. The first priority of combination of scarf with dress is color-matching. Sometimes contrasts can be intentionally made for matching dress and scarf.

图5-8 围巾的花型与服装花型相仿，风格调和

Figure 5-8 The pattern of the scarf is similar to that of the clothing, and the style is harmonious

1. 根据形体条件选择围巾
Scarf selecting by individual's figure

选择围巾搭配有两个要素：形体肤色、体型两个方面。

There are 2 rules for selecting scarf: complexion and figure:

肤色：肤色偏黄的女性应避开紫色和黄色，可选择奶白色、湖蓝色、中绿色的围巾，使自己的脸色在丝巾的衬托下更白净；如果肤色白皙，颈部修长，且身高适中，则各类色彩的围巾都可以取得不错的配饰效果。在图案上可选择大朵的花卉，宽大的格子，散点式或抽象色块的组合拼接，都可以使人看上去有朝气、睿智和干练之感。

Complexion: the tone of oriental women tends to appear yellow. Therefore, cream color, lake blue color, green color are preferred for clean and fair appearance while avoiding purple and yellow color; if the individual's figure is moderate with slim neck and fair complexion, then all types of colorful scarf are good. The pattern could be large floral, big plaid, dotted, or abstracted color pat. These could produce the effect of being active, clever and capable.

体型：身材苗条的青年女性可以选择有张力的橙色、柠檬黄色、果绿色，以及一些大花型面料的围巾，在人群中起到跳跃、醒目的作用；体形略为丰满，颈部不长的人，在佩戴围巾时就应尽量避免使用质感过于厚实的款式。即使是身材苗条但脸部较胖的女性，也避免选用过于跳跃的色彩，切忌大花大格的围巾，细条纹或小花型的围巾是不错的选择。

Figure: the slim young females can select expanding colors like orange, lemon,fuirt green, together with large floral patterned ones for the spectacular visual effects; if the figure is plump with short neck, the thick textured scarf should be avoided. The slim females with flat faces should avoid the hopping colors and big floral or plaid patterned scarf. The thin striped or small floral patterned scarf should be fittable.

2. 根据服装款式选择围巾
Scarf picking upon the dress type

不同服装款式所适合的围巾样式也不同。如为了搭配无领毛衫，可以选择一条色调柔和的、小碎花形的围巾；而为了搭配高领毛衫时，则以富有垂感的长纱巾为好。色彩纯度低的蓝、紫、墨绿、褐色等围巾较易与服饰搭配。

The appropriate scarf style varies with dress style. A gentle colored small floral scarf is suitable for round neck jersey; For polo neck jersey, drooping chiffon scarf is preferable. The scarf with colors like blue, purple, dark green, brown in low purity is easy to combine with dress.

3. 利用不同系扎方法塑造外观形象
Various apperances creating from different knotting methods

围巾色彩、面料、花型、款式的选择固然重要，围巾的系扎方法也非常有讲究。图5-9是大方巾、长围巾及小方巾的不同系扎手法，产生的视觉感截然不同。人们在日常生活中可以根据个人喜好选择。事实上，即使是同一条围巾，系扎的方法不同，就可以产生不同的外观效果。例如，同一款长巾，轻披于肩部，感觉可能是端庄、大方的，而将之系扎于腰间，感觉则可能是休闲而活泼的。

Though the choice of scarf color, fabric, pattern, style is important, knotting methods of scarf are also critical comparing with its color, fabric and style. The visual effects vary with knotting methods even it is the same scarf. For example Figure 5-9, if a long scarf covering on shoulders, the wearer would appear decent and graceful. The wearer would appear active and leisure if the long scarf fastend around waist. There are some fasten methods of small square scarf, long scarf and cape as illustrated. They may serve different preferences.

4. 根据场合选择围巾
Pick the right scarf for specific occasions

无论男女，正式场合使用的围巾都要庄重、大方。颜色可以兼顾个人爱好、整体风格及流行趋势，最好无图案，亦可选择典雅、庄重的图案。

On formal occasions, scarf should be decent and dignified regardless of male or female. The graceful pattern should be chosen and the colors could go with personal preferences, dress style and fashion trend.

五、领带、领结
E. Tie, bow tie

领带是上装领部的服饰件，系在衬衫领子上并在胸前打结，广义上包括领结。领带起源于欧洲，它通常与西装搭配使用，是人们尤其是男士日常生

图5-9 不同的围巾系扎手法
Figure 5-9 Different scarf tiets

活中最基本的服饰品。现代男子使用的领带基本沿袭19世纪末的条状款式，45°斜向裁剪，内夹衬布、里子绸，长宽有一定的标准，色彩、图案多种多样。

Tie is an accessory for collar. It is fastend around the collor, including bow tie. It is originated from Europe, normally matching with suit as the most fundamental accessories for males in daily life. The current male ties follow the strip shaped sytle of 19th century, 45° angled side cutting, cloth lining inside, with silk on the back. Certain standards are for length and width with various patterns and colors.

现代服饰搭配理念中，领带与领结是属于男子的专用服饰配件，且都是在比较正规的场合佩戴的。相比较而言，领带比领结的使用场合更多一些，一般穿着西服、衬衫都可以扎领带，而领结只有穿着礼服时才能使用。领带被称为西服的灵魂，有时一条精致的领带搭配，可以起到画龙点睛的作用。

In the modern fashion combination ideas, tie and bow tie belong to men's dress only, especially in formal occasions. Tie is widely used comparatively. Normally, it can go with either suits or shirts. Bow tie is only for tuxedo. Tie has been called as the soul of suit. Sometimes a refined tie pinpoints the suit.

1. 男子领带选用的注意事项
Notes for male tie selecting

（1）面料。

Fabric.

领带的选择主要考虑其面料、色彩、图案、款式等几个方面，领带首选的面料为真丝，真丝是制作领带最高档、最正宗的面料，除真丝之外，其他面料如尼龙、棉布、麻料、羊毛、皮革等，也可用于制作领带，但档次较低。还有一些在旅游区工艺品商店常见的如纸张、竹篾、珍珠等特殊材质制成的领带，大多不适合在正式场合使用。

The choice of tie mainly lies on its fabric, color, pattern, style and so on. The preferred fabric for ties is silk. Silk is the most high-grade and authentic fabric for making ties. Besides silk, other fabrics such as nylon, cotton, linen, wool, leather, etc., can also be made into tie, but the grade is lower. There are also some ties made of special materials such as paper, bamboo rafts and pearls that are common in craft shops in tourist areas. Most of them are not suitable for formal occasions.

（2）色彩。

Color.

领带的色彩有单色、多色之分，单色领带适用的场合较广，一些公务活动和隆重的社交场合都适用，一些中性的颜色如黑色、白色、灰色、蓝色、不同深浅的棕色、紫红色等最受欢迎，多色领带颜色选择上一般不要超过三种。需要注意的是，色彩过于艳丽的领带用途并不广泛，只有在非正式的社交、休闲时才适合使用（图5-10）。

The color of the tie is monochromatic and multi-color. The monochrome tie is suitable for a wide range of occasions. Some official activities and grand social occasions are applicable. Some neutral colors such as black, white, gray, blue, and different shades Brown, fuchsia and other popular, multi-color tie color choices generally do not exceed three. It should be noted that the use of ties with too bright colors is not widespread and can only be used when informal social and leisure (Figure 5-10).

图5-10　T台上的领带图案可选择的余地相对较大

Figure 5-10　The tie pattern on the runway has a relatively large choice

（3）纹样。

Decorative Patterns.

多色领带的花形一般有抽象纹样以及具象纹样两类，以抽象纹样花型的领带使用面较广。用于正式场合的领带，其图案应规则、传统，最常见的有斜条、横条、竖条、圆点、方格以及规则的碎花，它们多有一定的寓意。印有人物、动物、植物、花卉、房屋、景观、怪异神秘图案的领带，仅适用于非正式的场合。

The pattern of the multi-color tie generally has two types of abstract patterns and figurative patterns, and the tie with the abstract pattern is wider. For ties used in formal occasions, the pattern should be regular and traditional. The most common ones are diagonal strips, horizontal strips, vertical strips, dots, squares and regular floras, which have certain meanings. Ties with characters, animals, plants, flowers, houses, landscapes, weird and mysterious patterns are only available for informal occasions.

2. 领带的搭配、使用方法
Matching and use of tie

总体而言，领带与服装最简单的搭配方法就是衬衫和西装颜色搭配，领带和衬衫搭配，如果是白衬衫，领带直接去搭配西装颜色即可，使领带起到一个衔接衬衫和西装颜色的作用。即使是纹样极为复杂的领带，其色彩一般也会统一在一个色调里。按照西装—衬衫—领带三者的顺序，比较简便的配色法是：深—浅—深、浅—中—浅或深—中—浅的配色方法。如图5-9所示，T台上的领带图案可选择的空间较大，但其图案的选择依旧从属于服装，或纹样相仿或色彩协调，很好地起到衬托服装主体的作用。

Generally, the simplest way for matching tie and suit is its color coordinating, the colors of shirt and suit, tie and shirts. If the shirt is white, then match the color of tie directly with suit, that the color of tie bridging the shirt and suit. Even as for the complicate patterned tie, the colors can be categorized into one color section. According to the order of suit—shirt—tie, the easy strategy for color combinations is dark—ligth—dark, light—moderate—light, or dark—moderate—dark. Figure 5-9, the spaces for the spaces for tie patterns choosing in run way are relatively wider. The pattern still follows the suit, either mimicing or coordinating the suit color with good supportive effect.

只要色彩搭配统一和谐，不管哪种配色法，均能起到不错的效果。领带有宽窄之分，这主要受到时尚流行的左右。进行选择时，应注意最好使领带的宽度与自己身体的宽度成正比，而不要反差过大。领带还有箭头与平头之别。前者下端为倒三角形，适用于各种场合，比较传统；后者下端为平头，属于比较时髦的款式，多适用于非正式场合。除西服外，穿着其他的服饰如大衣、风衣、夹克、猎装、毛衣、短袖衬衫而不穿西装时，最好不要打领带。

The good visual effect can be fulfilled following the cooperation of colors regardless of which color combination strategy taken. The style of tie is influenced by fashion trend, varying in shapes like wide or thin. The ratio of the width of tie would better coperate with the width of individual's figure, not too much contrast. The bottoms of tie are classified as arrow shaped and flat. The previous one is the reversed triangle, suitable for most occasions, relatively classical; the latter one's bottom is flat, fashion styled, mostly for unformal occasions. Apart from suit, other clothes like coat, jacket, windcoat, hunting jacket, jersey and short sleeved shirts are better not to be with ties.

传统的服饰观念，领带是属于男士的饰物，女士一般不打领带。但现代时尚的发展，将领带也带入了女性服饰搭配的领域。在中性风格以及军装风格大行其道的时候，一些设计师巧妙地把原属男性的领带饰品引领进了女性时尚概念，营造出率性的女性形象。如图5-11所示，领带的选用恰到好处地展示了女性干练而率性的服饰气质。一些服饰搭配中，领带的配饰形象甚至完全颠覆了其在人们心目中的传统印象，显得大胆而前卫。

In classical fashion ideas, tie belongs to men's accessories. Women don't wear it. But along with the current fashion development, the tie is also used in the women's fashion. The tie is used into women's fashion ideas when neutral and military styles are introduced into female fashion areas by some designers. It creates the cool and handsome female images. In some fashion combination Figure 5-11, the tie has been used for bold and contemporary images, totally reversing the classical ideas.

图5-11 女服装中领带的使用
Figure 5-11　Use of a tie in a female garment

系领带的同时，要注意领带夹的使用，领带夹是为了使领带保持贴身、下垂的服饰用品。在正式场合把领带夹在衬衣襟上，这样领带会显得挺直，既不会被风吹起，在弯腰时也不会直垂向地面。领带夹在穿西服时使用，如仅单穿长袖衬衫时没必要使用领带夹，穿夹克时更无需使用领带夹。穿西服时使用领带夹，应将其别在特定的位置，即从上往下数，在衬衫的第四粒与第五粒纽扣，将领带夹别上，然后扣上西服上衣的扣子，从外面一般应看不见领带夹。因为按照妆饰礼仪的规定，领带夹这种饰物的主要用途是固定领带，如稍许外露，尚可，如果把它固定得太靠上，甚至直逼衬衫领扣，就显得过分张扬。

The use of tie clippers should be cared while being with the tie. The purpose of tie clipper is for the fit and drooping of the tie. In formal social circumstances, it is clipped on the shirt's edge for keeping the tie straight, not blown away by wind, not dropping to the ground when bending forward. Tie clipper is used in suit, not necessary only for wearing long sleeved shirt, not for jacket either. The position of the tie clipper should be noted when wearing the suit: it should be within the 4th and 5th shirt's button counting from upside down. And then button the suit. It should been seen from outside, as the purpose is for fastening upon the dress etiquette. It would appear too conspicuous if it is put upper, so close to the shirt button. It is fine just being revealed a little bit.

领带夹的材质，有镀金的、仿金的、K金的和白银的，有的领带夹上还镶有天然或人造的宝石加以装饰。从近年的时尚趋势分析，除特别正式的场合，男子着西服、戴领带已经很少同时使用领带夹了。

The materials for tie clippers are: plating gold, synthetic gold, K gold and silver, etc. some are decorated with natural or synthetic gems. Upon the current fashion trend, the tie clipper is rarely used with suit and tie, except some extremely serious occasion.

六、腰饰
F. Belt

1. 常见的腰饰及其风格
Common waist decoration and its style

腰饰是用于人体腰部位置的装饰品，由各种材料制作而成，兼具绑束衣服的实用功能以及装饰的美学功能。在古代，不论穿着官服、便服，腰间都要束上腰带，多以布帛及皮革制成，一些贵族还常常在革带上悬挂刀剑、印章之类的随身物件，天长日久，腰带便成了服装中必不可少的一种饰物。20世纪70～80年代，裤子款式普遍腰部肥大，因此要在腰带系扎，国人利用一些废弃的面料边角经过简单加工制成腰带，系住裙或裤，这类腰带人们更加通俗的称它为"裤带"，因其主要的目的在于实用，不涉及美观的概念。当时这类腰带男女皆用，现在已经很少有人使用了。在现代服饰中，只有一些女性的服装中还可以见到布帛类软质的腰带，如一些连衣裙、大衣等，但主要的目的是为了凸显腰身系腰带，是服装款式美观的需要，已基本脱离了实用的目的。另还有一些用纱、缎类制成的腰带，也主要是装饰之用。

Belt is the accessory for human waist made by various materials with the practical purpose and artistic decorational function. In ancient time, the belt was fastend on the waist regardless of the official dress or daily wear. They were mostly made by cloth or leather, on which some noble persons hung swords, stamples etc. Later it turns to be a neessory accessory in fashion. In 70th and 80th century, the pants' style was loose-waisted. Thus the fastening was needed. Chinese people used to manufacture the remnants of the fabrics into waistband to fasten skirt or pants. Its purpose was mainly for practical function, not related to visual ideas. In that time the waistband was popular among both men and women, but now it is rarely used. In current fashion accessories, it is still used in a few female dress as soft textured one, like the long dress or coat for visual purpose to highlight the waist. It is basically not related with practical function. Sometimes, satin and chiffon are used for waistband too, mainly for decoration.

皮革类材质制成的腰带在现代服饰中极为常见，不论男女皆可使用。除天然的各类动物皮革外，还有很多人造材质也被广泛应用。男子服饰中多使用一些高档的皮革制品，搭配一些款式比较正规的西裤穿着，而一般休闲的牛仔类裤装就很少使用了。男子腰带的款式比较单一，一般为深色皮革，带上没有装饰，带头多为金属制成。领带头的造型大致相同，至多在细节上略有差别。现代男子所用之腰带与古代男子所用款式有所类似，但不同的是现代男子腰带鲜有再悬挂其他配饰的。此类腰带女子服饰也常使用，不但材质、色彩多样，且款式也更为多样化，钉、镶、绣、编多种装饰手法皆可使用，还常常饰以各式挂缀。除皮革类腰带外，一些塑料、绳结、金属，甚至珠类饰品也被广泛运用于腰带的设计，在现代服饰艺术中，腰带已经成为服饰搭配的重要组成元素，其实用的功能已经大大削弱了。

Leather made belts are very popular in contemporary fashion regardless of males or females. A large amount of synthetic materials are used besides the natural ones. Some premium leather products are applied in the male fashion with formal suit pants, but rarely used in leisure wearing. The men belts' styles are simple, mostly in dark color with no decoration. The belt head is made by metal. The shapes are relatively identical except in some details. The styles of waistbands of contemporary men and ancient men are similar, the current people rarely hang any decorations on the belts. Now belt is frequently used in women fashion. Its color, style, material are so diversed. The techniques like pressing, setting, embroiding, knitting, etc are seen in the belt decoration with hanging. Apart from leather belts, some plastic, cordgrass, metal, even beads are largely seen in the belt design. In current fashion design, belt has turned to be one of the important accessories with its practical functions greatly reduced.

（1）腰封。

Waist sealing.

腰封与腰链是腰饰中两个较特殊的品种，它们一宽一窄，一粗一细，对服装起到了很好的装饰作用：

The waist seal and waist chain are two special varieties in the waist. They are wide and narrow, one thick and one thin, which plays a very good decorative role for the garment:

腰封原指"书腰纸",是包裹在图书封面中部的一条纸带,主要作用是装饰封面或补充封面的表现不足。现代服饰艺术中,腰封又被称为腰夹或胸衣等,是束于人体中部——腰部的一种服饰用品,有外穿和内穿两种。无肩、无袖、修身式外穿式腰封适合配合礼服穿用,内穿式高弹性的腰封非常适合生育后想恢复体型的以及腰部松弛或太粗的女性,它一般采用立体裁剪,有9~12条纵向的弹性压条,长度是从胸下围至臀部以上,长期穿着可有效地分散均匀脂肪,维护脊椎,抬高胸线,防止胃部扩大,控制食欲,美化腰线。

The waist seal originally refers to the "book waist paper", which is a paper tape wrapped in the middle of the book cover. The main function is to decorate the cover or supplement the cover. In modern costume art, waist seal is also called waist clip or chest coat, etc. It is a kind of clothing articles of use tied in the middle of human body—waist. There are two kinds: outer wear and inner wear. Shoulderless, sleeveless, or figure-displaying outside wears fit with the formal dresses while internal wear with high elastic waist sealing is very ideal to restore body and the waist for fertile women with thick, loose figures space. It usually uses the draping with 9~12 longitudinal elastic layering, extending from above under the chest circumference and hip. Long、term wearing can effectively dispersed evenly fat, maintaining spine, raise the chest line, prevent stomach expand, control appetite, and beautify the waist line.

时尚界所说腰封,一般指系在人体腰部的宽腰带,材质有很多,通常材料为皮、帆布等。腰封的紧收作用,有利于表现女性纤细的腰肢,便于与各式服装搭配,尤其与各种连衣裙搭配效果尤佳。腰封风格夸张,往往以水晶、珠片或铆钉作为装饰,带有帅性、坚毅的气质倾向。这种夸张的腰带能与任何服装搭配出彩:有时女性十足的连衣裙与之搭配,可显出柔中带刚的气质;而与帅气的短打军装或粗犷的牛仔裤相搭配则显得酷味十足。

The waistband of the fashion industry is more often referred as the wide waistband attached to the waist of the upper body. There are many materials, usually made of leather, canvas and so on. The tightening effect of the girdle is beneficial to the performance of women's slender waist, which is easy to match with all kinds of clothing, especially with various dresses. The style of the girdle is exaggerated, often decorated with crystals, beads or rivets, with a handsome, determined temperament. This exaggerated belt can be paired with any outfit: sometimes a feminine dress is paired with it to show the temperament of a soft and medium-sized temperament; it is cool with a pair of handsome short-sleeved uniforms or rugged jeans.

(2)腰链。

Waist chain.

以各类材质串链而成的腰链,腰链既细又长,显得清爽飘逸。腰链品种丰富,大多为珍珠腰链、金属腰链、丝带腰链等,这些腰链充满着简约柔美的风格,细长的腰链末端延伸出的流苏吊坠在走路时摇曳生姿。这些腰链的最佳"搭档"就是牛仔裤和各种裙装,这些裤或裙装的垂感可以更好地衬托出腰链的摇坠之感;裙装与纤细的金属腰链的搭配;牛仔裤与晶莹的珍珠宽腰链的搭配;或者连衣长裙或长风衣搭配以珍珠、金属、丝带腰链,都会衬托出着装者曼妙风姿;有时甚至可以是数条不同风格的腰链进行搭配,强烈的风格撞击反而起到了不同凡响的视觉冲击力。

The waist chain is made of various kinds of materials. The waist chain is both thin and long, and it looks fresh and elegant. The waist chain is also rich in variety, mostly pearl waist chain, metal waist chain, ribbon waist chain, etc. These waist chains are full of simple and soft style, and the tassel pendant extended from the end of the slender waist chain sways when walking. The best "partner" of these waist chains is jeans and various skirts. The draping of these trousers or skirts can better bring out the feeling of the waist chain. The skirt and the slim metal waist chain pairing, jeans with crystal pearl wide waist chain; or long skirt dressing, or long trench coat with pearl, metal, ribbon waist chain, will all bring out the graceful elegance of the wearer. Sometimes even a few different styles of the waist chain can be matched, leading to an extraordinary visual impact by the strong styles' conflict.

（3）腰带的选配要符合个人的身材、气质条件的要求，一般说来腰部较粗的人不适合色彩艳丽、款式夸张的腰饰，以免凸显腰部的不足；而即使腰部较细但盆骨较宽的人也不适合过于夸张的腰带，一些带有花饰的腰带也不适用；细细的腰链只有小腹平坦的人才能穿出它的美感，且搭配的服装裁剪须顺畅合体。

The selection of the belt should meet the requirements of the individual's body and temperament conditions. Generally speaking, the person with thicker waist is not suitable to wear colorful and exaggerated waist chains, so as not to highlight the lack of the waist People with thin waist, but wider pelvis is neither suitable to wear exaggerated ones, as well as some belts with floral ornaments; As for thin waist chains, only the people with flat belly can wear and show their beauty, and the matching clothing must be smoothly combined.

2. 腰饰的搭配要点

The key points of waist accessories matching

时至今日，腰饰已经成为一种时尚，细看国际大大小小时装秀，设计师已经离不开腰饰了。作为配饰的一部分，腰饰能很好地起收身勾勒线条的作用。因其处于人体上下服饰的衔接之处，腰饰在服饰搭配时往往具有承上启下的功效（图5-12）。腰饰与服饰的搭配多在色彩及风格上形成呼应：

Up to now, belt is a part of fashion. No fashion shows is without belt recently. As part of the accessories, belt can shape the line of waist, bridging not only the upper and lower body, but the integrated image of the dress Figure 5-12. The color and style coordniate with dress:

（1）色彩。

Color.

在服饰搭配时，腰饰的色彩可以与服装相融合、

图5-12　腰带起到了承上启下的视觉效果

Figure 5-12　The belt provides the visual effect of connecting the top and bottom

达到色彩和谐的视觉效果，但也可以有意拉开色彩差距。在图5-11中，不同款式的腰饰与服饰融合为一个不可分割的整体，或醒目或含蓄，起到了独特的装饰效果。

The color of belt can merge with dress to create the harmonious visual effect when do the combination. Sometimes the color contrast can be intentionally made. Figure 5-11, different styled belts merge with dress for unique decoration effects: conspicuous or introverted.

（2）风格。

Style.

腰饰的配饰作用以风格搭配最为关键，在选服饰配套的腰饰时首先应对服饰进行准确的风格定位，同时根据风格要求选择腰饰的款式、材质、色彩。如皮革制成的腰饰具有硬朗帅气之感；柔软的丝绸或雪纺制成的腰饰，给人以温暖亲切的感觉，富有女性化的温柔气质；皮革如搭配柔和的花朵形态，则可以中和皮革硬朗感，在帅气中又不失女性的气韵……见图5-13。腰饰的选择取决于服装的总体风格，一般而言：软而薄的丝质腰饰比较轻和飘逸，所以在服饰选配时应避免把它们与重和闷的衣服拴在一起，另外，那种层叠感很强的裙子也应尽量避免，以免让人产生累赘之感；硬而挺的皮质腰饰宜与具有干练感的服装搭配在一起，方具有协调之感。

The style of belt is critical in combination. The style of dress should be accurately set before selecting the style, color material of belt. Some cascading silk belts by chiffon or silk, create the strong feminine, warm and easy atmosphere in sweet colors and traditional embroiding floral decorations. As it appears visually light and flowing, the heavy and stuff clothes should be avoided to match with it Figure 5-13. Besides, the cascading skirts should not be matched either for not being redudant. Girdle and waist chain are two special waist accessories. One is wide and thick while the other is narrow and thin. Both of them can achieve good dress decorating effects.

图5-13　材质的差异使腰带展示不同的风格

Figure 5-13　The different materials make the belt show different styles

七、花饰品
G. Floral accessory

簪花即在鬟发或冠帽上插戴花朵，这是古代的一种装饰习俗。簪花的习俗在秦汉时期已有之。簪花的妇女形象在唐代周昉的绘画《簪花仕女图》中有形象的表现，数个唐代贵妇身披轻纱，她们的发髻上都戴有一朵特大的花朵。古代妇女所戴的花朵以色彩鲜艳的居多，尤以红花最受欢迎。妇女簪戴的除了鲜花之外，还有假花，如以通草、丝绒、色纸、珠宝等。现代女性很少使用鲜花为饰，一些以各类材料制成的假花往往成为绝好的发部装饰，或是装饰于发辫，或是点缀于发髻，或是将活泼俏皮的花束轻别散发上，妆出花漾精灵般的容颜，夸张的花饰，衬托出女性妩媚的气质。

Hair flower means clipping the flower on the hair pin or hat. It is a traditional decoration custom starting from Qin Han dynasty. There was a famous paintng' ladies with pinned flowers' by Zhou Fang in Tang dynasty, vividly reflecting that image. A couple of noble ladies all had an extra large flower on their hair with light chiffon covering their bodies. The bright color of flowers was preferred by ancient women, especially red colored. Apart from the fresh flower, artificial flowers were also worn. They were made by grass, velvet, colored paper, jewerly, etc. Modern women rarely put on fresh real flowers. Some artificial flowers by various materials were perfect for hair decoration, or on the braid, or hair pin, or hair bun, or clipping the flower bunch on the hair. The charming feminine atmosphere was reflected through the exaggerating flower accessories and fairy styled make up.

图5-14中，与服装同质的花饰品点缀于帽子，甚至与上装相融合，成为服饰形象中不可或缺的部分。

Fig 5-14, the same textured floral accessroies on the hat, coordinate the dress as an necessary part of the dress.

八、伞
H. Umbrella

伞本是雨天为了遮蔽雨水的生活用品，现代的伞类制品除了具有避雨、遮阳的功效外，也同时兼具了美观的服饰配件功效。如传统的油纸伞，虽已不再占据现代防雨用具的重要地位，但外观十分漂亮，搭配一些女性化风格的服装十分出彩（图5-15）。

图5-14 与服装相协调的花饰品，为服装增色不少
Figure 5-14 Floral accessories in harmony with the garment add color to the garment

图5-15 纸伞很好的衬托出服装的柔美情怀
Figure 5-15 The paper umbrella is a good foil to the soft beauty of the costume

Umbrella is the daily consuming product for rainny day. The modern umbrella is also for decoration purpose apart from shielding the rain and sunlight. The traditional oiled paper umberlla serves no longer for the current rain shielding, nor the important rain protection tools. But with its nice look, it helps to highlight the dress when combining with (Figure 5-15).

反之如果是一身运动风格的装束使用这样的伞就不太适合了；还有一些透明塑料材质制成的伞，虽不具备遮阳的功效，但在有微雨的春秋季，用以搭配比较飘逸、色彩清淡的衣裙，则可以营造出一种柔柔的浪漫情怀；再者，带有卡通图案的短T恤搭配以印制着儿童绘画的伞，在休闲的时节无论是避雨还是遮阳，都能够吸引不少注目的眼光。

On the opposite, umbrella would not suitable for the sports wear; some transparent umbrella in plastic materials could produce the romantic and tender atmosphere with delicate colored flowing dress in the drizzle weather; what is more, the cartoon patterned T shirt matching with the kids painted umbrella, could be attractive in the leisure days no matter it is for rain or sun shielding.

九、手表
I. Watch

1. 手表的分类
The material and color of the watch belt should be considered

手表或称为腕表，是指戴在手腕上，用以计时并显示时间的仪器。手表通常是用皮革、橡胶、尼龙布、不锈钢等材料制成表带，将显示时间的"表盘"束在手腕上。按照不同标准，手表的分类方式是不同的：按使用的能源分，手表可以分为全自动、半自动、石英、机械、人体动能等不同能源；按所用材质分，可以分为钢、陶瓷、钨钢等不同材质；按风格分，可以分为运动式、时装式等不同的风格。

Watch is also called wrist watch, a device for time display wearing on the wrist. The watch belts are usually leather, rubber, nylon, stainless steel, etc. They band the watch plate on the wrist. The classification of watch differs from standards. They can be classified as automatic, semi automatic, quartz, mechanical, eco drive, etc. according to the energy. Or by materials, they can be classified as steel, porcelain, tungsten steel, etc. or by styles, they can be classified as sports, fashion, etc.

2. 手表的搭配要点
Key points of the watch

手表的选用一般需要从人的体型出发，同时结合手表及其表带的材质、色彩，表的风格特征综合进行考虑：

Generally speaking, the selection of the watch should start from the body shape of the person. Meanwhile, the material and color of the watch and its band should be combined to consider the style and features of the watch comprehensively:

（1）根据体型选择。

By individual figure.

体型往往会决定一个人的气质，所以考虑体型也就是考虑到气质。体型高大强健者应该选择大表盘的手表，在造型与风格上略显粗犷的那种，甚至可以选择军表这样的另类表型。而体型较瘦小者应该选择表盘较薄小一些的表款。一般体型者容易选择，偶尔的大表款可以增加人的强悍气势，而小表款则可以显得谦逊内敛一些。

An individual's temperament is determined by its figure. Considering its figure equals considering the temperament. A tall and strong figured person should choose large sized watch plate, because its shape and style appear to be tough, even some special types like military watch can be considered. Small figured persons should choose the thin plated watch. Slecting watches for moderate sized persons is easy. Occasional wearing large sized watch look tough while wearing small sized watch appear modest.

（2）根据表风格进行选择。

Selecting watches upon style.

服装一般总是比手表更多，因此，在选择与服装搭配的手表时，可以以手表为出发点，搭配合适的服装。

根据设计的不同，手表也呈现出不同的风格，如，有的手表表盘偏薄，显得斯文且大方，这类超薄型的手表适合在商务谈判、日常上班的场合中与职业装和正装搭配出现；相对较厚而多功能的运动表，因为自身的运动型设计元素，与高品质的运动时装搭配会带来更多的时尚感。高雅的表款可以适合搭配运动型的服装，不仅在运动场合上可以佩戴，与太空感服装以及狂野的毛皮服装也可以搭衬；一些装饰有亮钻、花卉等具有甜美风格的手表，则一般多为女性在搭配时装时使用。

The amount of clothes are often more than the amount of watches. Therefore, watch can be chosen first, then pick the coping attires for the combination.

With different design, the watch presents different style. For example: elegant style reflected by thin watch plate, fitting for business occasion, or daily work circumstances, usually goes well with either serious dress or professional attire; while thick and multi functional sports watch can create the fashionable atmosphere while combining with the high quality sports fashion clothes due to its own sports elements. The elegant watch is suitable not only for sports apparel in sports occasion, but also fit for wild fur dress or space styled clothes. Some charming styled watches set with shining gem or floral mostly are also fit for women's fashion.

（3）根据表及表带的材质选择。

Selecting watches by belts materials.

表及表带的材质多种多样。精钢表盘尤其受到男性的喜爱，在商务场合出现也最为频繁；此外，还有以质地上好的皮革制作而成的手表，这类皮革表，需要佩戴者细心呵护，因此皮革手表最适合优雅的人士佩戴；塑料制作或装饰的手表，往往带有更强的时尚意味；因此，手表的搭配并没有太多特殊的规定或习惯，这些文化主要来自西方国家，只要佩戴大方得体即可。

The materials for watch and belts are various. Men are especially fond of steel watch plate, which is most popular on business occasions; besides some premium leather made watch requires the wearer's good care. Thus, this type of watch is suitable for elegant persons. Some watches are made by plastic for decoration purpose, which has strong fashion sense.

Generally, there are not many rules about the watch matchng. The watch culture is originated from western countries. Just wearing the watch properly is fine.

十、眼镜
J. Glasses

眼镜也属于被纳入时尚范围的生活实用品，流行因素的介入使眼镜这一原本为纠正视力而发明的物件具有了时代的气息。眼镜基本可以分为两大类，框架眼镜（有形眼镜）和隐形眼镜，在时尚的氛围下不论是哪一类眼镜都具有了美观与实用的双重功效。

Among the daily consuming products, glasses are categorized into fashion area. Fashion elements are added to the device, which is originally invented for sight adjustment. There are 2 basic types of the glasses: framed glasses (Shaped glasses) and contact lens. Both of them have the dual practical and decorational functions in fashion area.

框架眼镜与服饰的搭配可以从眼镜的外形、眼镜的颜色等方面进行考虑。如以协调为总原则，将眼镜（包括镜架及镜片）的色彩与服装的色彩统一在一种色调中：如果服装是红色调，眼镜就选择接近红的颜色；服装是白色调，眼镜就选择接近白色的颜色等，如图5-16所示。或是以对比为宗旨，选择与服装色彩形成强烈对比的眼镜：如服装颜色是冷色调，眼镜颜色就选择暖色调，或反之；如服装颜色是红色，眼镜就选取蓝色；服装是紫色，眼镜就选取黄色等。再者以眼镜为点缀：用醒目的眼镜颜色点缀大面积、大体积的服装颜色搭配，起到万绿丛中一点红的效果。作为时尚舞台上的重要一员，眼镜为服装起到了添光增彩的作用。如图5-16所示，时装展示舞台上的眼镜。有色隐形眼镜的问世，有色的镜片改变了人们与生俱来的瞳孔色彩，又为时尚增添了一个亮丽的元素。有色隐形眼镜作为服饰的搭配元素，或是在色彩上与服饰产生色彩的共鸣，或是反其道而行之——拉开色彩差别，在对比中展现生动（图5-17）。

Shapes, colors of the framed glasses need to be considered when matching with clothes. If the general rule is about harmony, the color of the glasses (lens and frames are included) and the attire should coordinate. For instance, if the color of the dress is red toned, the color of glasses should be close to red; if the color tone of clothes is white, then the glasses' color should be close to it, as in Figure 5-16. The contrast rule is also applicable for choosing the contrasting colored glasses to attire. For instance, if the color tone of the clothes is cold, then the glasses' color can be warm color toned. Or on the opposite, if the clothes' color is red, then the glasses' color could be blue; if the clothes' color is purple, then the glasses' color could be yellow, etc; besides, the glasses can serve as decoration purpose. Its distinguishing color can match the large volumed color of the clothes for the dramatic prominent effect (Figure5-17).

十一、首饰
K. Jewelries (Discussed with Pictures)

1. 首饰的分类及特征
Classification and characteristics of jewelry

与现代装饰于人体各个部位的饰品概念有所差异，首饰在古代主要是指用于头部装饰的饰品，典型的如发笄、发簪、发钗、簪花、花钿、胜等。在现代的服饰搭配概念中，首饰的概念涵盖了发饰、足饰、手饰、颈饰、耳饰等多个方面。本书以现代服饰搭配的基本划分概念叙述服饰配件与服饰搭配之间的关系。

Differed from the ancient jewlries concept that only for hair decorating, the contemporary jewelries are for all parts of the body. The typical jewlries before are hairpin, hairsticks, hairforks, hairflower, floral sticks, etc. while currently, the concept of jewlries covers the hair, feet, hand, neck, ear accessories, etc. The basic classifications of the jewelries matching with fashion will be discussed.

（1）发饰

Hair accessories.

发钗的品种繁多，仅文献的记载就有金钗、银钗、铜钗、翡翠钗、宝钗、珊瑚钗、玳瑁钗、琉璃钗、琥珀钗等多种。

There are various kinds of hair chains, such as

图5-16　眼镜色彩与服装协调
Figure 5-16　color of glasses coordinates dress

图5-17　时尚的眼镜为服装增色
Figure 5-17　Fashionable eyeglasses add color to clothing

gold chains, silver chains, copper chains, emerald chains, coral chains, tortoise chains, glazed chains and amber chains.

发钗和发簪是女性发部不可缺少的装饰，其材料不仅仅局限于金银、玉石等贵重材料，很多新型合成材料也被广泛运用。尤其是夏季，一款具有复古风情的服饰，精致的盘发装饰以小巧的发钗或是发簪，步履之间摇曳生姿。

The materials are not only limited to gold, silver, jade these precious ones. Many newly synthetic materials are also widely used. Especially in summer, the retro styled dress with refined hairstick or hairpin on the upbound hair is fabulous in walking movement.

簪花即在鬓发或冠帽上插戴花朵，这是古代的一种装饰习俗。古代妇女所戴的花朵以色彩鲜艳的居多，尤以红花最受欢迎。妇女簪戴的除了鲜花之外，还有假花，如以通草、丝绒、色纸、珠宝等。现代女性很少使用鲜花为饰，一些以各类材料制成的假花往往成为绝好的发部装饰，或是装饰于发辫，或是点缀于发髻，或是将活泼俏皮的花束轻别散发上，妆出花漾精灵般的容颜，夸张的花饰，衬托出女性妩媚的气质。

现代服饰艺术中，发饰的作用不仅仅是简单的用于头发的固定以及装饰，它已经成为服饰形象的一个部分，设计师们会根据服饰的需要，创造出独特的发饰。需要注意的是，夸张的发饰有时较服装更易成为视觉瞩目的焦点，服饰搭配时应根据实际需要正确处理发饰与服装之间的主次关系，以达到渲染整体的效果。图5-18趣味性的巧克力外形发饰，色泽闪耀，与服装颇具叛逆的风格形成对比。

Hairflower is the flower on hair or on hat, a traditional customs before. The flowers they wore were mostly bright colored, among which red was the most popular. Apart from fresh flowers, the artificial flowers are used which made by grass, velvet, colored paper, jewleries, etc. Modern women rarely put on fresh flowers. Some artificial flowers by various materials are perfect for hair decoration, or on the braid, or hair

图5-18　巧克力外形发饰
Figure 5-18　Chocolate-shaped hair accessory

pin, or hair bun, or clipping the flower bunch on the hair. The charming feminine atmosphere is reflected through the exaggerating flower accessories and fairy styled make up.

In the contemparory fashion art, the hair accessory is not only for the fastening or decoration of hair, it is a part of fashion. The unique hair accessories will be designed upon needed. What needs to be noted is that the exaggerating hair accessories tend to be more attractive than the dress. The priority should be adjust for the dress and the hair accessories for the Integrated effects. As figure 5-18 shows: chocolate-shaped hair accessory with bright color contrasts sharply with clothing's rebellious style.

现代社会比较常见的首饰有颈饰、手饰、耳饰、足饰等品类。

Neck, hands, ear, foot accessories are commonly seen in modern society.

串饰、念珠、项链、项圈等都属于颈饰之类；手镯、戒指、指甲装饰等属于手饰范畴；鞋、袜、脚链等属于足饰品，耳饰包括耳环、耳坠、穿耳等多个类别。

Stringed ornaments, prayer beads, necklace, necklet, etc belong to the nect accessories; bracelet, rings, nail ornament belong to hand accessories; shoes, socks, ankelet belong to the foot accessories, while ear accessories include earings, eardrop, earpiece, etc.

（2）颈饰。

Neck accessories.

串饰是颈饰最常见的形式，多用各种材料穿组而成，常见材料如金属、骨、玉、陶、石、水晶、玛瑙、竹、木等，用以穿起串饰的绳子材质也非常丰富，丝带、金属、皮缕皆可为之，颈饰的使用没有人员年龄的限制。以竹、木、陶、水晶等组成的串饰往往体积较大，多具有自然的气质，用以搭配休闲或具有民族感的服饰较为适合，且使用者的年龄也不宜偏大（图5-19）。而一些贵重的金属、玉石、玛瑙制成的串饰，多具有高贵典雅之感，更适合年龄大一些的女性使用。严格说来，念珠也属于串饰的一个种类，由于其具有一定的特殊性故单列叙述：念珠又称为"佛珠"或"数珠"，原本是佛教用品，通常用多枚珠子穿串而成。念珠中的珠子数量计一百零八颗，据说念珠用一百零八颗是为了醒人间"百八烦恼"。现代社会不少男女常在颈部挂一串念珠作为装饰，以求凡事顺心之意，但凡挂了念珠的人，颈部就不要再佩戴其他饰品。

The material of cord is also rich, ranging from silk ribbon, metal, skin ray. There is no age limitation for wearing. The bamboo, wood, porcelain, crystal ones are normally large sized, better with leisure or tribal styled dress for young wearers (Figure 5-19). Some precious ones made by metal, jade, agat are more suitable for senior women. Prayer beads are categorized into neck accessories. For its special purpose, it will discussed separately: prayer beads are also called' Buddha beads' or 'counting beads' for they are used to be the religious device, normally made by quite a few beads. The amount of the beads is usually about 108 for the reasons that the human troubles numbered around 108. Now quite a few young people like to wear prayer beads for praying all as wish. The other ornaments are not suggested to wear if the prayer beads are worn.

图5-19　搭配休闲服的项链
Figure 5-19　Necklace matching casual wear

（3）项链。

The necklace.

项链是在串饰基础上演变的一种颈饰，通常由链索、链坠和一个搭勾或搭扣组成。考虑到人体的形体生理条件差别，选择项链时，颈部修长的女性长或短的项链都可以选择，但是颈部较短的女性最好不要选用粗短的项链，V型开口的领子配合细细的带有坠子的项链有视觉的拉伸感；高领的毛衣外面不要佩戴短小的项链，长一些的链子随性地绕上两圈或在链子的下端打一个结，效果会十分的和谐……按照服装的风格倾向，项链的选择可在色彩、造型等方面与服装相协调，使得服饰形象更加熠熠生辉。图5-20中的各服饰形象，项链均采用与服装相似的统色调，且造型也与服装的纹样相匹配。

It is a neck ornament that evolved on the basis of string ornaments, usually consisting of a chain, a pendant and a hook or buckle. Considering the difference in physical and physiological conditions of the human body, when choosing a necklace, women with long neck lengths or short necklaces can choose, but women with shorter necks should not use thick and short necklaces, V-shaped open collars. The thin necklace with the pendant has a visual stretch; don't wear a short necklace outside the high-necked sweater The longer chain is wrapped around two times or at the lower end of the chain. The effect will be very harmonious…According to the style of the clothing, the choice of the necklace can be coordinated with the clothing in terms of color and shape, which makes the costume image more brilliant. Figure 5-20 The image of each costume, the necklaces are similar to the uniform color of the clothing, and the shape also matches the pattern of the clothing.

我国还有给儿童带锁的传统，在项链或串饰上坠上一个金属或玉石的锁（锁片），表达了人们对于孩子殷切的希望与祝福。现代社会这类装饰的使用者依旧以孩童居多，也不乏一些时尚的女性用之搭配具有民族风情的服饰，在古朴中透出一丝顽皮。

There is also a tradition of locking children in our country. A metal or jade lock (locking piece) is placed on a necklace or string ornament, expressing people's earnest hopes and blessings for children. In modern society, the users of such decoration are still mostly children, and there are also some fashionable women who use the costumes with ethnic customs, revealing a trace of naughty in the quaint.

（4）手饰。

Hands accessories.

手饰类最常见的是手镯，手镯在古代被称为"腕环"，是根据其形状以及装饰的部位定名的。手镯的形制有几个大的类型：圆筒状、圆环状、由两个半圆形环合并而成、以各种材质制成的饰牌连缀而成等，手镯有开口的，有不开口的，也有以链扣或搭勾进行连接的。现代所称的手链也应为手镯的一种。手镯只能戴在右手，如果手镯有节，中间没有宝石，应戴得宽松一些；如果是中心饰有宝石的手镯，则需戴得紧一些，不要垂在手上。

Bracelets are the most common types of bracelets, known in ancient times as "wrist rings," according to their shape and decorative parts. There are several large types of bracelets in shape:

图5-20　与服装相协调的各式项链

Figure 5-20　Various necklaces in accordance with the garment

a cylinder, a circular ring, a combination of two semicircle rings, a patchwork of decorative plates made of various materials, etc. The bracelets have openings, of which some do not open, and some are linked by chains or hooks. The modern bracelet should also be a bracelet. The bracelet can only be worn on the right hand. If the bracelet is knotted and there is no gem in the middle, it should be loose. If it's a jewel-encrusted bracelet, keep it tight and don't hang over your hands.

指甲的装饰也属于首饰，方法有染甲、蓄甲、戴义甲等，染甲使用的染料主要是凤仙花，指甲花。

The decoration of fingernails is also a kind of hands' accessory mainly including dyeing nail, storing nail, wearing yijia, etc. The dyes used in dyeing nail are mainly henna flowers, henna flowers.

戒指是人们套在手指上的环状装饰品，在古代也称指环。常见的指环材质主要有玉、骨、铜、金、银、各种宝石等。戒指多戴在左手上，戴在不同的手指有不同的含义，戴在中指表示未婚，戴在无名指表示已婚，而戴在小指上则表示独身之意。现在有不少时尚人士为了达到标新立异的目的还有戴在食指以及拇指上的，前者据说是带有求偶的含义，而后者则是得益于我国清代扳指的启示，扳指原本是游牧民族满族在射箭打猎时保护手指之用。现代社会戒指多为女性所使用之饰品，但婚戒则不然，为男女皆用，佩戴于左手无名指。由于一些广告的渲染，钻戒在现代生活中几乎已是结婚必购之品，但真正的婚戒应为素金，即使有钻石为饰，也是较小的装饰，甚至是镶嵌于戒指内圈，且一定要是指环而不可为有断口之设计，以象征婚姻的圆满。钻石戒指实为订婚戒指，通常可以戴在无名指上，对于手臂以及手指细长的女性而言，一枚精巧的钻戒足以显示出典雅的气质。

Ring is the round-shaped decoration for figures. The common materials are jade, bone, copper, gold, silver and gems. They are worn on the left hand, with different meanings applied if worn on different fingers: wearing on the middle finger means unmarried; on ring finger means married, on little finger means single. Some pioneer fashion people put rings on index finger or thumb. It was said that an index finger means proposing. Wearing ring on thumb is inspired from the fingerstall in Qing dynasty. The fingerstall was originated from the Man tribes protecting their fingers when hunting. Now rings are mostly worn by women. But wedding ring is for both man and woman on the left ring finger. Due to the advertisement, diamond ring almost turns to be the necessity for marriage. But the real wedding ring should be pure gold. Even with the diamond on it, the diamond should be small sized or set inside the ring. Besides, the ring should be designed without openings, representing the perfection of marriage. Diamond ring is actually for engagement, normally on the ring finger. For the women with slender arm and fingers, a refined diamond ring is enough for reflecting the elegant style.

时尚舞台上的手饰品并不强调材质的昂贵，而在于款式的新颖及与服装的契合程度（图5-21）。

Hand ornaments on the fashion stage do not emphasize the high cost of the material, but the novelty of the style and the degree of matching with the garment (Figure 5-21).

图5-21 时尚舞台上的手饰
Figure 5-21　Hand ornaments on the fashion stage

（5）耳饰。

Ear accessories.

我国古代崇尚穿耳，并喜欢在穿孔的耳垂上悬挂各种饰品。耳饰古代又称珥、珰。大部分耳饰都是金属的，有些可能是石头、木或其他相似的硬物料。佩戴在耳垂上的耳饰。造型丰富，佩戴主要以妇女为主，个别男子也有佩戴。佩戴的方式通常有3种：穿挂于耳孔；以簧片夹住耳垂；或以螺丝钉固定。一般用金银制成，也有镶嵌珠玉或悬挂珠玉镶成的坠饰。

现代人常用的耳饰有耳环、耳坠等。耳环是一种环状的装饰品；耳坠是在耳环的基础上演变而来的一种饰品，它的上半部分是圆形的耳环，下半部分再悬挂一枚或一组坠子，故名耳坠；还有一种紧贴耳垂的耳部装饰——耳钉，因其小巧而不张扬之感，深为众多女性所喜爱。除金属、玉石等传统材料外，很多新型的材料也被大量运用于现代耳饰品的设计制作，甚至借助科技的力量，将保健功能与耳饰、颈饰等结合，研制出既有装饰效果又有保健功能的饰品。相对而言，耳饰所占的面积比例比较小，但依旧需要在色彩、材质乃至造型等方面与服装相配套，如图5-22所示。

耳饰的佩戴因人而异，可以从人的脸型、颈部的长短、耳饰的色彩、材质、服装的色调，乃至季节的变化等多个方面考虑加以选择。

Ancient Chinese cared about ear piercing. They liked to put accessories on the pierced earlobe. Most of the earrings are metal, some are probably stone, wood or solid hard materials. The forms are various, especially for women. But some men also wear earrings. The wearing methods are hanging on the pieced hold; clipping on the earlobe, or fastening by screws. They are commonly made by gold and silver, or sometimes set with jades or pearls or with gem-set drops. The modern ear accessories are earrings, eardrops, etc. Earing is a circle shaped accessory. Eardrop is developed from earing, its upper part is round ring, while the lower part is one or a group of drops, from which the name came. Another form of ear accessory is the ear nail which closely stick to the earlobe, which is popular among women for its refined and modest style. Apart from metal, jade the traditional materials, some updated mat-

图5-22　耳饰的色彩造型与服装相呼应

Figure 5-22　The color styling of the earrings echoes the clothing

erals are largely used into the earing design. Even the new technology is applied into for the health function combined into the earrings and neck accessories. Ear accessories should match the dress. In some cases, earrings represent certain customs, religious ideas, social status, wealth, etc, Figure 5-21.

Earing' wearing differ from persons. They may be selected upon the face's shape, neck length, the color and material of earrings, the dress' color, even the seasons, etc.

（6）足饰。

Foot accessories.

足饰在我国古代主要是指鞋制品。而现代服饰搭配艺术中，足饰品的概念就要广泛很多。因现代社会对于女子着装的宽容，现代女子足部、脚踝，乃至大腿都有了展示的机会。因此，现代足饰的概念除鞋袜以外，还可以包括脚链、染足指甲等多种装饰手段。脚链其实是手链使用的一种扩展，佩戴于人体的脚踝部位进行装饰；而染足部指甲则是染手指甲的延伸，只是现代社会的染甲已经不再是使用天然的指甲花、凤仙花等材料了，而是色泽更为鲜亮、色彩种类繁多的各类人造指甲油。

In ancient China, foot accessory refers to shoes. The concept of it expands a lot in the modern fashion art. Since the much more tolerance for the women's appearnce currently, the part of foot, ankle, even thigh are allowed be exposed. Thus, the concept for foot accessory is not only limited into shoes and socks, the decorations like anklelaces, toenails dying,etc are also included. According to the classifacion of fashion matching. Anklelace is the expansion of the bracelet binding around the ankle. Now the diverse vivid synthetic colored dying polish is used for nails replacing the natural nail flowers.

2. 首饰搭配的要素
Notes for jewlery selection

在选择各类首饰品搭配时，考虑的因素除了首饰品本身的设计与造型因素外，还包括流行的色彩、首饰的材质等多个因素，如色彩的对比、材质的混搭、天然材料与非天然材料的肌理组合、首饰的品质感等。选择首饰与服装进行搭配要服从于服装的整体风格，服装穿着的TOP原则——即时间time、目的object、地点place，在此处同样适用。饰品的搭配无须多。用一两件精巧的装饰、点缀即可，而多于三件饰品则会显得庸俗。因为饰品只是起点缀作用，用于调节着装，使之与自己所要展现的美相融。在选择首饰品时不可忽视的因素如下：

In the selection of various types of jewelry, the factors considered include the popular design of the color and the material of the jewelry, such as the contrast of colors, the mix of materials, the natural materials, the texture combination of natural materials, the quality of jewelry, and so on. Choosing jewellery and clothing should be subject to the overall style of the clothing. The TOP principle of clothing wear—the time, the purpose object, and the place are also applicable here. There is no need to match the different styles of accessories, just one or two pieces of delicate decoration and embellishment. More than three pieces of jewelry will make appearance vulgar. Because the jewelry is only the starting point, it is used to adjust the dress to blend with the beauty that you want to show. Factors that cannot be ignored in choosing jewelry:

（1）首饰材质。

Jewlery material.

首饰的材料是服饰搭配时需要考虑的重要因素。首饰品的材质从金银等贵金属到竹木等。但不是所有的材质都适合于正式的场合使用，如银饰品则通常被认为是日常用品，正式或特殊的场合戴用显得不妥；塑料或竹木材质的饰品可以搭配休闲的服饰穿用，不能出现在晚宴等场合。

The material of the jewelry is an important factor to consider when dressing. Jewelry materials range from precious metals such as gold and silver to bamboo and wood. But not all materials are suitable for formal occasions. For example, silver jewelry is usually considered as a daily necessities. It is not suitable for formal or special occasions. Plastic or bamboo wood accessories can be worn with casual wear. but can not appear in a dinner or other formal occasions.

不论是华丽的晚礼服还是简洁的职业装，选择首饰品时都要切记，一次装扮只表明一种风格，在首饰品的选择时要有重点，如果耳环的款式已经比较夸张，项链的选择则以简洁为好，出于同样道理，戒指、手链、手镯可不戴；如果不想整体的服饰形象过于累赘，最好一次不要佩戴一个以上大型且夺目的戒指，可以佩戴2～3枚小戒指。成套项链、手链、手镯、戒指，在服饰搭配时不一次都戴上，可以错开搭配。项链与戒指、耳环与手镯组合是最佳搭配，还可以再加上一枚结婚戒指。在服饰的整体搭配时，不同材质的首饰品相互搭配，有时能够起到很好的效果，如银及金的项链会使暗色系的着装显露生机，但不要把银手镯与金项链搭配；如金项链与银耳环、钻石项链与金耳环之类的搭配是不适合的。

Whether it's a gorgeous evening dress or a simple business dress, remember that when choosing pieces of jewelries, one match only shows one style. In the selection of jewelry, there must be a focus. If the style of the earrings is already exaggerated, the choice of the necklace is simple with no ring, or bracelet、bangle for the same reason. If you don't want the overall costume image to be too cumbersome, don't wear more than one large and eye-catching ring once, but you can wear 2~3 small rings. Complete sets of necklaces, bracelets, bracelets, and rings are not worn once in the mix of clothing, and can be staggered. The combination of necklace and ring, earrings and bracelet is the best match, plus a wedding ring. In the overall matching of clothing, jewelry of different materials match each other, sometimes it can play a very good effect. For example, the necklace of silver and gold will make the dark color dress show life, but don't match the silver bracelet with the gold necklace; Gold necklaces with silver earrings, diamond necklaces and gold earrings are not suitable.

（2）年龄因素。

Age factor.

不同年龄段的女性，适合的首饰品也是不同的。一般年龄较小的时尚女性可以选择一些外形夸张、色彩鲜艳的首饰品，且材质不必一定是天然的；但年纪大一些的女性，选择的首饰品则最好是天然材质且品质感较好。

The jewleries different for different aged women. Some exaggerating shaped, bright colored jewleries are fine with young and fashionable women, with the materials not necessarily to be natural. Senior women better choose the natural materialed ones in high quality.

（3）个人喜好。

Personal preference.

服饰与首饰的搭配可根据个人的喜好，进行一些创新式的佩戴，如一条镶嵌了钻石或宝石的手链，可以将它穿在外套上，既起到了胸针的作用，而且给人耳目一新的感觉；如将长长的珠链或金属项链佩戴在腰间，充作腰链使用，或是将之绕在靴子上，这样的"视觉"感受一定与众不同；腰链在颈部绕上两圈，"夸张"的项链会使你在人群中分外引人注目；小小的戒指不一定要戴在指尖，穿在项链上、用作丝巾扣未尝不可；具有特色的耳坠动手装上一个别针可以充作胸针，效果一定新奇有趣。其实这样的创意有很多，主要在于设计者如何激发自己的想象力，加入了创意的服饰搭配组合意味着个性和别具一格，会让人有意想不到的惊喜。

The innovative wearing can be made for the combination of fashion with jewleries. For example, a diamond ring or gem bracelet, if worn on the outfit, it can serve as the brooch with shining effects; the long pearl chain or the metal necklace can be worn around the waist as waistlace, or bind around the boots to achieve unique "visual" effects. If binding the waistlace around the neck for 2 circles, it would be very distinguishing in the "exaggerating" style. The small sized ring does not have to be worn on the finger. It can be put through the necklace as a pendant, or used as scarf pin. The specialized eardrop can be worn as brooch if the clip added manually for the fun effects. Actually there are quite a number of these innovations depending on the imagination capabilities of the designers. The surprising effects can be achieved through thse innovative combination.

（4）使用的场合。

Wearing occasions.

首饰佩戴要与场合相匹配。如在一些隆重的晚会场合，如晚会、婚宴或特殊的聚会，一般可佩戴亮色系列、造型华丽独特的珠宝。如红宝石戒指，翠绿色翡翠耳环，造型突出的钻石套装等。而在白天较普通的工作环境里，不宜佩戴大颗亮色系列的珠宝，而佩戴颜色素静、造型典雅简单的冷色系列宝石首饰，较为适宜，如蓝宝石吊坠、紫晶戒指等。除此之外，佩戴首饰时还应综合考虑佩戴者的肤色、脸型、服装、发型等特征。

Jewelry should be worn to match the occasion. For example, in some grand evening occasions, such as parties, wedding banquets or special gatherings, you can wear bright colors and gorgeous jewelry. such as ruby ring with diamonds, emerald green jade earrings, diamond set with outstanding shape. In the ordinary working environment during the day, it is not easy to wear large bright color jewels, and it is more suitable to wear cool color series gemstone jewelry with elegant color and simple style, such as sapphire pendant and amethyst ring. In addition, the wearer should also consider the skin color, face, clothing, hairstyle and other characteristics of the wearer.

处于职场的女性在工作时佩戴的首饰不能仅仅以喜好为选择标准，一般应注意：不戴有碍于工作的首饰、不戴炫耀财力的首饰、不戴突出个人性别特征的首饰；佩戴数量以少为佳；最好项链、耳环、戒指等同色同质，风格划一。

Jewelry worn by women in the workplace can't just choose the preference according to their preference. Generally, they should pay attention to: not wearing jewelry that hinders work, jewelry without showing off financial resources, jewelry without highlighting individual gender characteristics; Less is better, so the best necklaces, earrings, and rings are better of the same color and style.

（5）与服装的搭配。

With the outfit.

首饰的穿戴基本是以与服装搭配和谐为宜，应该与服装在风格、色彩、款式，乃至价值上协调一致，以起到点缀衬托的效果为要，忌过于夸张、醒目的效果。具体地说，"风格一致"的服饰与配件搭配，在选择搭配豪放、粗犷风格服装的首饰时，以造型和色彩等方面表现具有热情奔放、粗大圆润、光亮鲜艳的首饰为宜；在选择职业装的搭配首饰品时，饰品要尽量简洁、大气，色调要与服饰相协调；在选择晚宴场合与礼服相搭配的首饰品时，则以明亮的材质与色彩、且款式夸张一些的饰品让其出彩。在色彩上搭配协调一致，首饰的色彩和服装的色彩既可以是同类色相配，也可以在总体协调中以小对比点缀。如选择服装中某一色作为首饰主色调或素色的服装，配以鲜艳、漂亮、多色的首饰，或艳丽的服装配以素色的首饰为好等。

The wear of jewellery is basically in harmony with the clothing. It should be coordinated with the style, color, style, and even value of the clothing, so as to play the effect of the setting, avoid excessive exaggeration and eye-catching effect. Specifically, when choosing jewelry with bold and rough style clothing, the "consistent style" of clothing and accessories is suitable for styling and color, with enthusiasm, bold and bright, bright and bright jewelry; When fitting the jewellery, the jewelry should be as simple and atmospheric as possible, and the color should be coordinated with the costumes. When choosing the jewellery that matches the dress at the dinner occasion, the accessories with bright materials and colors, should be selected to display exaggerated styles. As for coordinating in color, the color of the jewelry and the color of the clothing can be matched with the same color, or with a small contrast in the overall coordination. If you choose a certain color in the clothing as the main color of the jewelry or plain clothing, it is accompanied by bright, beautiful, multi-colored jewelry, or gorgeous clothing with plain jewelry as well.

首饰和服装搭配的协调性还表现在，两者其中有一为主导。一般情况下，首饰的风格要服从于服饰的风格，在搭配中寻找一种平衡之感：如简单的着装可以搭配相对夸张繁复的饰品；整体多层次的穿着或比较花哨的服装可能只需要一件精致的首饰加以提亮点缀即可。如果搭配领子较高的服装，一条长长的项链就能够拉长人们的视觉感受；大开领的服装则要搭配短链，但领口本身已经有闪亮珠片作为装饰的，就无需再画蛇添足使用项链；长袖不宜再使用手链或手镯，但是可以佩戴戒指；项链的长度不宜超过衣服的长度等，关键还是要在视觉上保持一种平衡感。如果是首饰品发布会之类的特殊场合，服装处于从属地位，款式应尽量简洁，以突出首饰为原则。如果搭配时首饰与服装二者平分秋色，那就失去了搭配的意义。

The coordination of jewelry and clothing collocation is also shown in the fact that one of them is dominant. Generally speaking, the style of jewelry should be subject to the style of clothing, and a sense of balance should be found in the collocation. Whole and multilayer dress or the flowery dress perhaps only needs a delicate jewelry to lift brightness. A long necklace can lengthen a person's visual perception if worn with a high collar. The garments with large open collars should be matched with short chains, but the neckline itself has already been decorated with shiny beads, so there is no need to add a touch to the necklace. Long sleeves should no longer use bracelets or brace chains, but rings should be worn. The length of the necklace should not exceed the length of the clothes and so on. The key is to maintain a sense of balance visually. If it is a special occasion such as the first jewelry conference, the clothing is in a subordinate position, the style should be as simple as possible, to highlight the jewelry as a principle. If jewelry and clothing are evenly matched, the meaning of matching is lost.

第二节　服饰品与服装搭配的关系及协调

Section 2　Relationship and coordination between clothing items and clothing collocation

服饰品是服装的点缀和补充，服饰品与服装搭配得好，可起到画龙点睛的功效，衬托出个人独特的气质，反之则会破坏服装的整体美感。服饰品与服装的表现力是由它们的材料和本身的特征所决定的。服装材料一般为各种纤维织物，温暖、柔软、亲和。服饰品所用材料比较广泛，根据其种类的不同，既有质地硬的珠宝、金属等，也有质地轻柔的纤维材质。要实现其整体和谐，在搭配时应考虑服饰品与服装的和谐关系，从而创造出更加丰满的视觉感受。

Dress accessory is the ornament and complement of dress. If dress and dress accessory can be matched well, they can have the effect that makes the finishing point distinguished, gives individual distinctive temperament, otherwise, they can destroy the integral aesthetic feeling of dress conversely. The performance of clothing accessory and clothing itself is determined by their materials and their own characteristics. Clothing materials are generally a variety of fiber fabrics, warm, soft, affinity. The dress accessory materials can be quite extensive, according to their different types、They could be hard materials like jewelry, or metal or soft materials like gentle fibers. In order to achieve its overall harmony, the harmonious relationship between clothing and clothing should be taken into account when matching clothing and its accessories, so as to create a fuller visual experience.

从美学角度而言，一切装饰都不能脱离整体的概念，其装饰效果应达到对立与统一。以首饰为例，一件华丽的首饰的佩戴，在颜色沉着、面料考究的服装上相互衬托，能起到"画龙点睛"的作用。反之，将其佩戴在一件色彩艳丽、花纹图案复杂的衣裙上，则是"画蛇添足"，非但不能很好地发挥其点缀与装饰的效果，还会破坏原本很美的服装效果。

From the aesthetic point of view, all adornment cannot leave integral concept, its adornment effect should be opposite and unified. Take jewelry for example: a gorgeous piece of jewelry, wearing in color calm, fabrics fastidious clothing to foil each other, can play the role of "make the finishing point". On the contrary, wearing it on a colorful and decorative skirt, it is "painting the snake and adding feet", which not only cannot give full play to the effect of its ornament and decoration, but also can destroy the original beautiful clothing effect.

因此，配件的使用与服装之间是相互作用、相辅相成的，只有当两者协调和统一时，它们的美感才会充分展现出来。此处所指的"统一"为大统一的概念，即不论服饰配件与服饰之间在色彩、款式或造型上和谐抑或对比，只要在服饰形象完整的前提下，表现出了服饰美的特性，都可以称为"协调统一"。配件与服装的协调统一一般还要注意以下几个问题：

Therefore, the use of accessories and clothing are interactive and complementary. Only when they are coordinated and unified, their beauty will be fully displayed. The "unity" here refers to the concept of great unity, that is, no matter the color, style or modeling between clothing accessories and clothing harmony or contrast, they can be called "coordination unity" so long as they can display clothing beauty on the basis of keeping the dress image integrity. In the coordination and unification of accessories and clothing, attention should be paid to the following problems:

一、风格上相呼应
A. Style echoes

配件的选择是以服装的风格造型作为前提和依据的。选择与服饰相搭配的各类配件，首先应确定服饰主体的基本风格，而后根据实际情况考虑搭配的效果，服饰配件的选择要依据服装的风格来进行。如职业女性，由于职业的原因，选择服饰配件的限制较多，但可以选择适合自己气质和风格的饰品，突破职业装色彩的单纯性，可以在胸前和发际，以及项链上搭配一些色彩生动的有色宝石，在职业装的庄重严肃之外，透射出女性的生机和美丽。

The choice of accessories is based on the premise and basis of the style of the clothing. Choosing all kinds of accessories that match the costumes, we should first determine the basic style of the main body of the clothing, and then consider the effect of the matching according to the actual situation. The choice of the accessories should be based on the style of the clothing. For example, professional women, due to professional reasons, choose clothing accessories more restrictions, but you can choose the jewelry that suits your temperament and style, to break through the simplicity of professional color, you can mixing some on the chest, hair, or necklace. By doing so, they can transmit the vitality and beauty of women by mixing some colored gemstones in addition to the seriousness of the professional wear.

一般情况下，服饰配件的选择强调与服装之间的协调，如礼服的款式风格精致华贵，则要求配件的风格也应具有雍容的晚宴气质，便装的款式构成较为简洁大方，则配件的风格也要随意和自然等。风格呼应并不意味着服饰与配件的风格必然具有相似性，具有混搭意味的服饰组合，配件与服饰之间存在一定的对比，作为客体的配件反而使得服饰更为突出，这样所达到的统一关系也属于风格呼应的一种形式。

Under normal circumstances, the choice of clothing accessories emphasizes the coordination with the clothing. For example, if the style of the dress is exquisite and luxurious, the style of the accessories should also have a graceful dinner temperament; if the style of the casual wear is simple and generous, the style of the accessories should also be casual and natural. The style echo does not mean that the style of clothing and accessories must have similarities. There is a mix of clothing and accessories. There is a certain contrast between accessories and clothing. As a guest accessory, the clothing is more prominent, so that the unified relationship is achieved. It is also a form of style echo.

二、体积上对比
B. The contrast of volume

服饰配件与服装之间是局部与整体的关系，把握好局部与整体的大小比例关系是处理好配件与服饰搭配的关键性因素。配件是服装的从属性装饰，但并不是一味以减少其在整体服饰形象中所占据的体积比为前提。一件独具特色、精致漂亮的配件可以为服装增色不少，妥善运用各类服饰配件，是服饰搭配艺术中必须重视的一个问题。在配件和服装的关系中，服装是主体，配件则为客体，服装与配件之间的主从关系极为微妙，服饰与配件二者之间存在的主客体关系是始终贯穿于服饰搭配过程中的。一方面，服饰的主体关系不容忽视，另一方面配件的客体关系有时还会与主体产生倒置。服装与配件的主客体倒置，不能简单地理解为一味地去追求配件客体的作用，而是在一种新型的配件与服装的关系基础上，力图达到神形统一的效果。其实，适当的突出配件的客体作用，目的是为了更好地强调服装的主体地位。同时，配件与服饰的主客体倒置要避免配件与服装脱离太远，而达到一种既突出客体却又不改变其从属地位，弱化主体却又不和主体相脱离的状态。

The relationship of accessory and fashion is just like the part to integration. It is the key to handle well about the size ratio of the fashion and accessory. Accessory is the supportive decoration, but it doesn't mean that to simply decrease its volume to the entire image. A unique and refined accessory can promote the dress. It is crucial to handle well about different types of accessories in the fashion combination. About the relationship of fashion and accessories, fashion leads the accessories. The priority about them is complicated, which will be through all the matching process. On one side, the leading role of dress can't be ignored, on the other side, their priorities can be reversed. The reversed priority of fashion and accessories can't be simply understood to pursue the function of accessories, but based on their updated relationship to reach the physically and mentally consistent effect. In fact, some leaning priority added on the accessories enables the promotion of the leading position of the fashion. At the meantime, the reversed priority of fashion and accessories should avoid the divergence from the dress, to achieve the remotely attached status while promoting the accessories without reversing its supportive positions.

三、肌理上相对比
C. Texture contrast

制作服饰配件的材料种类很多，服装与配件组合可根据不同需求心理，审美情趣做相应的变化。服装与配件之间肌理对比，最为突出的体现是在面料上。比如，当服装的面料较为细腻时，可选择质感粗犷而奔放的包袋；当服装面料较为厚重而凹凸不平时，则可选择一些肌理光润柔滑的包袋，与服装面料造成鲜明对比。总之，从服饰的整体肌理效果来看，两者之间既可相互对比也可相互补充，既可互相衬托又可相互协调，在搭配变化中产生出一种特有的视觉美感。

There are various materials for fashion accessories. The combination vary with psychological status, and personal tastes. The contrast of fashion and accessoriesis mainly displays on the fabric. For example, as the attire's fabric is fine and delicate, the rough and passionate textured bag can be selected; when the fabric of fashion is heavy and irregular surfaced, the smooth textured bag can be selected to create the strong contrast. The comprehensive texture consequence can produce a specially beautiful visual effects if they coordinate or contrast.

四、色彩的配合
D. The combination of color

色彩是整体服饰形象的第一视觉印象。服饰配件常常在整体的服饰色彩效果中起到"画龙点睛"的作用。当服装的色彩过于单调或沉闷时，便可利用鲜明而多变的色彩运用到配件中，来调整色彩感觉，而当服装的色彩显得有些强烈和刺激时，又可利用配件单纯而含蓄的色彩来缓和气氛。服饰形象色彩的处理要根据整体效果的需要，这样既可以迅速快捷的选择好颜色，又易取得色彩的高度协调。

Color is the first visual impression of fashion image. The accessory serves as pinpointing in the entire combination. When the color of dress is too dull or low, the brightly vivid colored accessories can be applied to adjust. As the dress color is strong and irritating, the pure and introvert colored accessories can be applied to ease the atmosphere. The adjustment of fashion color should cope with the integration of the comprehensive visual effect, which enables the efficient selection of color to achieve the color harmony.

Accessory does enjoy certain priority in the fashion combination. Its artistic value is undividable with attire. when wearing with attire, it is a part of the comprehensive image in order to support, reflect the individual's inner disposition.

图5-23为黑色裙装与服饰品搭配的范例：搭配小黑裙的服饰品非常容易选择，饰品最好简洁、精致，与包包、鞋子配色不要太多，可在金色和白色选择一种色系。色彩简约而不杂乱才能符合小黑裙简洁、精致的格调。图5-24为浅色套装与服饰搭配的范例：白色服装有轻盈感，搭配白色珍珠或者亮晶晶的饰品显得干净。裙子是宝蓝色，配饰可以适当地相呼应的鞋子、衣服的颜色。饰品以小件闪光饰品，如钻石耳坠或颈饰为主，方显优雅精致。整体要显得和谐，不要让太多的颜色出现。

Figure 5-23 shows an example of a black dress and accessories: the accessories with a small black dress are very easy to choose: the jewelry is best simple, exquisite and the bag; the color of the shoes is not too messy; you can choose one in gold and white color system. The color is simple and not messy to meet the simple and delicate style of the little black dress. Figure 5-24 shows an example of a light-colored suit and clothing: the white garment has a light feel, and it is clean with white pearls or sparkling

图5-23　黑色裙装与服饰品的搭配
Figure 5-23　Black dress and accessories

图5-24　浅色套装与服饰品的搭配
Figure 5-24　Light color suits and accessories

jewelry. The skirt is royal blue, and the accessories can properly match the color of the shoes and clothes. The jewelry is mainly composed of small pieces of glitter jewelry such as diamond earrings or neck ornaments. The whole should be harmonious. Don't let too many colors appear.

服饰配件虽然在服饰的整体效果中占有一定的位置，然而在审美实践中人们认识到，其艺术价值是与服装分不开的。服装和配件一经穿戴，便成为人们外表的一个组成部分，烘托、陪衬和反映着人们的内在气质。

Although the clothing accessories occupy a certain position in the overall effect of the clothing, in the aesthetic practice, people realize that the artistic value is inseparable from the clothing. Once worn, clothing and accessories become an integral part of people's appearance, setting off, foiling and reflecting people's inner temperament.

第六章 世界不同地域的历史、文化背景与其服饰配件的设计

Chapter 6　The history and cultural background of different regions of the world and the design of their accessories

课程名称：世界不同地域的历史、文化背景与其服饰配件的设计

教学目的：通过学习掌握世界各地域服饰的历史文化及其特色、服饰形制与色彩分析及了解影响现代服饰潮流。

教学重点：本章学习以掌握世界各地域服饰的历史文化及其特色；掌握服饰形制与色彩分析；应用于服饰创意设计上。

Course Name: History, cultural background and design of clothing accessories in different regions of the world

Teaching purposes: Through learning, students can learn the history and culture of the world's clothing, its characteristics, clothing shape and color analysis and understand the impact of modern clothing trends.

Teaching focus: This chapter aims to master the history and culture of the world's local costumes and their characteristics; master the clothing shape and color analysis. and its characteristics applied to the creative design of clothing.

课时参考：8课时。

Class time reference: 8 class hours.

世界上各个国家和民族的历史中都有服饰配件的记载，其装饰形式及装饰行为与各地区的生活环境、生活习俗息息相关。随着人们的观念和技术的积累也促使服饰配件的进步和发展。如首饰的应用、腰带的式样、背包的功能、鞋靴的变化等都有其历史文化背景的痕迹。人类世代相传的习俗形成了服饰配件特定的内涵，具有区域特征和传统形式。

There are records of clothing accessories in the history of various countries and nations in the world. Their decorative forms and decorative behaviors are closely related to the living environment and living customs of each region. With the accumulation of people's ideas and technology, it also promotes the advancement and development of apparel accessories. For example, the application of jewelry, the style of the belt, the function of the backpack, the changes of the shoes and boots, etc. All have traces of its historical and cultural background. The custom passed down from generation to generation forms the specific connotation of costume accessories, with regional characteristics and traditional forms.

服饰作为一种文化形态，贯穿了东西方近现代的历史。在漫长的人类发展过程中，中国与西方服饰文化因为不同的历史积淀，具有不同的发展方向，进而产生了宗教、审美、生活方式等各方面的差异，具体表现在服装上各有特色，风格迥然。因而服饰

总是以它独特的物质存在和精神表现，反映着它所依存的社会历史时期，服装、服饰品正反映了它的社会基础。

As a cultural form, clothing and decoration run through the modern history of the East and West. In the long process of human development, Chinese and Western clothing and decoration cultures have different development directions because of different historical accumulations, which in turn has produced differences in religion, aesthetics, and lifestyles. The style is bleak. Therefore, clothing and decoration always express with its unique material existence and spirituality, reflecting the social and historical period on which it depends. Clothing and costumes reflect its social foundation.

起源于黄河文明的中国服饰文化深受传统东方文化的影响，在服装款式的设计上不露曲线，强调线形和纹饰的抽象寓意表达，透露出一种婉约、端庄、优雅、含蓄的东方情调；而以地中海文明为基础的西方服饰文化深受基督教思想、艺术文化思潮、科学技术等方面影响，其审美视觉重视立体的造型，追求科学的内涵。而服装饰品款式上却依附于人体形态的表现，它随着服装曲线明显，呈现出从实用中来的优美、大胆与雅致。西方服饰文化以其独特的历史文化和科学技术，成为国际服饰文化体系发展的主流。同时中西又多有交流借鉴，因而服饰上的艺术表现也出现互融。

The Chinese costume culture originated from the Yellow River civilization is deeply influenced by the traditional oriental culture. The design of the clothing style is not curved, emphasizing the abstract expression of the line and the ornamentation, revealing a graceful, dignified, elegant and subtle oriental sentiment; The Western clothing culture based on the Mediterranean civilization is deeply influenced by Christian thought, art and culture, and science and technology. Its aesthetic vision attaches to the three-dimensional shape and pursues the connotation of science. The clothing decoration style is attached to the human body's performance. It shows the beauty, boldness and elegance from the practical with the obvious clothing curve. Western clothing culture has become the mainstream of the development of the international costume culture system with its unique historical culture and science and technology. At the same time, there are many exchanges between China and the West, so the artistic expression on the clothing also appears to be in harmony.

本章将各时代、各民族的服饰配件做比较分析，可以看到，服饰配件有它固有的特性，如从属性与整体性、社会性与民族性、象征性与审美性等决定了服饰配件在服饰艺术中的内涵与作用。

This chapter compares and analyzes the accessories of various eras and nationalities. We can see that the accessories have their inherent characteristics, such as attributes and integrity, sociality and nationality, symbolism and aesthetics, determining the connotation and role of clothing accessories in the art of clothing.

第一节　亚洲及其四国的服饰配件特色

Section 1　Clothing accessories in Asia and its four countries

一般来说，亚洲服装外形是"修长"的，这是由于东方人的身材较为矮小，服装造型"修长"，在感官上产生"错觉"缘故，从而在比例上达到完美、和谐。自然修长的服装能使男性显得清秀，使女性显得窈窕。同时，平顺的服装外形与中国人脸部较柔和的轮廓线条相称。

Generally the shape of Asian clothing is "slender". This is because the Oriental are relatively short in size, and the clothing style is "slender", which creates an "illusion" in the senses, thus achieving perfection and harmony in proportion. Naturally slender clothing can make men look pretty and make women look elegant. At the same time, the smooth appearance of the garment is commensurate with the soft contours of the Chinese face.

从结构特征看，中式服装采用中国传统的平面直线裁剪方法，无论袍、衫、襦、褂，通常只有袖底缝和侧摆相连的一条结构线，无起肩和袖窿部分，整件衣服可以平铺于地，结构简单舒展。

From the structural characteristics, Chinese clothing adopts the traditional Chinese method of flat linear cutting. No matter for the robe, shirt, or crepe, there is usually only one structural line connecting the bottom of the sleeve and the side hem, without shoulder and sleeve parts, and the whole piece of clothing can be laid flat on the ground with a simple structure.

南亚，如印度尼西亚，印尼人的日常服装十分简朴轻便，印尼女子的服装比较特别，她们传统性的上衣长而宽敞，对襟长袖，但是没有衣领，衣服质料多半采用白色有花纹薄纱，纽扣用金色大粒的铜扣，也有用合金制成，也有用镶钻石的金纽做成。

In South Asia, like Indonesia, Indonesian daily clothing is very simple and light. Indonesian women's clothing is special. Their traditional tops are long and spacious. They have long sleeves, but no collars. Most of the clothes are made of white patterned tulle. Buckle with a large gold buckle, also made of alloy, but also made of diamond-plated gold buttons.

爪哇族和巴厘族的女性，上身穿着简单缝制的衣服，下身则穿着称为"纱龙"的漂亮长裙，质料为木棉或化学纤维；男性穿着轻快的衬衫型上衣，以及长裤型的纱龙。而各地域不同服装的款式、结构的差异，给服饰配件的变化与搭配带来设计的"无限可能"。

The Javanese and Balinese women wear simple sewed clothes on their upper body and a beautiful long dress called "Sauton" on the lower body. The material is kapok or chemical fiber; men wear a light shirt top and trousers. The differenccs in styles and structures of different garments in different regions bring "infinite possibilities" to the design and change of clothing accessories.

一、中国的文化
A. Chinese culture and its accessories' features

1. 文化背景
Cultural background

（1）天人合一的文化。

The culture of harmony between man and nature.

古代中国素有"衣冠礼仪之邦"的美誉，它是中国深厚人文传统的重要组成部分。服饰所表现的文化心态，是情感的纽带，它一端牵着日常的生命活动，一端连着用理性无法表述的美的世界。在古人眼里，人世是天道的展现，所以人们的服饰也应该要合于天理。所以无论是在服饰的形制上，还是颜色、纹样上，都蕴含有深刻的"天人合一"的理念（图6-1）。

Ancient China is known as the "state of clothing and ceremonies" and it is an important part of Chinese's profound humanistic tradition. The cultural mentality expressed by clothing and decoration is the bond of emotion. It carries the daily life activities at one end and the beautiful world that cannot be expressed by rationality at the other. In the eyes of the ancients, the world is a manifestation of heaven, so people's clothing should also be in harmony with heaven. Therefore, whether it is in the shape of clothing, color, or pattern, it contains a profound concept of "Heaven and Man" (Figure 6-1).

①衣裳❶。

Clthoes.

中国人上衣下裳的穿法就是对天地之别的认识，

---

❶《周易·系辞下》有云：黄帝尧舜"垂衣裳而天下治"。有记载中出现"衣裳"二字，衣裳制是华夏文明中服饰礼仪规格最高的形式。

服饰配件设计与应用

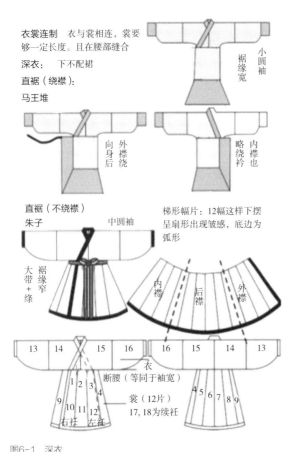

图6-1 深衣
Figure 6-1 Deep clothing

这一秩序不可颠倒。上衣像天，下裳像地，代表"天地之法"；帝王的上衣多为玄色，玄为黑色；下裳多为纁（xūn）色，纁为绛红色。冕服采用了两种颜色，上以象征未明之天，下以象征黄昏之地。蕴含"天人合一"的理念。

The wearing method of the Chinese people's tops and bottoms is the understanding of the world. This order cannot be reversed. The top is like the sky, the bottom is like the land, representing the "law of heaven and earth"; The emperor's top is mostly black, while the lower is mostly blush color, The robes are made in two colors, symbolizing the unidentified day and symbolizing the place of the evening. It contains the concept of "Heaven and Man".

②深衣❶。

Deep clothing.

深衣是最能体现华夏文化精神的服饰，有着丰富的哲学内涵和人文气息。

深衣是将上衣与下裳分别裁剪后缝合；上衣用布四幅，象征一年四季，下裳用布十二幅，象征一年十二月；袖口宽大，象征天道圆融，领口直角相交，象征地道方正，以应"天圆地方"；背后一条直缝贯通上下，象征人道正直，表示人要正直向上。

The deep coat is the very clothing that reflects the spirit of Chinese culture. It has rich philosophical connotation and humanistic atmosphere.

The deep coat is cut and stitched separately from the top and the lower skirt; the top is covered with four pieces of cloth, which symbolizes the seasons, and the 12 pieces of cloth are used for the next year, symbolizing the year of December; the cuffs are wide, symbolizing the harmony of the heavens and the neck, and the neckline intersects at right angles. Symbolizes the authenticity of the square. There is a straight line at the back from top to bottom, symbolizing human integrity, indicating that people must be upright.

（2）服饰中的"礼乐"文化。

"Ritual music"'s culture in clothing and accessories.

"礼"是统治阶级规定的秩序，其本质内涵是等级观念和等级制度；在深受"礼"文化影响的阶级社会中，有着森严的等级制度。各阶层的成员，从衣食住行到穿衣戴帽，都有严格的等级规定，不可随便逾越。而服饰则成为"昭名分，辨等威"的重要工具。中国古代服饰就是中国礼制社会的一种标志。服饰作为一种文化形态，已贯穿了中国古代各个时期的历史。

"Ritual" is the order stipulated by the ruling class. Its essential connotation is hierarchical concept and hierarchical system. In the class society deeply influenced by the "ritual" culture, it has a strict hierarchy.

---

❶ 深衣属于汉服，古时上下通行之衣为深衣，代表时代特征的服装亦为深衣，深衣实可为古服之特征。言古服者，应先及之。何谓深衣，《礼记·深衣》孔氏正义曰，"所以称深衣者，以余服则，上衣下裳不相连，此深衣衣裳相连，被体深邃，故谓之深衣"。

Members of all walks of life, from clothing and food to dressing and wearing hats, have strict grading rules and cannot be passed casually. And the clothing become an important tool to "distinguishing the reputation and ranks". Ancient Chinese clothing are a symbol of Chinese ritual society. As a cultural form, clothing have run through the history of ancient China.

从服饰的演变，可以看出历史的变迁、经济的发展和中国文化审美意识的嬗变。服饰从历史上看，无论是商的"威严庄重"，周的"秩序井然"，战国的"清新"，汉的"凝重"，还是六朝的"清瘦"，唐的"丰满华丽"，宋的"理性美"，元的"粗壮豪放"，明的"敦厚繁丽"，清的"纤巧"，无不体现出中国古人的对服饰审美设计倾向和思想内涵。而审美设计倾向、审美意识，它必然根植于特定的时代。因此，中国服饰的颜色、纹样、配饰等特征，便是"礼"文化的鲜明物化形式。

From the evolution of costumes, we can see the changes of history, the development of economy and the transformation of Chinese cultural aesthetic consciousness. From the historical point of view, whether it is the "dignity and magnificence" of Shang dynasty, the "neatly ordered" of Zhou dynasty , the "freshness" of the Warring States dynasty, the "denseness" of the Han dynasty, or the "slimness" of the Six dynasties, the "plumpness and gorgeousness" of the Tang Dynasty, "Rational beauty" of Song dynasty, "rough and bold" of Yuan's dynasty, "durable and prosperous" of Ming dynasty, "slim" of Qing dynasty all reflect the Chinese ancients' aesthetic design tendency and ideological connotation. Aesthetic design tendency, and aesthetic consciousness, must be rooted in a specific era. Therefore, the characteristics of the colors, patterns and accessories of Chinese clothing are the vivid and materialized forms of the "ritual" culture.

①服饰颜色体现尊卑。

The color of the clothing reflects the honor.

a. 黄——帝王之尊。

Yellow—Emperor's Respect.

b. 阴阳五行与颜色——五行：木、火、土、金、水。

Yin and Yang Five Elements and Colors—Five Elements: Wood, Fire, Earth, Gold, Water.

c. 五方：东、南、中、西、北。

Five parties: East, South, Middle, West, North.

d. 五色：青、赤、黄、白、黑（图6-2）。

Five colors: blue, red, yellow, white, black (Figure 6-2).

周——红色为高级服色。

Zhou—Red was a high color.

秦——黑色为最高级服色，帝王百官都穿黑色衣服。

Qin—Black was the most advanced color, and the emperor is wearing black clothes.

汉——黄色逐渐成为为最高级服色，皇帝穿黄色衣服。

Han—Yellow gradually became the most advanced color, and the emperor wore yellow clothes.

唐至清——除皇帝以外，一律不许穿黄。

Tang dynasty to Qing dynasty—No one was allowed to wear "yellow" except the emperor.

紫绯绿——达官显贵。

Purple green—distinguished nobles.

魏初，文帝曹丕制定九品官位制度，"以紫绯绿三色为九品之别"。这一制度此后历代相沿杂而用之，直到元明。唐、宋官服服色，三品以上紫色；五品以上，绯色；六品、七品为绿色；八品、九品为青色。明朝因皇帝姓朱，遂以朱为正色；百官公服自南北朝以来紫色为贵，因《论语》有"恶紫之夺朱也"，紫色从官服中废除不用。一至四品，用绯色；五至七品，用青色；八至九品，用绿色（图6-2）。

In the early days of the Wei dynasty, Emperor Wen of Cao Pi formulated the system of the nine-positioned official ranks: "the three colors of purple, green and yellow to distinguish the 9 ranks." This system has been used in various generations until the Yuan dynasty and Ming dynasty. Tang and Song officials wear color, includes: purple for the officials ranking

3rd level or above, twilight for the ones ranking 5th or above, green for the ones ranking 6th and 7th level, and cyan for the ones ranking 8th and 9th level. In the Ming dynasty, the emperor surnamed Zhu, and Zhu was the right color; since the Northern and Southern dynasties, the purple color represented noble, because the "The Analects of Confucius" had "the evil purple to win Zhu", and the purple was abolished from the official uniform. The officials ranking 4th and above wear "twilight"; the ones ranking 5th to 7th wear "cyan", and the ones ranking 8th and 9th wear "green" (Figure 6-2).

龙、华虫、黼、黻八章在衣上；其余四种藻、火、宗彝、粉米在裳上（图6-3）。

12 stamps—The unique pattern of the emperor: Twelve stamps refer to the twelve patterns representing the emperor, in which the day, the moon, the stars, the mountains, the dragons, the worms, the scorpions, and the eight chapters are on the clothes; the remaining four Algae, fire, ancestral, and rice are on the skirt (Figure 6-3).

图6-3 日，月，星辰，照临；山，稳重；龙，应变；华虫，文丽；宗彝，忠孝；藻，洁净；火，光明；粉米，滋养；黼，决断，黻，明辨。
Figure 6-3 Day, month, star, shine; mountain, steady; dragon, strain; Chinese insect, Wenli;Zongtang, Zhongxiao; algae, clean; fire, light; powder rice, nourish;determination, correct judgement

a. 龙纹——九五之尊。

Dragon pattern—the noblest.

最典型的要数龙纹，皇帝衣服用五爪龙纹样装饰，即龙袍，而其他人一般情况下是不能穿着龙纹服装的，只有少数高官可以穿三或四爪龙纹服装，称为蟒袍。

The most typical one is the dragon pattern. The emperor's clothes are decorated with a five-claw dragon pattern, that is, the robe, while others generally cannot wear dragon clothing. Only a few senior

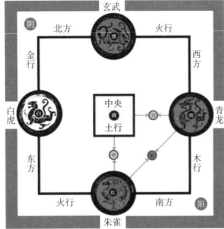

图6-2 五行色与尊卑体现（秦朝皇袍）
Figure 6-2 Five-line color and the embodiment of seniority (imperial robe of qin dynasty)

②服饰纹样彰显等威。

Clothing patterns reflect ranks.

十二章——皇帝的特有纹样：十二章，指的是代表帝德的十二种图案，其中日、月、星辰、山、

officials can wear three or four-claw dragon clothing. It is called python robe.

b. 官服中的纹样。

Pattern in official clothes.

唐代女皇武则天曾赐百官绣袍，以文官绣禽，武官绣兽。明朝对此加以仿效，开始在官服前胸、后背处分别装饰一块补子来区分文武官员的品级。清代胸前绣"补子"的做法却直接取自前代明朝。

In the Tang dynasty, Empress Wu Zetian gave the officers embroidered robe, the civil officers' were embroidered poultry, and the military officers' were embroidered beast. The Ming dynasty emulated this and began to decorate a section of the front and back of the official uniform to distinguish the ranks of civil and military officials. The embroidered chest in the Qing dynasty was directly taken from the previous Ming dynasty.

c. 补子。

Complement.

补子的寓意：文官儒雅娴静，官服以禽鸟为补子图案纹样，以彰显其贤德。武官勇武剽悍，威风凛凛，以猛兽为官服补子图案，以彰显其威仪（图6-4、图6-5）。

The meaning of the complement: the civil officers were elegant and quiet, and the official clothes are decorated with birds to show their virtues. The military officer were brave and arrogant, and the prestige is sturdy, with the beast as the official costume to complement the pattern to highlight its prestige (Figure 6-4、Figure 6-5).

d. 平民百姓纹样——吉祥。

Common people's patterns—auspicious.

不仅官服，民间服饰也常以各种具有吉祥寓意的图案妆点，希望以此可以带来好运（图6-6、图6-7）。

Not only official uniforms, folk clothing were often decorated with a variety of patterns with auspicious meaning, hoping to bring good luck (Figure 6-6、Figure 6-7).

图6-4 明早期六品文官鹭鸶纹缂丝方补
Figure 6-4 Eqret pattern complement using silk tapestry with cut design for the sixth ranking civil officials in the early ming Dyrasty

图6-5 文一品官补子——仙鹤（传世实物）
Figure 6-5 Crane pattern complement for the first ranking civil officials (the real one handed down from arcient times)

图6-6 忍冬纹：东汉末期开始出现，南北朝时最流行，因忍冬越冬而不死，有"延年益寿"的吉祥之意
Figure 6-6 Honeysuckle pattern: It began to appear in the late Eastern Han dynasty. It was the most popular in the Northern and Southern dynasties. Because it was wintering, it was not dead, and it had the auspicious meaning of "longevity"

图6-7 牡丹纹
Figure 6-7 Pattern of poney

二、服饰形制与色彩分析
B. The analysis of shape system and color of clothing and accessories

天子、诸侯穿衮服，戴冕；大夫穿裨衣，戴冕；士人以素积为裳，戴白鹿皮做的皮弁；平民则只能穿布衣。

The emperor and his male relatives wear the luxurious clothes, and the hat; the officers wear robes and the hat; the scholars use the plains as the slings, and the hat in white deer skins; while the civilians can only wear the cotton clothes.

纵观几千年的历史，汉族的服饰，在式样上主要有上衣下裳和衣裳相连两种基本的形式，大襟右衽是其服装始终保留的鲜明特点。

Throughout the history of thousands of years, the clothes of the Han people mainly have two basic forms: the tops and the bottom or connected clothes. The big right side is the distinctive features of their clothing.

在配饰上，早期为玉器。金钗、翠翘、系在绶带末端的金坠子等等都是贵族饰物，持笏佩玉也是贵族服饰的一种风尚（具体分类请参考上一章）。

At the early age, the accessories were jade gold or gem hair fork, and the golden pendants at the end of the sash are all aristocratic ornaments. Holding or wearing jade is also a style of aristocratic clothing (see the section in the previous chapter for details).

三、服饰配件特征
C. Clothing accessories features

新石器时代晚期的良渚文化皆已出现了由玉璜❶（玉牌饰）、管状珠组合而成的玉项饰，其中玉璜通常被串连在玉佩中部的显著位置上，以充当组合的主体。此期佩戴玉项饰应主要为了装饰及象征财富、权力之用，尚未具备礼制方面的作用（图6-8）。

In the late Neolithic period, the Liangzhu culture has appeared a jade ornament composed of jade and tubular beads. The jade is usually connected in a prominent position in the middle of the jade to serve as the main body of the combination. Wearing jade ornaments in this issue should be mainly for decoration and symbolize wealth and power without playing a role in ritual system (Figure 6-8).

图6-8　良渚文化神人兽面纹玉璜　上海震旦博物馆
Figure 6-8　Liangzhu Culture styled jade ornament with God and Human animal face, Shanghai Aurora Museum

1. 商周至秦汉服饰
The clothing and accessories from Shang Zhou Dynasty to Qin Han Dynasty

从商代到西周，是中国奴隶社会的兴盛时期，也是区分等级的上衣下裳形制和冠服制度以及服章制度，逐步建立的时期。

From the Shang Dynasty to the Western Zhou dynasty, it was the prosperous period of the Chinese slave society, and it was also a period of gradual establishment of the grading system and the clothing system.

衣料主要是皮、革、丝、麻，其中丝、麻织物占重要地位。商代人亦能织造极薄的绸和提花几何纹的锦、绮。奴隶主和贵族，平时已穿色彩华美的丝绸衣服。色彩以暖色为多，尤其以黄红为主，间

---
❶ 玉璜：六器之中的玉璜、玉琮、玉璧、玉圭等四种玉器，历史最悠久，早在新石器时代就已出现。在良渚文化中，玉璜是一种礼仪性的挂饰。每当进行宗教礼仪活动时，巫师就戴上它，它经常与玉管、玉串组合成一串精美的挂饰，显示出巫师神秘的身份。

第六章 世界不同地域的历史、文化背景与其服饰配件的设计

有棕色和褐色，但并不等于不存在蓝、绿等冷色（图6-9）。

The clothing materials are mainly leather, silk and hemp, of which silk and linen fabrics play an important role. Shang dynasty people can also weave brocades and silks of extremely thin silk and jacquard geometric patterns. The slave owners and nobles usually wear colorful silk clothes. The color is mostly warm, especially yellow-red, with brown and brown, but it does not mean that there are no cool colors such as blue and green (Figure 6-9).

周代首饰大致沿袭商代服制而略有变化。这个时期的服装还没有纽扣，一般腰间系带，并在腰上还挂有玉制饰物，腰带主要有两种：一种是丝织的为"大带"或称为"绅带"、另一种是皮革制成的为"革带"，这种服装为矩领，领、袖、襟、裾均有缘饰，肩上有披肩，腰系绦带，并在右侧挂玉佩（图6-10）。

The jewellery of the Zhou dynasty has changed slightly depending on the commercial service system. There are no buttons in the clothing of this period. Generally, there are belts on the waist, and jade ornaments are hung on the waist. There are two kinds of belts: one is silky or the other is called "big belt" or "strap". The other is made of leather, which is a leather belt. The type of clothing has edge decoration for collars, sleeves, and cymbals. The shoulders are covered with shawls, the waist is strapped, and the jade is hung on the right side (Figure 6-10).

西周时期，可谓组玉佩发展的鼎盛时期，此期标准的组玉佩开始出现，且数量之多，形制之复杂，居历代之冠（图6-11）。

During the Western Zhou dynasty, it was the heyday of the development of the group, and the standard group of jade began to appear in this period, The number and the complication of jade ornaments make the development reach the peak of the previous generation (Figure 6-11).

2. 春秋战国时期服饰
The clothes and accessories in Spring and Autumn Warring States

春秋战国时期的衣着，上层人物的宽博，下层社会的窄小，已趋迥然。由于上衣下裳并不方便，需要以衣物紧密包裹才不至于暴露身体，深衣因为"被体深邃"，因而得名。深衣的衣缘形式主要有两种：一为交领曲裾式（图6-12），另一为交领直裾式（图6-13）。

图6-9 东周 窄袖织纹衣穿戴展示图（根据出土铜人服饰复原）
Figure 6-9 Eastern Zhou dynasty Narrow sleeve textured wear display (recovery from unearthed bronze costume)

图6-10 东周王室和各路诸侯佩挂玉饰
Figure 6-10 The jade ornaments wearing by the Eastern Zhou dynasty royal family and the various princes

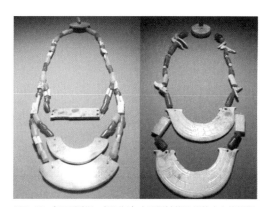

图6-11 （西周时期·组玉佩）山西运城绛县倗伯夫人毕姬墓
Figure 6-11 (Western Zhou dynasty·Jade Groups) Shanxi yuncheng jixian county ladies tomb

The clothes of the Spring and Autumn Period and the Warring States Period, the broad-mindedness of the upper-level figures, and the narrowness of the lower-level society have become obvious. Because the top of the shirt is not convenient, it needs to be tightly wrapped in clothes so as not to expose the body. The deep clothes are named because they are "deeply squatted". There are two main types of clothing in the deep clothes: one is to lead the QuJv (Figure 6-12), and the other is to cross the straight (Figure 6-13).

春秋晚期，玉佩的佩戴方式及形制都发生了巨大的变化，这和东周时旧制度逐步瓦解、"礼崩乐坏"的历史潮流是合拍的。在佩戴方式上，此期组玉佩均不再套于颈部，而是系在腰间的革带，并垂至下肢。

在形制上，构件中的玉璜被珩类器所替代，且新增加了玉璧、玉环、玉管等器类（图6-14～图6-17）。

In the late spring and autumn, the wearing style and shape of the grouped jade had undergone tremendous changes. This was in line with the historical trend of the old system gradually disintegrating in the Eastern Zhou dynasty and the "courtesy and music collapses". In the way of wearing, this period of jade was no longer the one placed on the neck, but the leather belt around the waist, and down to the lower limbs. In terms of shape, the jade in the component was replaced, and the jade ring, jade tube, and other instruments were newly added (Figure 6-14~Figure 6-17).

图6-12 战国 楚曲裾式单衣图（参照长沙楚墓《人物龙凤帛画》绘制）
Figure 6-12 Warring States Chuqu Ju style single figure (refer to Changsha Chu Tomb "People Long Feng painting" draw)

图6-13 楚国贵妇交领、右衽直裾单衣 湖北江陵墓出土实物
Figure 6-13 The Chu State's lady is handed over, and the right hand is straight. The Hubei Jiangling Tomb is unearthed.

图6-14 玉璧
Figure 6-14 Jade ring

图6-15 玉环
Figure 6-15 Jade Brace

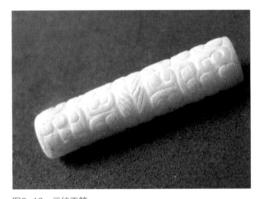

图6-16 云纹玉管
Figure 6-16 Cloud patterned jade tube

图6-17 青玉勾云纹圆环 战国
Figure 6-17 Sapphire Hook Cloud patterned Ring Warring State periods

3. 秦汉服饰

clothing and accessories in Qin Han dynasty

秦统一六国后不久，刘氏汉朝建立。汉朝建立初期，用儒家礼教思想教育人们。西汉男女服装，沿袭深衣形式。西汉时典型的女子深衣，以长沙马王堆出土实物最为精美。有直裾（直襟）和曲裾（三角斜襟式）两种。

Shortly after Qin unified the six countries, the Liu Han dynasty was established. In the early days of the Han dynasty, people were educated with Confucian ethics. Western Han men and women clothing, followed the form of a deep coat. The typical women's deep clothes in the Western Han dynasty were the most exquisitely unearthed from the Mawangdui in Changsha. There were two types: straight (straight) and curved (triangular).

曲裾（图6-18）：下裳部分面积加大。穿上身，静立时衣面悬垂自然贴体，走动时则裙裳部分膨大如伞，不束缚脚步。这种斜领连襟合成锐角的曲裾衣，是战国深衣的变例。

QuJv (Figure 6-18): The area of the lower skirt was increased. Put on the body, when standing still, the clothes were hanging over the natural body. When walking, the skirts were partially swollen like umbrellas, and they were not bound. This kind of oblique collar and the combination of the sharp angle of the QuJv clothing was a variant of the Warring States deep coat.

直裾（图6-19）：在西汉以前就已出现，但不能作为正式的礼服。原因是古代裤字皆无裤裆，仅有两条裤腿套到膝部，用带子系于腰间。所以，外要穿着曲裾深衣。

Straight (Figure 6-19): It appeared before the Western Han dynasty, but it could not be used as a formal dress. The reason was that the ancient pants had no connection parts, only two trousers were placed on the knees, and the belt was tied to the waist. Therefore, you must wear a QuJv deep coat.

襦裙：是中国妇女服装中最主要的形式之一。自战国直至清代，前后二千多年，尽管长短宽窄时有变化，但基本形制始终保持着最初的样式（图6-19）。

Skirt: It is one of the most important forms of Chinese women's clothing. From the Warring States to the Qing dynasty, for more than two thousand years, although the length and width have changed, the basic form has always maintained the original style (Figure 6-19).

图6-18 曲裾
Figure 6-18 QuJv clothing

图6-19 汉代妇女的襦裙图
Figure 6-19 Han dynasty women's skirt map

汉代首饰，以玉见称，而汉代的玉器继承了战国时代玉器的传统，并有所变化和发展。礼仪性的玉器（所谓"瑞玉"）较前减少，组成佩饰的各种佩玉在种类上趋于简化，用于丧葬的玉器显著增加，玉制的日用品和装饰品也有较大的发展（图6-20）。

Jewelry of the Han dynasty, was known as jade, and the jade of the Han dynasty inherited the tradition of jade articles in the Warring States period, with changes and development. The ceremonial jade (the so-called "Ruiyu") was reduced, the variety of jade that constitutes the ornament is simplified in terms of types, the jade used for funeral is significantly increased, and the jade articles and decorations are also greatly developed (Figure 6-20).

在雕琢工艺方面，圆雕、高浮雕、透雕的玉器和镶玉器物较前增多。纹饰的风格由以抽象为主转向以写实为主（图6-21）。汉代的玉器种类很多，按社会功能和用途的不同，可以分为日用品、装饰品、艺术品、辟邪用玉、礼仪用玉和丧葬用玉。

In the carving process, the round carvings,

high reliefs, openwork jade and jade objects were increasing. The style of the ornamentation shifted from abstraction to realism (Figure 6-21). There were are many kinds of jade articles in the Han dynasty. According to different social functions and uses, they can be divided into daily necessities, decorations, art works, jade for evil spirits, jade for ceremonies, and jade for funeral.

图6-20 东汉玉梳，美国大都会博物馆藏
Figure 6-20 Eastern Han dynasty jade comb, the American Metropolitan Museum

图6-21 汉 玉璧
Figure 6-21 Han dynasty Jade ring

汉代男子的首饰一般只用笄。妇女除笄以外还用钗和擿。汉钗的形状比较简单，是将一根金属丝弯曲为两股而成。此外，汉代妇女的发饰还有金胜、华胜、三子钗等，皆绾于头部正面额上的发中。汉代妇女还戴耳珰。这时的珰多作腰鼓形，一端较粗，常凸起呈半球状。戴的时候以细端塞入耳垂的穿孔中，粗端留在耳垂前部（参考第三章第三节）。

Men's jewelry in the Han dynasty was generally only hair pins. Women used sputum and "Zan" in addition to "Ji". The shape is relatively simple, which is made by bending one wire into two strands. In addition, the Han dynasty women's hair accessories also included Jin Sheng, Hua Sheng, three branched fork, etc., all in the hair on the front of the head. Han dynasty women also wore ear rings. At this time, the scorpion was mostly in the shape of a waist drum, with the one end thicker, and the bulge often hemispherical. When worn, the end was inserted into the perforation of the earlobe, and the thick end was left in the front of the earlobe (refer to chapter Ⅲ section 3).

4. 魏晋南北朝服饰
Clothing and accessories in Wei and Jin Dynasties

魏晋南北朝在历史上的文化、艺术的传承与被中原本土文化征服。佛教在这一时期得到最广泛的传播，宗教性质的工艺美术在这个时期得到发展和人们以全新的眼光来看待现实世界相关。那时的服饰作品出现了许多清新之处（图6-22）。

The Wei, Jin, Southern and Northern dynasties were inherited in the history of culture and art and conquered by the Central Plains native culture. Buddhism was the most widely disseminated during this period, and religious arts and crafts developed during this period and people looked at the real world with a new perspective. At that time, the costumes displayed a lot of fresh things (Figure 6-22).

在服装上有襦裙的发展、袍服的普及和裤褶流行（图6-23）。

In the clothing, there were the development of aprons, the popularity of gowns and the popularity of pleats (Figure 6-23).

魏晋南北朝时期的首饰与配饰，根据文物发现主要有替钗、步摇、梳篦、指环、耳坠、玉双璃鸡心佩（图6-24）、金花饰片等。

Jewelry and accessories during the Wei, Jin and Southern and Northern dynasties, according to the cultural relics, mainly found that there were substitutes, stepping, combing, ring, earrings, jade and double-colored chicken hearts (Figure 6-24), gold flower ornaments and so on.

第六章　世界不同地域的历史、文化背景与其服饰配件的设计

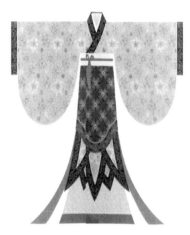

图6-22　杂裾垂髾女服展示图（根据传世帛画及壁画复原绘制）

Figure 6-22　Display of the mix sided coveted women's clothing (drawn according to the handed down paintings and murals)

图6-23　北魏彩绘陶文武士俑

Figure 6-23　Painted pottery warriors in the Northern Wei dynasty

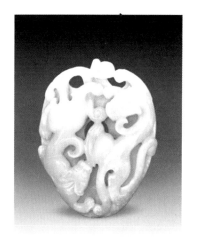

图6-24　魏晋南北朝　玉双螭鸡心佩

Figure 6-24　Jade double heart Wei, Jin and Northern and Southern dynasties

5. 隋唐至清朝服饰

Sui and Tang dynasties to Qing dynasty clothing and accessories

唐代衣冠服饰承上启下，丰富多彩，是我国古代服饰发展的重要时期唐代，也是首饰发展的一个高峰期。服装样式的新变化：

In the Tang dynasty, the clothing of the Tang dynasty was dazzled and colorful. It was an important period for the development of ancient Chinese costumes in the Tang dynasty, and it was also a peak period for the development of jewelry. New changes in clothing style:

唐——半臂、大袖衫、回鹘装。

Tang dynasty—half-arm, big-sleeved shirt, back to wear.

宋——褙子、袆（huī）衣。

Song dynasty—Bei zi, huī clothing.

辽金元——左衽、辫线袄、织金锦袍。

Liao, Jin and Yuan dynasty—ZuoRen, braided threads coat, woven gold robes.

明——比甲、水田衣。

Ming dynesty—Sleeveless over-dress、paddy field Clothes.

清——长袍马褂。

Qing dynasty—long robes.

服装上的配饰：

Accessories on clothing:

唐——披帛。

Tang dynasty—hanging silk.

宋——玉环绶。

Song dynasty—jade ring ribbon.

元——瓦楞帽、姑姑冠。

Yuan dynasty—corrugated cap, aunt crown.

明——宫绦、凤冠霞帔。

Ming dynasty—royal sash, Phoenix crown and cloud shoulder wrap.

清——纬帽、瓜皮帽。

Qing dynasty—weft hat, melon cap.

隋唐女装有时装性，往往由宫廷妇女服装发展到民间，又往往受西北民族影响而别具一格。

Sui and Tang women's fashion, often developed from the court women's clothing to the folk ones, and was often influenced by the Northwestern people and unique.

半臂：有对襟、套头、翻领或无领式样，袖长齐肘，身长及腰，以小带子当胸结住。因领口宽大，

211

穿时袒露上胸（图6-25）。

Half-armed: There were confrontation, hedging, lapel or collarless styles. Sleeves were elbow long and waist long. The straps were tied with a small strap. Because the neckline was wide, it was exposed to the chest when worn (Figure 6-25).

襦裙：在隋代及初唐时期，妇女的短襦都用小袖，下着曳地长裙，裙腰高系，一般都在腰部以上，有的甚至系在腋下，并以丝带系扎，给人一种俏丽修长的感觉。

Skirt: In the Sui dynasty and the early Tang dynasty, women's short squats were all made up of small sleeves, with long skirts, and high waists, usually above the waist, and some even under the armpits, tied with ribbons, giving people A pretty slender feeling.

披帛：当时还流行长巾子，一端固定在半臂的胸带上，再披搭肩上，旋绕于手臂间，名曰披帛。

The hanging silk: At the time, a long towel was also popular. One end was fixed on the chest strap of the half-arm, and then draped over the shoulder, swirling around the arm, from which the name was called.

披帛多以丝绸裁制，上面印画纹样，花色和披戴方式很多。披帛会随女子行动时而飘舞，非常优美（图6-26）。

The hanging silk was mostly made of silk, with patterns printed on it, and a lot of colors and patterns. It would dance with the woman's action beautifully (Figure 6-26).

回鹘（hú）装：回鹘（中国西北地区的少数民族）女装的基本款是连衣长裙，翻领、窄袖，衣身比较宽松，下长曳地，腰际束带。一般在翻领和袖口上都有凤衔折枝花的纹饰。颜色以暖色调为主，尤喜用红色。材料大多用质地厚实的织锦，领、袖均镶有较宽阔的织金锦花边（图6-27）。

Hui Hu clothing (the ethnic minorities in Northwest China), the basic style of women's wear was a long dress, lapels, narrow sleeves, loose body, long stretch, waist belt. Generally, there were embossed flowers on the lapels and cuffs. The color was mainly warm, especially in red. Most of the materials were made of thick brocade, and the collar and sleeves were all set with wide woven gold lace (Figure 6-27).

女子在穿这种服装时要梳椎状的回鹘髻，上饰珠玉，簪钗双插，戴金凤冠，穿笏头履。

When women wore this kind of clothing, they should comb the vertebrae back, adorned with beads and jade, double-inserted, wearing a golden phoenix crown, and a hoe.

图6-25 高髻、锦半臂、柿蒂绫长裙妇女（西安三彩釉陶俑）
Figure 6-25 Women with sorghum, brocade, and persimmon skirts (Xi'an Sancai glazed pottery)

图6-26 隋朝时期的短襦、长裙、披帛女服穿戴展示图
Figure 6-26 Short-shouldered dresses, long skirts, and cloaked women's wears in the Sui dynasty

图6-27 回鹘装贵妇
Figure 6-27 Hui Hu dressed noble woman

第六章 世界不同地域的历史、文化背景与其服饰配件的设计

从唐代金银器首饰看，主要以忍冬、莲花（图6-28）、石榴、蔓枝结成的花朵。编结方式是枝条向内或向外对卷，形成一个石榴形、桃形。图形内外繁简适度地点缀一些花枝、花叶纹。显得清丽素雅，结曲轻巧，疏密相间，主次分明（图6-29）。

The Tang dynasty gold and silver jewelry, mainly patterned the flowers of honeysuckle, lotus (Figure 6-28), pomegranate, and vine branches. The braiding method was that the branches were rolled inward or outward to form a pomegranate shape and a peach shape. The inside and outside of the graphic were moderately and appropriately decorated with some flower branches and mosaics. It looked clear and elegant, light and knotted, and densely spaced, with a clear distinction between the primary and the secondary elements (Figure 6-29).

6. 宋辽夏金元服饰
Song Liaoxia Jinyuan dynasty clothing and accessories

在这个时期，服饰开始崇尚俭朴，重视沿袭传统，朴素和理性成为宋朝服饰的主要特征。以宋朝为例，当时流行一种称为褙子（图6-30）的外衣。宋代襦裙（图6-31）的样式和唐代的襦裙大体相同。身上的装饰并不复杂，除披帛以外，只在腰间正中部位佩的飘带上增加一个玉制圆环饰物，它的作用，主要是为了压住裙幅，使其在走路或活动时不致会随风飘舞而影响美观，称"玉环绶"。这种结环加玉佩方式，一直影响到明清。

In this period, the costumes began to admire the simplicity, attach importance to the tradition, and simplicity and rationality became the main features of the Song dynasty costumes. Take the Song dynasty as an example. At that time, a coat called BeiZi (Figure 6-30) was popular. The style of the Song dynasty skirt (Figure 6-31) was roughly the same as that of the Tang dynasty. The decoration on the body was not complicated. In addition to the cloak, only a jade ring ornament was added to the streamer in the middle of the waist. Its function was mainly to press the skirt so that it did not walk or move. It would affect the beauty with the wind and dance, called "Yuhuan." This kind of knot and jade method had always affected the Ming and Qing dynasties.

图6-28 唐莲花纹金梳
Figure 6-28 Tang dynasty, Lotus flower gold comb

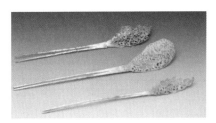

图6-29 唐代银镏金簪
Figure 6-29 Silver jubilee in the Tang dynasty

图6-30 褙子
Figure 6-30 BeiZi

图6-31 右衽交领小袖上襦，长裙系于腰部以上，佩戴披帛，仍有唐代遗韵
Figure 6-31 The right-handed collar is on the small sleeves, the long skirt is tied above the waist, wearing a cloak, and there is still the Tang dynasty styles

服饰配件设计与应用

宋代首饰主要有头饰、耳饰、颈饰、腕饰、腰饰、带饰等。可见，宋人有佩戴坠饰的风气，且作品多以玉为之。辽、金、元时期，有相当成熟的制玉业，它的制玉工艺已达到了一定的水平。

The jewelry of the Song dynasty mainly includes headwear, earrings, neck ornaments, wrist ornaments, waist ornaments and belt ornaments. It can be seen that the Song people have the atmosphere of wearing pendants, and the works are mostly made of jade. During the Liao, Jin and Yuan periods, there was a fairly mature jade industry, and its jade craftsmanship had reached a certain level.

金人与元人皆有于帽顶加饰件的习俗。辽、金、西夏、元时期多有鎏金银冠和金步摇钗头饰、耳饰、银项圈、项饰、臂饰、革带、佩饰等（图6-32、图6-33）。

Both the Jin and the Yuan have the customs of the top and the ornaments. In the Liao, Jin, Xixia and Yuan dynasties, there were many gold and silver crowns and gold stepping head ornaments, ear ornaments, silver collars, ornaments, arm ornaments, leather belts, and accessories (Figure 6-32、Figure 6-33).

7. 明代服饰
Ming dynasty clothing

明代重视推广植棉，棉布得到普及，普通百姓的衣着也得到了改善。明代男子的便服，多用袍衫，其制为大襟、右衽、宽袖，下长过膝。

袍衫上的纹样，多寓有吉祥之意，比较常见的团云和蝙蝠中间，嵌一团型"寿"字，意为"五蝠捧寿"（图6-34）。装饰图案为宝相花是一种抽象的纹样，通常以莲花、忍冬或牡丹花为基本形象，经变形、夸张，并穿插一些枝叶和花苞，组成一种既工整端庄，又活泼奔放的装饰图案（图6-35）。这种服饰纹样在当时深受欢迎。

In the Ming dynasty, emphasis was placed on the promotion of cotton planting, cotton cloth was popularized, and the clothing of ordinary people was improved. The men's casual clothes of the Ming

图6-32 辽 高翅鎏金银冠 内蒙古赤峰木盟奈曼旗青龙山镇陈国公主驸马合葬墓出土
Figure 6-32 Liao Gold and Silver Crown with high wings from the unearthed tomb of Princess Chen and her husband in QingLongshan Town, Naiman Banner, chifengmu Alliance, Inner Mongolia.

图6-33 南宋官窑博物馆黄金制品 五花头凤鸟纹金簪
Figure 6-33 Gold Products of the Southern Song dynasty Guan Kiln Museum

图6-34 五蝠捧寿纹大襟袍展示图及戴四方平定巾、穿大襟袍的男子
Figure 6-34 A large robes with five bats celebrating longevity and a man wearing a flat towel and a big robes

第六章 世界不同地域的历史、文化背景与其服饰配件的设计

dynasty used robes, which were made of big squats, right squats, wide sleeves, and long knees.

The pattern on the robes is more auspicious. The more common cluster clouds and bats are embedded in a group of "longevity" characters, meaning "five bats holding life" (Figure 6-34). The decorative pattern is an abstract pattern, usually with lotus, honeysuckle or peony as the basic image. It is deformed, exaggerated, and interspersed with branches and flower buds to form a decorative pattern that is both dignified and lively (Figure 6-35). This style of clothing was very popular at the time.

在首饰的制作上，有宫廷首饰的种类和造型特点，工艺的细巧严谨，如皇冠和金钗、金簪等（图6-36），这些与世俗民众生活紧密相连的民间首饰也大大发展（图6-37）。民间首饰则表现出自由健康的精神面貌和浓厚的生活气息，求善，求真，求美和迎祥祈福的心理。

In the production of jewellery, there are the types and styling characteristics of court jewellery, and the craftsmanship is exquisite and rigorous, such as the crown and the golden plaque, the golden plaque, etc (Figure 6-36). These folk jewellery closely connected with the secular people's life have also developed greatly (Figure 6-37). Folk jewellery shows a free and healthy spirit and a strong sense of life, pursuing good, seeking truth, seeking beauty and welcoming the spirit of blessing.

8. 清代服饰
Qing dynasty clothing and accessories

清朝是中国封建社会的后期，是由满族人所建立的王朝，文化发展趋于保守。男子服装有袍、褂、袄、衫、裤等。袍挂是主要的礼服，长袍多开衩，官吏开双衩，皇族开四衩，袖口装有箭袖，便于骑马射箭，形似马蹄故称马蹄袖，不开衩之袍称一裹圆，为平时百姓之服（图6-38）。

The Qing dynasty was the late stage of China's feudal society. It was a dynasty established by the Manchu and its cultural development tends to be conservative. Men's clothing has robe, sash, shirt, pants and so on. The robes are the main dresses, the robes are more open, the bureaucrats open the shackles, the royals open four squats, the cuffs are equipped with arrow sleeves, which are convenient for horse-riding archery, and are shaped like horseshoes, so called horseshoe sleeves. Usually the clothes of the people (Figure 6-38).

图6-35 缠枝宝相花纹织锦袍展示图，面料是织"宝相花"纹样的织金锦
Figure 6-35 The picture of the woven flower robe, the fabric is the weaving gold brocade of the weaving "Baoxiang flower" pattern

图6-36 明 定陵出土镶嵌红蓝宝石艺术品
Figure 6-36 Ming dynasty Dingling unearthed inlaid red sapphire artwork

图6-37 明 明间嵌宝玉金簪银簪
Figure 6-37 Ming dynasty embedded in the jade gold and silver hair pin

图6-38 清 男子服装
Figure 6-38 Qing dynasty men's clothing

满族妇女着"旗装",梳旗髻(俗称两把头),穿"花盆底"旗鞋。至于后世流传的所谓旗袍,长期主要用于宫廷和王室。

清代后期,旗袍也为汉族中的贵妇所仿用(图6-39)。

Manchu women wear "Qi clothes", comb the Qi hair styles (commonly known as two handful hair), and wear "flower pot bottom" flag shoes. As for the so-called cheongsam that has been circulated in the later generations, it has been used mainly for the court and the royal family.

In the late Qing dynasty, cheongsam was also used by the ladies of the Han nationality (Figure 6-39).

清朝金银首饰一改唐宋以来或丰满富丽、生机勃勃,或清秀典雅、一曲恬淡的风格,而越来越多地趋于华丽、浓艳,宫廷气息也越来越浓厚。造型的雍容华贵,宝石镶嵌的色彩斑斓(图6-40),特别是那满眼皆是的龙凤图案,象征着不可企及的皇权(图6-41)。这一切都和明清两代整个宫廷装饰艺术的总体风格和谐一致,但与贴近世俗生活的宋元金银器制品迥然不同。

Since the Tang and Song dynasties, the gold and silver jewellery of the Qing dynasty has become rich and vigorous, full of vitality, or elegant and elegant, and a faint style, and more and more tend to be gorgeous and rich, and the court atmosphere is getting

图6-39 清 女子服装
Figure 6-39 Qing dynasty Women's clothes

图6-40 清代串珠簪(大都会博物馆)
Figure 6-40 The Beads of the Qing dynasty (Metropolitan Museum)

图6-41 清代 点翠发簪
Figure 6-41 The Qing dynasty green colored hair pin

stronger and stronger. The shape is graceful and luxurious, and the gemstones are intricately colored (Figure 6-40), especially the dragon and phoenix pattern that is full of eyes, symbolizing the imperfect royal power (Figure 6-41). All of this is in harmony with the overall style of the entire court decoration art of the Ming and Qing dynasties, but it is very different from the Song and Yuan gold and silver products that are close to the secular life.

第二节 日本的文化及其服饰配件特色
Section 2 Japanese culture and its clothing and accessories features

一、文化背景
A. Cultural background

日本是亚洲东北部的岛国，日本独特的地理条件和悠久的历史，孕育了日本的文化。日本文化是传承于汉唐宋明的汉民族古典文化。

Japan is an island country in northeastern Asia. Japan's unique geographical conditions and long history have given birth to Japan's culture. Japanese culture is a classical culture of the Han nationality that is inherited from the Han, Tang, and Song dynasties.

从古至今，日本文化的发展有其独有的特点，有许多既不同于中国，又不同于西方的发展规律。日本服装——和服是日本的民族服饰。江户时代以前称吴服，语出《古事记》《日本书纪》《松窗梦语》，在称为和服之前，日本的服装被称为"着物"，而日本古代所称的"吴服"是"着物"的一种。

From ancient times to the present, the development of Japanese culture has its own unique characteristics, and there are many development laws that are different from China and different from the West. Japanese clothing—kimono is a national costume of Japan. Before the Edo era, it was called Wufu, and the words "The Ancients" "Japanese Book" and "Song Window Dream". Before being called a kimono, Japanese clothing was called "attached thing", and what the ancient Japanese called "Wu suit" is a kind of "object".

二、服饰形制与色彩分析
B. The analysis of clothing shape system and color

日本传统的和服只有两种尺码：男物和女物，因此，只有着衣者挑选合适自己的衣服穿，鲜有量体裁衣者。

There are only two sizes of traditional Japanese kimonos: men and women, so only the wearer chooses the right clothes for them, and there are few tailors.

吴服，它是在依照中国唐代服装的基础上，经过1000多年的演变形成的，和服属于平面裁剪，几乎全部由直线构成，即以直线创造和服的美感。和服裁剪几乎没有曲线，只是在领窝处开有一个20厘米的口子，绱领时将多余的部分叠在一起。如将和服拆开，人们可以看到，用以制作和服的面料，仍然是一个完整的长方形。日本人将他们对艺术的感觉表现在了和服上。

Wufu, which was formed on the basis of Chinese Tang dynasty clothing, after more than 1,000 years of evolution, the kimono belongs to plane cutting, almost all composed of straight lines, that is, the beauty of kimono is created by straight lines. The kimono cuts almost no curve, except that there is a 20 cm hole in the collar and the excess is stacked on the collar. If the kimono is taken apart, one can see that the fabric used to make the kimono is still a complete rectangle. The Japanese showed their feelings about art in a kimono.

每个日本人一生必须有三套和服：

第一套和服：是3~7岁时，孩子参加儿童节时穿着。3岁、5岁和7岁是小朋友特别幸运的3个年纪，每年的11月15号是日本的"753"节（图6-42）。

Every Japanese must have three sets of kimonos in his life: the first set is 3~7 years old when children are wearing them on Children's Day. 3 years old, 5 years old and 7 years old are three years of extraordinarily lucky for children. On November 15th of each year, it is Japan's "753" festival (Figure 6-42).

第二套和服：20岁成人节的时候，女孩子们已经出落得落落大方，靓丽妩媚。成人日是日本的节日（图6-43），目的为向全国本年度年满20岁的青年男女表示祝福。

The second set of kimonos: At the age of 20, the girls have been beautiful and charming. Adult Day is a Japanese holiday (Figure 6-43), which aims to express blessings to young men and women who are 20 years old this year.

第三套和服：结婚时作为礼服（图6-44）。花嫁就是日语的新娘的意思，花嫁衣裳也就是新娘服饰。花嫁衣裳是传统婚礼中必不可少的服饰，以白色和红色锦绣的和服为主。

The third set of kimonos: as a dress when you get married (Figure 6-44). Flower marriage is the meaning of the Japanese bride, and the wedding dress is the bride's dress. Flower wedding clothes are essential in traditional weddings, mainly in white and red kimonos.

三、和服的种类
C. The categories of Kimono

留袖和服特点：和服下摆有整体对花图案，衣袖及上半身成一色，带家纹。女性参加婚礼和正式的仪式，典礼等时穿的礼服，主要分为黑留袖和色留袖。

Sleeved kimono features: The kimono hem has an overall floral pattern, and the sleeves and upper body are in a single color with a home pattern. The kimonos women wear to attend weddings and formal ceremonies, ceremonies, etc., are mainly divided into "black sleeves" and color sleeves.

1. 黑留袖和服
Black sleeve Kimono

已婚女性的第一礼服，以黑色为底色，染有五个花纹，在和服前身下摆两端印有图案的，称为"黑留袖"（图6-45）。

The first dress of a married woman, with a black background, is dyed with five patterns, and the pattern on both ends of the kimono's front hem is called "black sleeves" (Figure 6-45).

图6-42　孩童时期和服
Figure 6-42　Childhood kimono

图6-43　成人日和服
Figure 6-43　Adult Day Kimono

图6-44　结婚和服
Figure 6-44　Wedding kimono

2. 色留袖和服
Color sleeve Kimono

在黑色之外的其他颜色的面料上印有三个或一个花纹，且底边有图案的，称为"色留"。底边图案整体对花（图6-46）。

Three or one pattern is printed on fabrics of other colors than black, and the hem is patterned, called "color retention." The hem pattern is overall to the flower (Figure 6-46).

3. 访问和服
Kimono for visiting occasions

整体上染上图案的和服从底边，左前袖，左肩到领子展开后是一幅图画，开学仪式，朋友的宴会，晚会，茶会等场合都可以穿，穿着场合比较多，并且没有年龄和婚否的限制（图6-47）。

The whole pattern is woven from the hem, the left front sleeve, the left shoulder to the collar is a picture. The kimonos are worn on occasions like the opening ceremony, friends' banquets, parties, tea parties and so on with no age and marriage Limits (Figure 6-47).

4. 振袖和服
Vibrating sleeve kimono

是未婚小姐们的第一礼服，根据袖子长度又分为"大振袖""中振袖"和"小振袖"，其中穿得最多的是"中振袖"。主要用于成人仪式、毕业典礼、宴会、晚会、访友等场合。振袖和服花样绚丽色彩鲜艳展开就像一幅美丽的图画（图6-48）。

It is the first dress of the unmarried ladies. According to the length of the sleeve, it is divided into "big sleeves" "mid sleeves" and "small sleeves". The most worn ones are "mid sleeves". Mainly used for adult ceremonies, graduation ceremonies, banquets, parties, friends and other occasions. The colorful sleeves of the kimono and the kimono are like a beautiful picture (Figure 6-48).

5. 小纹和服
Small grain patterned kimono

衣服上染有碎小花纹（图6-49）。因为适用于练习穿着，所以一般作为日常的时髦服装，在约会和外出购物的场合，常常可以看到。小纹和服也是年青女性用于半正式晚会的礼服。

The clothes are dyed with small patterns (Figure 6-49). Because it is very suitable for practice wearing, it is often seen as a daily fashion outfit, in the occasion of dating and shopping. The small-grain kimono is also a dress for young women to use for a semi-formal party.

图6-45　黑留袖和服
Figure 6-45　Black sleeved kimono

图6-46　色留袖和服
Figure 6-46　Color sleeves kimono

图6-47　访问和服
Figure 6-47　Kimono for visiting occasions

6. 婚礼和服
Wedding Kimono

打褂是幕府时期侯国夫人的正式礼装，仅在节日、庆贺仪式上着用。未婚女性也可以着用。白无垢（图6-50）是幕府时期士族女子婚礼礼服，绫、纶子、羽二重质地。结婚当天只穿着白色打褂，因为白色象征纯洁，结婚几天后才换上色打褂（带夫家家纹的）。色打褂也是礼服的一种（图6-51）。

Wedding Kimono is the official dress of Mrs. Hou Guo in the Shogunate period. It is only used in festivals and celebrations. Unmarried women can also wear. White Kimono (Figure 6-50) is a sectarian period women's wedding dress, with silk, feather two texture. On the wedding day only white coat could be worn, because white symbolizes purity. After a few days of marriage the dress could be changed to long over-dress (with husband's family lines). Color gown is also a kind of dress (Figure 6-51)

四、和服的配饰
D. The accessoria for Kimono

1. 袋带
Bag belt

宽八寸（日本的1寸=3.03cm），正面有花纹，底面是素色，是日本最流行的带子。其中一种织入锦线或金线的带子可与礼装搭配，其他染有轻快图案的带子则用于时装服。穿着和服会用到腰带，将其打成各种结以使其稳固地穿着在身上而不需要用到任何纽扣。带的款式种类很多，男的会用角带，小孩会用兵儿带，而女的则会用袋带、名古屋带、半巾带等（图6-52）。

图6-48　振袖和服
Figure 6-48　vibrating sleeve Kimono

图6-49　小纹和服
Figure 6-49　Small grain patterned kimono

图6-50　白无垢
Figure 6-50　white Kimono

图6-51　色打褂
Figure 6-51　Long over-dress

图6-52　名古屋带（八寸带）、半巾带
Figure 6-52　Nagoya belt (8inch belt) and semi-towel belt

It is eight inches wide (1 inch = 3.03cm in Japan), with a pattern on the front and a plain color on the bottom. It is the most popular belt in Japan. One of the straps woven into the brocade or gold thread can be paired with the attire, while other straps with a light pattern are used in the outfit. Wearing a kimono will use a belt and tie it into a variety of knots to make it securely worn on the body without the need for any buttons. There are many types of belts, males will use the corner belt, children will use the belt, while women will use the bag belt, Nagoya belt and so on (Figure 6-52).

2. 结
Knot

日本的腰带在背后打有不同的花结，象征不同的意义，表达着本人的信仰和祈愿。据统计，日本带的普通结法就有289种。和服的带结是和服整装中一个重要的组成部分，种类很多，有孩子结、少女结、婚嫁结等（图6-53）。

Japan's belts have different knots in the back, symbolizing different meanings and expressing their beliefs and wishes. According to statistics, there are 289 common methods in Japan. The knot of the kimono is an important part of the uniform of the kimono. There are many types, such as children's knots, girl knots, and marriage knots (Figure 6-53).

3. 带扬
Dai Yang

作用除了在制作带结是能够固定和包覆带枕外，而且在装饰上也严格要求与和服、和服带配套。带扬的材料一般选用纺绸、绫、绉织物等，上面装饰扎染纹样、刺绣图案或者是色无地（图6-54）。

In addition to the function of fixing and raping the belt pillow in the production of the belt knot, Dai Yang is also strictly required to match the kimono and kimono belt in the decoration. The materials with the yang are generally made of spun silk, crepe, enamel fabric, etc., and the top is decorated with tie-dyed patterns, embroidered patterns or no color (Figure 6-54).

4. 头饰
Headwear

和服搭配的发饰有梳（图6-55）、簪（图6-56）、丝带等，选择发饰要注意使用的场合，穿振袖之类的华丽和服，选择发饰也就要花哨些，若是穿丧服则要避免戴珊瑚、翡翠之类的发饰。艺伎类常用华丽的风珠花簪、龟甲笄等。

The kimono is equipped with combs (Figure 6-55), hairpin (Figure 6-56), ribbons, etc., hair accessories should be chosen according to different occasions. For example, if you wear gorgeous kimonos such as sleeves, you have to choose hair fancy accessories. If you wear mourning clothes, you have to avoid wearing hair accessories such as corals and jade. The geisha is often used in gorgeous flower beads, tortoise shells, etc.

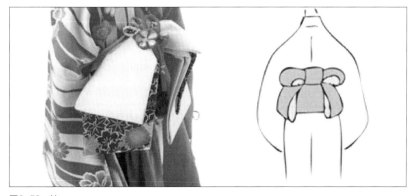

图6-53 结
Figure 6-53 Knot

图6-54 带扬
Figure 6-54 Dai Yang

图6-55 梳子
Figure 6-55 Comb

图6-56 簪，和风，古风发饰，华丽花簪
Figure 6-56 Hairpin, Harmonious wind, antique hair decoration, beautiful flower hairpin

第三节 韩国的文化及其服饰配件特色
Section 3 Korean culture and its clothing and accessories' features

一、文化背景
A. Cultural background

韩国位于北面和中国大陆接壤、东西南三面环海的韩半岛上。朝鲜半岛向南北延伸并形成东面高西面低的东高西低的地形。这样地理位置对韩国的交通、气候、产业、文化等有很大的影响。

South Korea is located on the Korean Peninsula, bordering on the mainland of China and surrounded by the East and West. The Korean peninsula extends north-south and forms an east-high, low-lying terrain with a low east side and a high west side. This geographical location has a great impact on South Korea's transportation, climate, industry, and culture.

韩国文化的最大特征之一是"混性"，把很多东西杂糅在一起形成的文化。

西方社会用100年走完的路，在韩国只用了短短30年就实现了。他们受到中国文化的影响；也受日本及西方文化的影响。这是韩国文化固有的混性所致。这在服饰表现上得到了充分的体现。

One of the greatest characteristics of Korean culture is the "mixedness", a culture that combines many things together.

The way in which Western society has completed in 100 years has been achieved in Korea in just 30 years. They are influenced by Chinese culture; they are also influenced by Japanese and Western culture. This is due to the inherent incompatibility of Korean culture. This is fully reflected in the performance of the apparel.

二、服饰配件形制与色彩分析
B. The analysis of the shape system and color of the clothing an accessories

端庄与华丽的完美结合的韩服，是一种传统的朝鲜族服装。韩国服饰最初主要是受中国唐代服饰的影响。对此，史书中就有记载："服制礼仪，生活起居，奚同中国"。

The hanbok, the perfect combination of dignity and grandeur, is a traditional Korean costume. Korean costumes were originally mainly influenced by Chinese Tang dynasty costumes. In this regard, there are recorded in the history books: "Styles of rituals and ways of living are like the ones in China".

韩服的个性发展开始于李氏朝鲜中期。从那以后，韩服特别是女装，逐渐向高腰、襦裙发展，同中国服饰的区别逐渐增大。但官服、朝服等重要礼服，仍一直延续着较多的中国特色。朝鲜战争期间，西式服装进入韩国。然而，韩国也积极推销韩服，并设计出了合乎时代、容易穿着的式样，使韩服重新恢复了活力（图6-57）。

The personality development of Hanbok began in the middle of Lee's North Korea. Since then, hanbok, especially women's wear, has gradually developed into a high waist and a skirt, and the difference from Chinese clothing has gradually increased. However, important dresses such as official uniforms and kimonos continue to have more Chinese characteristics. During the Korean War, Western-style clothing entered Korea. However, South Korea also actively promoted hanbok and designed a style that was easy to wear and made the hanbok regain its vitality (Figure 6-57).

韩服的特色是设计简单和衣服上并没有口袋。韩服的结构自成一格，上衣自肩至袖头的笔直线条同领子、底边、袖肚的曲线，构成曲线与直线的组合，没有多余的装饰，体现了"白衣民族"的古老袍服的特点。韩服的线条兼具曲线与直线之美，尤其是女士韩服的短上衣和长裙上薄下厚，端庄、娴雅。

The hanbok is characterized by simple design and no pockets on the clothes. The structure of Hanbok is self-contained. The straight line from the shoulder to the sleeve of the shirt is the same as the curve of the collar, the hem and the sleeve belly. It forms a combination of curves and straight lines. There is no extra decoration, which reflects the ancient robe of the "white nationality" Features. The lines of Hanbok combine the beauty of curves and straight lines, especially the short blouses and long skirts of women's hanbok, which are dignified and elegant.

三、服饰配件的特征
C. The features of accessories

1. 香囊
Sachet

香囊又名香袋、花囊。香囊与荷包并非同一种物件。它是古代劳动妇女创造的一种民间刺绣工艺品。香囊用五色丝线缠成的。它是用彩色丝线在彩绸上绣制出各种内涵古老神奇、博大精深的图案纹饰，缝制成形状各异、大小不等的小绣囊，内装多种浓烈芳香气味的中草药研制的细末。古代的香囊是用来提神的，也有用香料来做的，因其香适合很多人的喜欢，后逐步改为纯香料（图6-58）。

Sachet is also known as sachet and flower pouch. Sachets and purses are not the same type of

图6-57　韩国传统服装
Figure 6-57　The traditional clothes of Korean

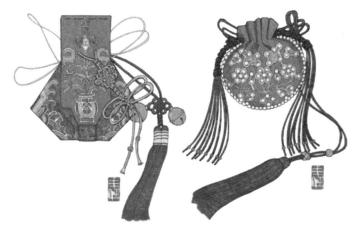

图6-58　韩国传统服饰插画（由于韩服没有口袋，所以无论男女老幼，都会佩戴一个类似于香囊的袋子）
Figure 6-58　Korean traditional clothing illustration (because hanbok has no pockets, both men, women and children will wear a bag similar to a sachet)

object. It is a folk embroidery craft created by ancient working women. The sachet is wrapped in a five-color silk thread. It is embroidered with colored silk thread on the colored silk with various ancient and magical and profound patterns. It is stitched into small embroidered sacs of various shapes and sizes, and the fineness of Chinese herbal medicine with various strong aromatic smells is built. Ancient sachets are used for refreshing and also with spices. Because their fragrance is suitable for many people, they are gradually changed to pure spices (Figure 6-58).

2. 假发与簪

Fake hair and wigs

韩国传统服饰插画用在假发（加髢）上的装饰，古代朝鲜常使用如金，银，铜，珊瑚，琥珀，珍珠，玉，玛瑙等作为装饰（图6-59）。

Korean traditional costume illustrations Used in the decoration of wigs (the crown), ancient North Korea often used gold, silver, copper, coral, amber, pearl, jade, agate and so on as decoration (Figure 6-59).

3. 头饰——发簪（图6-60）

Headwear—hairpin (Figure 6-60)

图6-59 假发与簪
Figure 6-59 Wigs and hairpins

图6-60 韩国已婚妇女使用的头饰
Figure 6-60 Headwear used by Korean married women

第六章　世界不同地域的历史、文化背景与其服饰配件的设计

4. 男子梁冠
Men's crown

金冠，或称祭冠、梁冠，与朝服搭配穿着（有一定等级的官员佩戴）。金冠上的线条，表示的是品级高低，数目越多级别越高（图6-61）。

The golden crown, or ceremonial crown, is worn with the jersey (a certain level of official wear). The lines on the golden crown indicate the level of the grade. The higher the number, the higher the level (Figure 6-61).

5. 腰饰（图6-62）
Waist ornament (Figure 6-62)

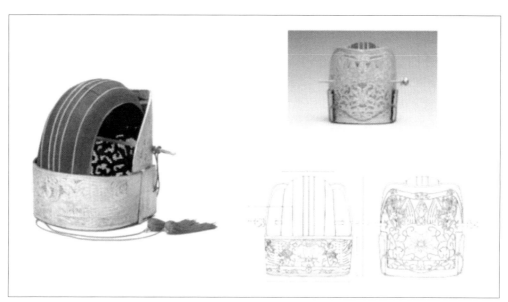

图6-61　男子梁冠
Figure 6-61　Men's crown

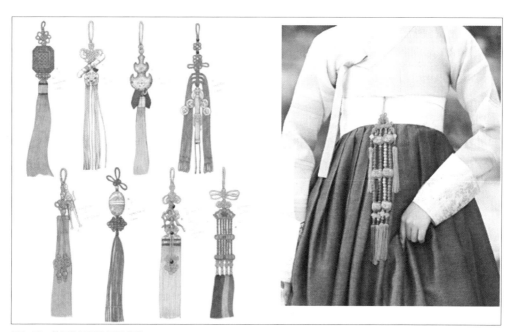

图6-62　韩国传统服饰插画 腰饰
Figure 6-62　Korean traditional costume illustration

第四节 印度的文化及其服饰配件特色
Section 4　Indian culture and its clothing and accessories' features

一、文化背景
A. Cultural background

印度共和国位于亚洲南部，南亚次大陆中心。社会重男轻女，穆斯林女子一般很少抛头露面，更不与陌生人随便交往，也不可同异性握手。妇女多在前额中央点上吉祥痣，喜欢佩带项链、胸饰、耳环、鼻圈、戒指、脚镯等饰物。

The Republic of India is located in the south of Asia, the center of the South Asian subcontinent. The society is patriarchal, Muslim women generally rarely show their faces, and do not casually interact with strangers, nor can they shake hands with the opposite sex. Women are more auspicious at the central point of the forehead, preferring to wear necklaces, chest ornaments, earrings, nose rings, rings, anklets and other accessories.

二、服饰配件形制与色彩分析
B. The analysis of the shape system and color of clothing and accessories

纱丽（图6-63）是印度女性最具代表性的服饰配件。他们以自己文化的创造力，诠释着他们对于生活、审美、色彩的理解和追求。印度的妇女大多体态丰腴，她们穿着纱丽，半遮半露，显得绰约美丽，一如神秘的印度。

Sari (Figure 6-63) is the most representative clothing accessory for Indian women. They use their creativity to interpret their understanding and pursuit of life, aesthetics and color. Most of the women in India are plump. They are dressed in sari, half-covered and half-faced, and look beautiful, just like the mysterious India.

同时，印度是世界上最大的珠宝首饰出口国。当地首饰多以24K黄金为主，造型极富印度民族特色，做工精湛，以古典、华美、复杂和夸张闻名于世。首饰是女性身份的象征。印度的女性首饰种类繁多，有耳环（图6-64）、项链（图6-65、图6-66）、戒指、手链、手镯、发饰（图6-67）、额饰、胸饰（图6-68）、脚链等。印度人认为，"首饰是女性生活的一半"，女子应当充分地利用首饰打扮自己。即使家境清贫的妇女，也要倾其所有，佩戴一些低廉的金属或塑料首饰。

India is the world's largest exporter of jewellery. The local jewellery is mainly made up of 24K gold. The shape is rich in Indian national characteristics, and the workmanship is exquisite. It is famous for its classical, gorgeous, complicated and exaggerated. Jewelry is a symbol of female identity. There are a wide variety of women's jewelry in India, including earrings (Figure 6-64), necklaces (Figure 6-65、Figure 6-66), rings, bracelets, bracelets, hair accessories (Figure 6-67), foreheads, chest ornaments (Figure 6-68), and anklets. Indians believe that "jewelry is half of women's life" and women should make full use of jewelry to dress themselves up. Even women who are poor in their families should be punished and wear some cheap metal or plastic jewelry.

银质脚铃，是大多数印度新娘必备的装饰物。在马哈拉施特拉邦，穿有黑色珠子的吉祥金链，是已婚妇女的标志；在西孟加拉邦，戴在头发中缝的星环，是女子出嫁的标志。除了婚礼，在生孩子、过生日、朋友聚会等各种大型的社交场合，女性都会把自己打扮得珠光宝气。长辈给晚辈的见面礼一般也是首饰。在印度，成年男性至少人均一枚戒指，有的手腕上还戴着手镯或拴着红线，以保佑平安。

Silver footbells are a must-have for most Indian brides. In Maharashtra, the auspicious gold chain with black beads is a symbol of married women; in West Bengal, the star-rings worn in the hair are the symbol of women's marriage. In addition to weddings, women will dress themselves up in various large social occasions such as having children, birthdays, and gatherings of friends. The elders' meeting gifts for the younger generation are also jewelry. In India, adult males have at least one ring per person, and some wear bracelets or red lines on their wrists to protect their peace.

印度珠宝常出现的主题是印度的国鸟孔雀。在印度教的体系之中，孔雀是湿婆神的大儿子，六头四臂的战神俱摩罗的坐骑，1963年，印度政府正式把蓝孔雀定为国鸟。印度人将珠宝的光芒聚焦到这种性灵美丽的鸟儿身上，无论是孔雀造型还是羽翎上的图腾，都成为设计师传统珠宝设计中不可或缺的灵感元素。

The theme of Indian jewelry is often the national bird peacock of India. In the Hindu system, the peacock is the eldest son of Shiva, the six-armed and four-armed mount of the gods of Moro. In 1963, the Indian government officially designated the blue peacock as a national bird. Indians focus the jewels on this beautiful bird. Whether it is a peacock or a totem on a feather, it is an indispensable element of the designer's traditional jewelry design.

他们购买珠宝不是以一件为单位，而是以一个成套的系列为基础。几千年来，每逢节日庆典，印度女子身上就会装点以成套的珠宝于行走间叮当作响，各式珠宝承载着不同的吉祥寓意随身晃动，与眉心的一点朱红交相辉映，形成了独有的印度传统文化。

They buy jewelry not on a single unit, but on a complete set. For thousands of years, every festival celebration, the Indian woman will wear a set of jewels to sing in the walk, all kinds of jewels carry different auspicious meanings to shake with the eyebrows, and form a unique Indian traditional culture.

这种珠宝套装的历史起源于王室对于身份的追求，他们希望用庄严而浑然一体的整体美感来彰显自己的财富和权力。后转化成一种华丽而奢侈的传统，并且逐渐成了印度最普遍的风俗。

The history of this jewellery set originated from the royal family's pursuit of identity, and they hope to demonstrate their wealth and power with a solemn and seamless overall aesthetic. It was later transformed into a gorgeous and extravagant tradition and gradually became the most common custom in India.

在印度人的心目中，金色是最辉煌灿烂的颜色，也是最具能量和质感的高贵色彩。目前印度全年黄金需求居世界首位。除了纯金首饰，印度目前流行的首饰品种还包括黄金镶嵌的钻石、红宝石、绿宝石、蓝宝石、黄玉、紫晶、石榴石、猫眼等。

In the minds of Indians, gold is the most brilliant color, and it is also the noblest color with the most energy and texture. At present, India's annual gold demand ranks first in the world. In addition to pure gold jewelry, India's current popular jewelry varieties include gold inlaid diamonds, rubies, emeralds, sapphires, topaz, amethyst, garnet, cat eyes and so on.

印度珠宝还拥有最夸张奢华的设计，就像满街色彩浓艳的纱丽。从大型摇曳的垂坠式耳环，到质地厚重的黄金手镯，加粗放大的经典符号，成就了印度珠宝奢侈华丽的恢宏气势。

Indian jewellery also has the most exaggerated luxury design, like the colorful sari in the street. From large swaying drop earrings to thick gold bracelets and boldly enlarged classic symbols, the Indian jewels are luxurious and magnificent.

印度珠宝中，钻石难见精雕细琢的痕迹，原石的形状被信手拈来，独一无二的天然光泽，竟然比多面人工切割后的尖锐效果更让人神往。另外一些广为人知的原始工艺也是如今印度珠宝中最常用的工艺，这些岁月悠长的制作方法更为印度珠宝平添了几分浓厚的文化感和历史气息。

图6-63　印度传统纱丽

Figure 6-63　Indian traditional Sari

In Indian jewellery, diamonds are hard to see and trace, and the shape of the original stone is hand-picked. The unique natural luster is even more fascinating than the sharp effect of multi-faceted artificial cutting. Other well-known original crafts are also the most commonly used crafts in Indian jewellery today. These long-lasting craftsmanship add a touch of cultural and historical flair to Indian jewellery.

Figure 6-64　Indian jewlery earing

Figure 6-65　Indian jewlery necklace

Figure 6-66　Indian jewlery neckbrace

Figure 6-67　Indian jewlery hair ornaments

Figure 6-68　Indian jewlery chest ornaments

第五节　欧洲服饰配件特色

Section 5　The features of European clothing and accessories

一、文化背景

A. Cultural background

欧洲服饰配件是西方文化的一个组成部分。在古代的奴隶社会，西方服装早期形式比较单纯简朴，性别区分不明显，以半成品披挂缠绕式为主。如埃及女子的筒形衣，波斯人的坎迪斯、克里特岛女子的紧身上衣与钟形裙、希腊的希顿和希玛纯、罗马的托加和丘尼卡都是这一时期的服装的代表，而且彼此之间有着本质上的联系。希腊的希顿、罗马的托加与埃及的多莱帕里、巴比伦的卷衣可说是一脉相承，透其表看其里，就以希腊文化为代表的希顿服饰而言它所流露出纯真的美感，在简洁中透露出人体最纯朴的美。体现出古希腊人对自然的热爱和对人体美的关注。

European clothing accessories are an integral part of Western culture. In the ancient slave society, the early forms of Western clothing were relatively simple, and the gender distinction was not obvious. Such as the Egyptian women's tubular clothing, the Persian Candis, the Cretan women's tights and bell-shaped skirts, Greece's Heaton and Shima Chun, Rome's Toga and Tunica are all this period representatives of clothing, and they are intrinsically linked to each other. Greece's Heaton, Rome's Toga and Egypt's Dole Pari, Babylon's reel can be said to be in the same vein, and it is pure in the case of Greek style represented by Greek culture. Beauty, in the simplicity reveals the most simple beauty of the human body. It reflects the ancient Greeks' love of nature and their concern for human beauty.

西方服饰配饰从诞生到发展的每一个阶段，都蕴涵着浓厚的西方文明和经济发展的历史，这些首饰的不同类型、材料及加工工艺特点都包含着社会和时代背景的信息。

From the birth to the development of Western clothing accessories, there is a strong history of Western civilization and economic development. The different types, materials and processing characteristics of these jewelry contain information about society and the background of the times.

二、服饰配件形制与色彩分析
B. The analysis of shape system and color of the clothing and accessories

1. 服装形制的表现
The presentation of clothing' shape system

西方服饰在形制上的演变过程幅度较大：从最初古希腊、古罗马的披裹式非成型类服饰发展到中世纪前期以半成型类的衣服为主的服饰。中世纪末期到文艺复兴时期，由于立体裁剪的出现，西方服饰慢慢走向了合身的体形型。而后，经过两次工业革命的洗礼，构筑式、体型型的服饰形态更趋完善。时至今日，现当代西方服饰主导了整个世界的服饰式样也已不言而喻。概括起来其形制的演变过程为：非成形——半成型——体型型。

The evolution of Western clothing in the form of a large extent: from the original ancient Greek, ancient Roman draped non-formed clothing to the clothing of the semi-formed clothes in the early Middle Ages. From the end of the Middle Ages to the Renaissance, due to the emergence of three-dimensional cutting, Western costumes gradually moved toward a fitted body shape. Then, after the baptism of the two industrial revolutions, the style of construction and body shape was more perfect. Today, the fact that modern and contemporary Western clothing dominates the world's clothing style is self-evident. To sum up, the evolution of its shape is: non-formed—semi-formed—body shape.

2. 色彩的运用
Use of color

西方服饰在古罗马时期就奠定了以白色寓意纯洁、紫色象征高贵的色彩偏爱，也就是说注重赋予色彩一定的情感寓意。直至今日西方女性的婚纱礼服仍以白色作为主要用色。而紫色用的紫草是最为贵重的染料，受当时染色技术水平的制约影响，紫色仅限于皇帝的衣装，民众禁用。到了中世纪在色彩上追求绚丽、奢华同时又弥漫着浓厚的宗教气息，以金、棕、黄、白等饰以紫红色条纹为装饰。欧洲文艺复兴以来，随着服饰奢华程度的升级，明亮的色彩受到人们欢迎，织锦缎和天鹅绒中还织进了闪闪发光的金银丝线。此时期各国的色彩喜爱不尽相同：法国人特别喜欢丁香色和蔷薇色，迷恋储蓄的天蓝和圣洁的白色；西班牙人崇尚高雅的玫瑰红和银灰色调；在英国，黑色被认为是神秘、高贵的色彩，特别是黑缎子和黑天鹅绒常是贵妇的首选。到了近代，西方服饰发展过程中的色彩浓烈与淡雅交替登场，用色十分丰富，也正是在这一时期奠定了将黑色尊为西方男子礼服用色的习惯，沿袭至今。

In the ancient Roman period, western costumes established a color preference with pure white and purple, which means that the color is given a certain emotional meaning. Until now, Western women's wedding dresses still use white as the main color. The purple comfrey is the most expensive dye, which is affected by the level of dyeing technology at that time. The purple color is limited to the emperor's clothing, and the people are disabled. In the Middle Ages, the pursuit of beauty and luxury in color was filled with a strong religious atmosphere. The gold, brown, yellow, white and other colors were decorated with purple-red stripes. Since the European Renaissance, with the upgrade of the luxury of clothing, bright colors have been welcomed, and the glittering gold and silver threads are also woven into the brocade and velvet. The colors of the countries are different during this period: the French especially like lilac and rose, the sky blue and the holy white that are obsessed with saving; the Spanish admire the elegant rose red and silver-gray tone; in the UK, black is considered mysterious and noble. The color, especially the black satin and black velvet is often the first choice for ladies. In the modern times, the color of the development of Western costumes was strong and elegant, and the color was very rich. It was during this period that it was the habit of using black as the color of western men's dresses.

因此，西方服饰在用色上受社会政治因素影响不十分明显，而将用色与个人特征、喜好联系较为紧密，着装者的肤色、心情、穿着场合等是西方人服饰用色考虑较多的因素。

Therefore, Western clothing is not obviously influenced by the social and political factors in terms of color, but the color is closely related to personal characteristics and preferences. The skin color, mood, and wearing occasions of the wearer are more considered by Westerners factor.

3. 服饰的装饰
The decoration of clothing and accessories

西方服饰被誉为"立体的雕塑"，因此西方服饰的装饰手法也在很大程度上起到加强服饰整体的立体造型作用。轮状领、膨胀的袖型、切口装饰、重叠的花边和花朵、晃动的流苏以及浆过的纱料和各部位的衬垫、紧身胸衣和裙撑，就连应用平面上的图案往往也追求一种不对称的流动效果（例如哥特时期的国家徽章图案）。层次丰富、虚实搭配、重叠穿插都是西方服饰的常用装饰手法，而它们是和西方服饰的立体造型相呼应的。

Western costumes are known as "three-dimensional sculptures", so the decorative techniques of Western costumes also play a large role in strengthening the overall three-dimensional shape of the costumes. Wheeled collars, expanded sleeves, cut-outs, overlapping laces and flowers, swaying tassels, and padded yarns and padding, bodice and skirts, even the pattern on the flat surface pursues an asymmetrical flow effect (such as the national badge pattern of the Gothic period). Rich layers, virtual and real, overlapping and interspersed are the common decorative techniques of Western clothing, and they are in line with the three-dimensional shape of Western clothing.

在首饰工艺上，较多地使用稀有金属和珐琅制造，并镶嵌珍珠、宝石，设计别致、精美而奢华，工艺水平高超，与服装搭配运用项链、耳饰、饰针等，会产生珠光宝气、绚烂多彩的效果。如拜占庭的奢华风格（图6-69～图6-72）。

In the jewellery process, it is made of rare metals and enamels, and is inlaid with pearls and gems. The design is chic, exquisite and luxurious, and the craftsmanship is superb. The use of necklaces, earrings, pins and so on with clothing will produce jewels. Gorgeous and colorful effects.Such as Byzantine luxury style (Figure 6-69~Figure 6-72).

由于追求缤纷多变的装饰性，形成独特的镶贴艺术，在男女宫廷服的大斗篷、帽饰以及鞋饰上都出现了镶贴光彩夺目的珠宝和华丽图案的刺绣。尤

第六章　世界不同地域的历史、文化背景与其服饰配件的设计

图6-69　拜占庭时期十字架吊坠
Figure 6-69　Byzantine cross pendant

图6-70　拜占庭时期珐琅彩圣徒像吊坠
Figure 6-70　Byzantine period enamel saints like pendants

图6-71　拜占庭时期服饰搭配
Figure 6-71　Byzantine clothing mix

图6-72　拜占庭时期奥托王朝圆碟形胸针
Figure 6-72　Byzantine Otto dynasty round dish brooch

其是巴洛克珠宝，风靡于17世纪的欧洲，堪称奢华的珠宝盛宴。巨大的宝石形成一种豪华气派的风格，奢华的钻石、珍珠以及各种珍贵宝石都会被用在巴洛克的珠宝设计上，正好符合宫廷奢侈、浮夸的气势（图6-73）。

　　Due to the pursuit of colorful and decorative diversity, and the formation of unique inlay art, embroidered with dazzling jewels and gorgeous patterns appeared on the big cloaks, hats and shoes of the men's and women's court uniforms. Especially the Baroque jewellery, popular in Europe and the 17th century Europe, is a luxurious jewellery feast. Huge gems form a luxurious style, luxurious diamonds, pearls and precious gems will be used in the design of Baroque jewellery, in line with the court's luxury, exaggerated momentum (Figure 6-73).

　　在服饰纹样方面，当时多流行花草纹样，意大利文艺复兴时期流行华丽的花卉图案，法国路易

231

十五时期（图6-74），受洛可可装饰风格的影响，流行再现S型或涡旋形的藤草和卷曲柔和的庭园花草纹样（图6-75）。

In terms of clothing patterns, the flowers and patterns were popular at that time, and the gorgeous flower patterns of the Italian Renaissance, the French Louis XV period (Figure 6-74), influenced by the Rococo decorative style, the popular reproduction of S-shaped or vortex-shaped vines And curled soft garden flowers and patterns (Figure 6-75).

洛可可装饰风格的配饰工艺更注重宝石的配色。在大小宝石的排列，以及金属花卉图案的穿插上，用大宝石边上绕上一圈又一圈的小宝石，绿配红，红配蓝，或是祖母绿搭配钻石和珍珠，极尽最绚烂的冲突色彩（图6-76）。这一时期的珠宝设计方式也一直延续至今。

The Rococo-decorated accessory craftsmanship focuses on the color of the gemstone. In the arrangement of large and small gemstones, and the interweaving of metal flower patterns, a large gemstone is used to circulate a small circle of gems, green with red, red with blue, or emerald with diamonds and pearls, displaying the color of the conflict (Figure 6-76). The way jewelry was designed during this period has continued to this day.

带蕾丝衬底的宽领带珍珠项链，也是洛可可时期皇室的经典项链风格，柔软的蕾丝和温润的珍珠加上精致的蝴蝶结装饰，直到今日仍在引领时尚界的潮流（图6-77）。

The Choker pearl necklace with lace backing is also the classic necklace style of the Rococo royal family. Soft lace and warm pearls combined with delicate bows are still leading the fashion world today (Figure 6-77).

图6-73 巴洛克珍珠
Figure 56-73 Baroque pearls

图6-74 玛丽·安托瓦内特皇后 洛可可装饰风格
Figure 6-74 Queen Marie Antoinette rococo decorating style

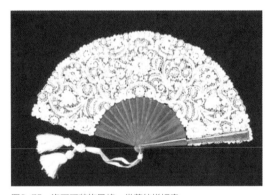

图6-75 洛可可装饰风格 卷草纹样折扇
Figure 6-75 Rococo decorating style rolling pattern folding fan

图6-76 洛可可时期珠宝安妮·费尔柴尔德·鲍勤肖像画
Figure 6-76 Jewelry in Rococo periods The portrait of Anne Fairchild Bowler

图6-77 宽领带在现代首饰设计中的应用 香奈儿2016款
Figure 6-77 Choker's application in modern jewelry design Chanel 2016

近代，有影响的流行图案花样有波普风格、后现代主义风格服饰。以星系（图6-78）、宇宙为主题的迪斯科花样（图6-79），利用几何错视原理设计的欧普图案以及用计算机设计的电子图案等（图6-80）。这些图案及纹饰也完全渗透于欧洲的首饰及饰品上。

In modern times, influential popular patterns have pop style and postmodern style clothing. A disco pattern based on the galaxy (Figure 6-78) and the universe (Figure 6-79), an Op pattern designed using the principle of geometric illusion, and an electronic pattern designed by a computer (Figure 6-80). These patterns are also completely penetrated into European jewelry and accessories.

三、影响现代服饰潮流分析
C. Affecting the trend analysis of modern clothing

1. 波普风格服饰
Pop style clothing, decoration

从服装设计上来说，波普风格并不是一种单纯的、一致性的风格，而是各种风格的混合，是形形色色、各种各样的，带有折衷主义的特点，它被认为是一个形式主义的设计风格（图6-81～图6-83）。

便宜、惹眼、大胆、艳俗、放肆、傲慢、消遣、年轻、快乐、眼花缭乱和惊世骇俗是人们对波普风格服装的评价。

图6-78　星云图案戒指
Figure 6-78　Nebula pattern ring

图6-79　法国艾迪儿珠宝迪斯科风格耳钉
Figure 6-79　French idee jewelry disco style stud earrings

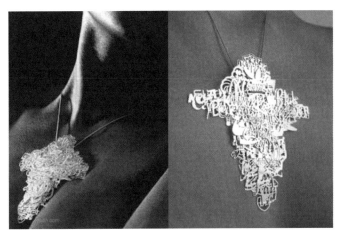

图6-80　美国《下一个》电脑打印技术／图案饰品
Figure 6-80　"The next Bling" from U.S.digtial printing / patterned jewelry

图6-81　几何元素波普风格服饰
Figure 6-81　Geometric elements pop style clothing, accessories

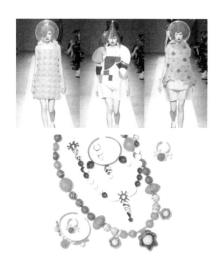

图6-82　年轻、快乐　波普风格服饰
Figure 6-82　Young, happy Pop style clothing, accessories

图6-83　怪诞风格服饰
Figure 6-83　Weird style clothing, accessories

From the point of view of fashion design, Pop style is not a simple, consistent style, but a mixture of various styles. It is a variety of styles, with various characteristics, with eclecticism. It is considered to be a Formalistic design style (Figure 6-81~ Figure 6-83).

Cheap, eye-catching, bold, gaudy, arrogant, pastime, young, happy, dazzling and shocking is the evaluation of pop style clothing.

2. 后现代主义风格服饰
Postmodern style clothing, decoration

后现代主义服饰的特点：风格泛化。

Characteristics of postmodernism clothing and decoration: style generalization.

图6-84　后现代主义服饰
Figure 6-84　Postmodernism costumes

第六章 世界不同地域的历史、文化背景与其服饰配件的设计

风格泛化其实就是一种拼凑艺术，无论材料还是技法，后现代主义服饰都在跨国界、跨时间的范围内吸收文化，后现代的信息是泛化的（图6-84）。将各种元素解构、拆散后重新组合搭配，设计师致力于追求全新的样式和表现形式、标新立异，给人们以视觉上的强烈冲击（图6-85、图6-86）。

Style generalization is actually a piece of art. Regardless of materials or techniques, postmodernism clothing and accessories absorb culture in cross-border and cross-time, and post-modern information is generalized (Figure 6-84). Deconstructing and disassembling various elements and recombining them. The designer is committed to pursuing new styles and expressions, making new changes and giving people a visually strong impact (Figure 6-85、Figure 6-86).

第六节 美洲服饰配件特色
Section 6 The features of american clothing and accessories

一、文化背景
A. Cultural background

美洲分为北美洲与南美洲，位于太平洋东岸、大西洋西岸。北美洲和南美洲，以巴拿马运河为界，总称亚美利加洲，简称美洲，美洲又被称为"新大陆"。印第安人，亦作Amerindian或Amerind，是对除因纽特人外的所有美洲原住民的总称（图6-87）。

图6-85　后现代主义建筑风格珠宝
Figure 6-85　Postmodern architectural style jewelry

图6-86　后现代主义未来风格饰品
Figure 6-86　Postmodernist future style jewelry

图6-87　印第安人装束
Figure 6-87　The clothing of Native Americans

The Americas are divided into North America and South America, on the east coast of the Pacific Ocean and on the west coast of the Atlantic Ocean.

North America and South America are bounded by the Panama Canal. They are collectively called Amerika, also referred to as the Americas, and the Americas is also known as the "New World." The Indians, also known as Amerindian or Amerind, are the general term for all Native Americans except the Inuit (Figure 6-87).

二、服饰配件形制与色彩分析
B. The analysis of shape system and color of clothing and accessories

1. 服饰配件形制的表现
The presentation of clothing shape system

印第安人他们出自亚洲最落后最贫穷的地方。但他们的风俗，表现在他们丰富多彩的节日，如有太阳节、阿拉西塔斯节、降魔节、民俗节、亡灵节等等这与古老的印第安人特有的宗教传统有关，图腾在印第安人中被推崇，各个部落每年都举行图腾祭拜的节日，演变至今，每一个部落的图腾都有各自的故事和传说，绚丽多彩的图腾为现代服饰设计增加了灵感。他们信仰"万物有灵论"的宗教，对大自然的一草一木都持敬畏态度。因此，印第安人的服饰艺术主要来源于自然，运用服饰的花纹都表现部族的崇尚和标识，极为美观。

Indians are from the most undeveloped and poorest places in Asia. But their customs are manifested in their colorful festivals, such as the Sun Festival, Alassitas Festival, Devil's Day, Folk Festival, Day of the Dead. This is related to the ancient Indians' unique religious traditions, totems among the Indians. It is admired that each tribe holds a totem worship festival every year. Since its evolution, each tribe's totem has its own stories and legends, and the colorful totems have added inspiration to modern costume design. They believe in the religion of "all things animistic" and respect the nature. Therefore, the Indian costume art is mainly derived from nature. The patterns of the clothings are all representative of the tribe's advocacy and logo, which is extremely beautiful.

2. 服饰配件色彩的运用
The application of color in the clothing and accessories

印第安人的服饰的色彩元素是运用高明度的色彩搭配，这会使他们的服装、配饰甚至建筑都变得引人注目，如红配绿的大胆配色，充分体现了这个民族的率性和任性，壮丽的大地色更突出了印第安人的淳朴和厚道。在印第安人生活的部落不会迷路，因为远远就可见到饱和度极高的色彩映入眼帘，循着色彩能安全到达他们的生活区，这就是印第安人特有的着装和生活方式。

The color elements of Indian clothing are made with high-quality color combinations, which will make their costumes, accessories and even buildings stand out. The bold color scheme of red and green fully reflects the temperament and willfulness of this nation. The magnificent earth color highlights the simplicity and kindness of the Indians. The tribes living in the Indians will not get lost, because the highly saturated colors will be seen in the distance, and they can safely reach their living areas. This is the unique dress and lifestyle of the Native Americans.

印第安色彩，可称为神秘色彩。有玛雅蓝（图6-88）这种颜色可保存持久鲜亮，目前有专家已找到原因。地衣红、姜黄色和已西木也红等，均彰显了印第安民族色彩元素的魅力和特征。玛雅人独钟爱蓝色，因为他们宗教祭祀中有一道必不可少的程序，是将被献祭的活人头部涂抹成蓝色，来祭祀雨神，祈祷一年的降雨。在玛雅人的壁画和其他生活物品上，这种蓝色的颜料也被广泛应用。

Indian colors can be called mysterious colors. There are Maya blue (Figure 6-88), this color can be preserved for long-lasting bright, and some experts have found the reason. Lichen red, ginger yellow and Hexi wood are also red, etc., all highlight the charm and characteristics of the Indian national color elements. The Mayas loved blue because they had an indispensable procedure in their religious rituals. They painted the heads of the sacrificed

第六章 世界不同地域的历史、文化背景与其服饰配件的设计

people in blue to sacrifice the Rain God and pray for a year of rain. This blue pigment is also widely used in Maya murals and other living objects.

有记载，美洲的印第安人将一种地衣煮沸，提取暖棕色。染料用来制作毛毯，用这种染料制成的衣物不但色泽鲜亮，而且具有防虫蛀的功效。例如莎丽服，姜黄色接近于大地的颜色，被大面积应用在衣服和饰品中（图6-89、图6-90）。印第安色彩元素中还有是充满活力的红色了，巴西木树密度大，弹性好，硬度强，抗腐性强抛光平滑，耐磨不裂，色泽光亮。里面的"红色"是印第安人获取染料的重要途径（图6-91）。

It is recorded that the American Indians will have a lichens boiled and extract the color of warm brown. Dyes are used to make blankets, and the clothes made from this dye are not only bright in color, but also have the effect of insect-proof frogs. For example, sari, ginger is close to the color of the earth, and is widely used in clothes and accessories (Figure 6-89、Figure 6-90). There is also a vibrant red color in the Indian color elements. The Brazilian wood tree has a high density, good elasticity, strong hardness, strong anti-corrosion and smooth polishing, non-cracking and bright color. The "red" inside is an important way for Indians to get dyes (Figure 6-91).

图6-88 玛雅丧葬面具和器物（玛雅蓝）
Figure 6-88 Mayan funeral masks and artifacts (Maya Blue)

图6-89 印第安手工串珠真皮牛皮流苏斜跨单肩包
Figure 6-89 Indian handmade beaded leather cowhide tassel crossbody shoulder bag

图6-90 美国印第安人国家博物馆-1
Figure 6-90 National museum of the american indians

图6-91 美国印第安人国家博物馆-2
Figure 6-91 National museum of the american indians

(1) 服装。

Clothing.

(2) 鞋子：靴款（图6-92）。

Shoes: Boots (Figure 6-92).

这款靴子选用上乘的墨绿色皮革作为鞋身，用绿色和姜黄的搭配，让整双鞋子显得颇有品味，浓浓的墨绿就像是热带雨林深处浓密的树叶，配上充满大地色彩的鞋带，更加充满了生机和活力，辅雕刻图案和挂件点缀，印第安色彩非常浓。

These boots are made of fine dark green leather as the shoes. With the combination of green and turmeric, the whole pair of shoes looks quite tasteful. The dark green is like the depth of the rainforest. The leaves, paired with earth-colored laces, are full of vitality, complemented by engravings and pendants, displaying the strong Indian color.

(3) 首饰（图6-93~图6-95）。

Jewlery (Figure 6-93~Figure 6-95).

(4) 图案与纹饰。

Patterns.

印第安人的图腾图案在原始社会中起着重要的作用，他们通过图腾标志，能得到图腾的认同，受到图腾的保护。如图腾标志最典型的就是图腾柱，是他们服饰的纹饰充分体现部落里崇拜的图腾或者信仰，有浓郁的民族特点。

The totem patterns of the Indians play an important role in the primitive society. They can get the identity of the totem through the totem mark and be protected by the totem. As shown in the picture, the totem pole is the most typical totem pole. The ornamentation of their costumes fully reflects the totems or beliefs worshipped in the tribe, and has strong national characteristics.

阿拉斯加印第安部族的服饰图案有形态逼真的鱼类、走兽和飞鸟（图6-96）等，特别是表现蓝鲸的生动形态。具体绘制印第安人的披肩和披毯，其图案别具匠心，不仅色彩搭配奇特，图案也表现了浓厚的生活气息，如羚羊、梅花鹿形象，呼之欲出。面具和纹面是印第安人的一种强烈的文化艺术表现。这是生存在圣塔仑河谷的印第安西瑞族妇女的纹面被认为是一种稀有的文化。她们文面的目的是为了得到幸福和爱情。十字形花纹是为了避邪；人形图

图6-92 美国缅因州土著指南靴子

Figure 6-92 Indigenous Guide Boots in maine, USA

图6-93 美国印第安人手镯

Figure 6-93 American indian bracelet

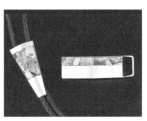

图6-94 美国印第安人项链

Figure 6-94 American indian necklace

图6-95 印第安民族风项链首饰

Figure 6-95 Indian national wind necklace jewelry

第六章　世界不同地域的历史、文化背景与其服饰配件的设计

案代表如钻石般明亮的美丽；文面只用黑、蓝、红和白4种颜色（图6-97）。已婚女子只能用蓝白两色，男人用黑色与红色，未婚姑娘也可以纹上花朵图案，如果年轻女子的脸颊纹上鱼尾花纹，那么就表示她成熟了。

The Alaska Indian tribe's costumes include realistic fish, beasts and birds (Figure 6-96), especially the vivid form of the blue whale. The shawls and rugs of the Indians are drawn in detail. The patterns are unique, not only the color and the unique colors, but also the expression of a strong sense of life, such as the image of an antelope and a sika deer. Masks and patterns are a strong cultural and artistic expression of the Indians. This is the rarity of the Indian Siri women who live in the Santarem Valley. The purpose of their texture is to get happiness and love. The cross-shaped pattern is to avoid evil; the human figure represents the beauty of diamonds; the lines are only black, blue, red and white (Figure 6-97). Married women can only use blue and white, men with black and red, while unmarried girls can also have flower patterns. If the young woman's cheeks show the fishtail pattern it means she is mature.

印第安人佩戴的饰品崇尚原始风格，其原料主要是高山大海赐予他们的贝壳和宝石。最早，印第安人佩戴饰物是为了祛邪或表示地位的区别，后来渐渐演变为一种生活的装点和对美的爱好，如今有些饰物已成为部分印第安部落的族徽或标记（图6-98～图6-100）。

The ornaments worn by the Indians admire the original style, and the raw materials are mainly the shells and gems from that the mountains and seas. At first, the Indians wore ornaments for the sake of defamation or representation of status. Later, they gradually evolved into a decoration of life and a hobby of beauty. Nowadays, some ornaments have become the emblem or mark of some Indian tribes (Figure 6-98~ Figure 6-100).

图6-98　印第安部落首领　项链
Figure 6-98　The native american's tribal leader's necklace

图6-96　印第安　雷鸟❶项链
Figure 6-96　Indian thunderbird necklace

图6-97　印第安项链
Figure 6-97　Indian necklace

图6-99　Zuni Rainbowman和Inlay Knifewing图案首饰珠宝设计
Figure 6-99　Zuni Rainbowman and Inlay Knifewing pattern Jewelry Design

---

❶ 印第安雷鸟：印第安神话是美国原住民的神话与故事，由于原住民神话深受萨满巫术文化影响，因此主要信仰与大自然的神灵相当接近，印第安人们不仅敬畏神明，也敬畏大自然中的一草一木，相信既使是植物，也拥有自己的灵魂，因此值得受到人的尊重。对于动物神灵的崇拜，也衍生出了图腾崇拜的信仰．

服饰配件设计与应用

图6-100　Mosaic Inlay　图案戒指
Figure 6-100　Mosaic Inlay patterned ring

第七节　非洲服饰配件特色
Section 7　The features of African clothing and accessories

一、文化背景
A. Cultural background

非洲位于东半球的西南部，地跨赤道南北，西北部的部分地区伸入西半球。

Africa is located in the southwestern part of the eastern hemisphere, spanning the north and south of the equator, and parts of the northwest extend into the western hemisphere.

由于风俗习惯、宗教信仰、地理环境、气候条件以及经济、文化等状况不同，他们对于服饰配件的审美要求、价值等的认同，差异很大。非洲的服饰配件分为两大部分：原始部落的朴素服饰和现代非洲的传统服饰。

Due to different clothing, religious beliefs, geographical conditions, climatic conditions, and economic and cultural conditions, their recognition of the aesthetic requirements and values of clothing accessories varies greatly. Africa's clothing accessories are divided into two parts: the primitive costumes of the original tribe and the traditional costumes of modern Africa.

原始部落比较封闭、流动性小，但他们有自己部落特定的标记原始部族是通过文身、装饰物与其他部落加以区分。他们装饰的标记丰富、奇特、明显，使人一看而知是哪个民族或部落的。如有的原始部落以鼻栓作为标记，有的则以发饰、胡须的装饰来表示，有的部落以某种图腾形象作为部落标，在他们的装饰物中，体现出来或在饰物上雕刻、刺绣方法表现出来，或绘制在兽皮、纺织品上，或用绳线编结出来。从服饰到日常用品中，处处可见这种现象，标志的重要作用被凸显出来（图6-101）。

图6-101　不同部落装饰上的差别
Figure 6-101　Different ornaments in different tribes

Primitive tribes are relatively closed and fluid, but they have their own tribal-specific marks. The original tribes are distinguished from other tribes by tattoos and decorations. The decorative marks are rich, peculiar and obvious, which makes it possible to see which ethnic or tribal one it is. Some primitive tribes are marked with nasal plugs, while others are decorated with hair ornaments and beards. Some tribes use some kind of totem image as a tribal standard, which is reflected in their decorations or engraved on ornaments. Embroidery methods are displayed, either on animal skins, textiles, or ropes. From clothing to daily necessities, this phenomenon can be seen everywhere, and the important role of the logo is highlighted (Figure 6-101).

二、服饰配件的形制与色彩分析
B. The analysis of shape system and color of clothing and accessories

1. 服饰配件的形制——护身与装饰
The shape of clothing accessories——body protection and decoration

一方面，非洲人对图腾非常崇拜，把许多动物

第六章　世界不同地域的历史、文化背景与其服饰配件的设计

视为神圣的；另一方面，他们认为繁衍后代十分重要，因此，对人体充满敬仰之情。

On the one hand, Africans admire totems and treat many animals as sacred; on the other hand, they believe that it is very important to reproduce and descend, so they are full of admiration for the human body.

非洲人在颈部佩戴装饰物品（图6-102），有赖以保护生命之意。他们认为颈部相接头部与躯干，是生命关键之所在。故必须在其上套以饰物，以超自然的魔力而保护之。图腾民族对于"图腾"是用来顶礼膜拜的偶像，是祈求得到庇护、得到帮助、辟邪、消灾、祛病，保佑人们幸福安康的"神物"。尤多选择图腾的一部分，如取动物的牙齿、角、贝壳、龟壳等悬于颈间，充当颈部的咒物。民族学家认为，原始民族这种佩戴项链的行为，并非是为了美貌，而是出于记数、记事的需要，最多也只是为了在同伴中比试谁猎取的动物多，随着时间的推移，他们逐渐形成的颈部佩戴装饰物品向审美性发展，并有赖以保护生命之意。

Africans wear decorative items on the neck (Figure 6-102), depending on the meaning of life. They believe that the neck joint and the trunk are the key to life. Therefore, it must be decorated with ornaments to protect it with supernatural magic. The totem people are the idols used to worship the "totems". They are the "magic objects" that pray for shelter, help, evil spirits, disaster relief, illness, and blessing people's happiness and well-being. Yudo chose a part of the totem, such as taking the animal's teeth, horns, shells, turtle shells, etc. hanging over the neck, and acting as a charm of the neck. According to ethnologists, the primitive people's behavior of wearing necklaces is not for beauty, but for the needs of counting and memorizing. At most, they only want to compare the animals in the companion who are hunting for more time. The gradually formed neck wears as decorative items develops towards aesthetic functions and relies on the protection of life.

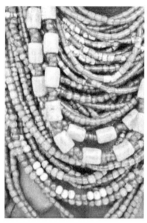

图6-102　护身符颈饰
Figure 6-102　Amulet Necklace

原始部落的人"不但很热心地搜集一切他认为可以做装饰品的东西，他还很耐心、很仔细地创制他的项链、手镯及其他饰物。他们实在是将他们所能收集的一切饰物都戴在身上，也是把身上可以戴装饰的部分都戴起装饰来的。"（摘自格罗塞，《艺术的起源》第62页）。自然界的花草、贝壳、石头等物原始人类充分利用装饰美化自己。同时他们会将不同色彩的贝壳按一定规律间隔排列，将处理过的纤维制成绳索编结起来，把兽皮按人体结构裁剪制作成鞋帽等，这些装饰物在实用的基础上都尽可能美观、漂亮。

The primitive tribes "are not only very enthusiastic in collecting everything that they think can be used as an ornament, but also they are very patient and very careful in creating his necklaces, bracelets and other accessories… They wear really all the accessories they can collect. They also decorated their body parts that can be worn" (Excerpt from Grosse, "*The Origin of Art*", p. 62). Natural flowers, shells, stones, etc. are all the items the primitives make full use to beautify themsleves. At the same time, they will arrange the shells of different colors at regular intervals, braid the treated fibers into ropes, cut the hides into human body structures to make shoes and hats, etc. These decorations are as practical and beautiful as possible.

在非洲最常见的颈饰层次多、夸张、引人注目，

241

头饰、腕饰也同样丰富多彩。但他们美感的实施有时会让人忍受痛苦（图6-103）。如有的部落小孩七八岁时开始在下唇和耳轮上穿孔栓塞，随着年龄的增大更换栓塞，直到定型为止，形成一种永久性的装饰。据说如果这个部族的人没有饰以栓塞则会令人难堪。

The most common neck ornaments in Africa are many, exaggerated and eye-catching, and the headwear and wrist ornaments are equally colorful. But the implementation of their beauty sometimes makes people suffer (Figure 6-103). If some tribal children start to embolize the lower lip and the ear wheel at the age of seven or eight, replace the embolism with age and form a permanent decoration until the shape is fixed. It is said that it would be embarrassing if the people of this tribe were not decorated with embolism.

2．服饰配件色彩装饰运用
The application of color in clothing and accessories

非洲的部落文化认为，各种色彩艳丽图案的传统是重要的，所以妇女将脸上涂抹出彩色图案是美丽的象征，在出席许多重要场合时，妇女除了佩戴各种装饰品外，还把自己的脸上画出色彩鲜艳的彩绘图案（图6-104）。

African tribal culture believes that the tradition of various colorful patterns is important, so women's color patterns on their faces are a symbol of beauty. When attending many important occasions, women also wear themselves in addition to various decorations. The face is painted with colorful painted patterns (Figure 6-104).

非洲人脸彩绘没有统一的样式要求，不过追求色彩艳丽是人脸彩绘的共同特点。非洲儿童从很小的年龄起就可以参与脸部彩绘活动，儿童们在进行各种文艺表演时通常学成年人的样子在脸上涂抹出色彩艳丽的彩绘图案。

There is no uniform style requirement for African face painting, but the pursuit of colorful colors is a common feature of face painting. African children can participate in facial painting activities from a very young age. Children often paint adults with colorful paintings on their faces when performing various cultural performances.

有时非洲人在进行传统舞蹈表演时会将身体和脸部全部涂上彩绘。非洲艺术家们表示，在身体上进行彩绘和演出前化妆一样，都是为了使表演者看上去更加美丽。根据部分非洲地区的传统，一些土著部落的武士也在自己的脸上图上彩绘，这些彩色图案象征着部落武士的勇敢和威严，在各种军事操练中武士们都会"盛妆"列队。

图6-103 痛苦的装饰
Figure 6-103 Painful ornament

图6-104 色彩、图案彩绘
Figure 6-104 Color and pattern painting

第六章 世界不同地域的历史、文化背景与其服饰配件的设计

Sometimes Africans paint their bodies and faces when performing traditional dance performances. African artists say that the same paintings on the body and the pre-show makeup are all designed to make the performers look more beautiful. According to the traditions of some African tribes, some indigenous tribe warriors also painted on their own faces. These colorful patterns symbolize the bravery and majesty of the tribal warriors. In various military exercises, the warriors will "dress".

（1）动物骨骼装饰。

Ornaments in animal bones.

原始民族的天敌之一是猛兽，但是当勇敢的捕猎者制服了这些猛兽后，将猛兽的齿、角、蹄、尾等部位的东西串饰起来佩戴于身（图6-105），同样起到了美化自身、展示勇猛的标志性作用。

One of the natural enemies of the primitive people is the beast, but when the brave hunters subdued these beasts, they decorated the teeth, horns, hooves, tails and other parts of the beast and wore them (Figure 6-105), which also beautified themselves, showing the iconic role of bravery.

In the original ceremonial activities, celebrations or religious activities, people's costumes are more prominent, and the form of decoration is more exaggerated than in everyday life (Figure 6-106).

总的来说，非洲的首饰是从实用转化过来的，非洲人佩戴首饰是为了增强自身的能力和美观，进而形成了自己独特的风格，至今还在影响着时尚界（图6-107）。

图6-106 非洲 国家地理杂志 1971年6月 柏柏尔女人穿她的珍贵的银饰品在一个朋友的婚礼
Figure 6-106 Africa National Geographic Magazine June 1971 Berber woman wears her precious silver jewelry at a friend's wedding

In general, African jewellery is transformed from practical use. Africans wear jewellery to enhance their abilities and aesthetics, and thus form their own unique style, which still affects the fashion world (Figure 6-107).

图6-105 猛兽骨骼与装饰
Figure 6-105 Beast bones and decorations

（2）宗教礼仪与装束。

Religious rituals and costumes.

在原始的礼仪活动、庆典活动或宗教活动中，人们的服饰装扮更为突出，装饰的形式也比日常生活中更加夸张（图6-106）。

图6-107 非洲首饰风格在现代服饰设计中应用
Figure 6-107 African jewelry style applied in modern apparel design

总之，从纵向来看，不同时期的文化、科技、工艺水平、政治、宗教及各方面对服装及配饰产生了深刻的影响;这种影响必然反映出艺术性、审美性、工艺性、装饰性等方面的变化。从横向看，不同的民族风情、民族习俗、地域环境、气候条件等因素，使不同民族、不同地域的服饰配饰具有各不相同的形式和内容。以上这些因素形成了丰富多彩、

243

精美迷人的服饰配件大家族。

In a word, from a vertical perspective, different periods of culture, technology, technological level, politics, religion and various aspects of clothing and accessories have had a profound impact. This kind of influence inevitably reflects the changes of art, aesthetics, craft and decoration. From the horizontal view, different ethnic customs, customs, regional environment, climate and other factors, so that different ethnic groups, different regions of clothing accessories with different forms and content. These factors have formed a colorful, beautiful and attractive clothing accessories large family.

每个时期的服饰配件都与社会的工业生产方式、社会政治因素等条件紧密相关。服饰配件也随着服装新的设计观念、新的风格、流行思潮以及层出不穷的新材料、新工艺而产生新的变化。现代服饰配件设计思维早已跨越了民族与国家的界限，超越了以往狭义的设计范畴。一切与人们日常生活息息相关的精神、文化、经济等因素，都被充实到设计当中。人们追求生活的富裕美满、心理上的丰硕感和满足感，新的思维方式给现代服饰配件的发展增添了新的气息和魅力。

Clothing accessories in each period are closely related to the industrial production mode and social and political factors of the society. Clothing accessories also produce new changes with the new design concept, new style, popular trend of thought and endless new materials and new technology. Modern costume accessories design thinking has already crossed the national and national boundaries, beyond the previous narrow design category. All the spiritual, cultural, economic and other factors that are closely related to People's Daily life are enriched in the design. People pursue the rich and happy life, the rich sense of psychology and satisfaction, the new way of thinking to the development of modern clothing accessories adds a new flavor and charm.

参考文献

[1] 陈东升,王秀艳. 新编服装配饰学[M]. 北京:中国轻工业出版社,2004.

[2] 王明葵. 服装品牌、价值及服用性之间关系的探讨[J]. 中国纤检,2011(10):66-67.

[3] 普列汉诺夫. 论艺术[M]. 上海:生活·读书·新知出版社,1999.

[4] 盛羽. 服饰品设计[M]. 郑州:河南美术出版社,2010.

[5] 陆晓云. 装饰艺术设计[M]. 北京:北京大学出版社,2011.

[6] 石海滨. 从比较视角看艺术本质审美价值理性[J]. 湖南社会科学,2005(5):142-144.

[7] 格尔兹. 文化的解释[M]. 纳日碧力戈等,译. 王铭铭校. 上海:上海人民出版社,1999.

[8] 何星亮. 象征的类型[J]. 民族研究,2003(1):39-47.

[9] 曾强,奚源,蔡晓艳. 服饰品设计教程[M]. 重庆:西南师范大学出版社,2014.

[10] 张嘉秋,车岩鑫. 服饰品设计[M]. 北京:中国传媒大学出版社,2012.

[11] 陈京松. 论戏曲的商品形态[J]. 广东艺术,2000(5):5-8.

[12] 杨小清."非客观性"艺术形态学的理解与分类问题[J]. 文艺研究,2001(4):41-47.

[13] 许星. 服饰配件艺术[M]. 北京:中国纺织出版社,2015.

[14] 彭永茂. 20世纪世界服装大师及品牌服饰[M]. 沈阳:辽宁美术出版社,2001.

[15] 涂途. 现代科学技术之花一枝术美学[M]. 沈阳:辽宁人民出版社,1987.

[16] 郑辉,潘力. 服装配件设计[M]. 沈阳:辽宁科学技术出版社,2009.

[17] 张珊,马颖. 前卫服装设计风格浅议[J]. 青年文学家,2015(20).

[18] 龙凤梅. 浅谈田园风格在服装设计中的应用[J]. 艺术科技,2012(4):56-56.

[19] 董怡. 巴洛克风格在现代男装中的运用状况研究[J]. 装饰,2009(4):114-115.

[20] 李芃,杨贤春. 洛可可风格在中西方服饰上的体现[J]. 湖南工业大学学报,2001,15(4):43-45.

[21] 李超德. 设计美学[M]. 合肥:安徽美术出版社,2004.

[22] 黄思华. 服饰技术美及其艺术表现的研究[J]. 服饰导刊,2017,6(1):43-46.

[23] 麦克格兰斯. 英国珠宝首饰制作基础教程[M]. 张晓燕,译. 上海:上海人民美术出版社,2009.

[24] 邹宁馨,傅永和,高伟. 现代首饰工艺与设计[M]. 北京:中国纺织出版社,2003.

[25] 伊丽莎白·奥利弗. 首饰设计[M]. 刘超,甘治欣,译. 北京:中国纺织出版社,2004.

[26] 郭新. 珠宝首饰设计[M]. 上海:上海人民美术出版社,2009.

[27] 陈征,郭守国. 珠宝首饰设计与鉴赏[M]. 北京:学林出版社,2008.

[28] 阿纳斯塔西娅·杨. 珠宝技术的工作合指南[M]. 英国:泰晤士河—哈德逊出版社,2010.

[29] 赵丹绮,王意婷. 玩金术[M]. 台北:兆星图书事业股份有限公司,2015.

[30] 扬水之. 中国古代金银首饰[M]. 北京: 故宫出版社, 2014.

[31] Jargstorfs. Glass In Jewelry[M]. USA: Schiffer Publishing Ltd, 1991.

[32] Yvonne Coffey. Glass Jewellery[M]. London: A & CBlack, 2009.

[33] 张婧婧. 时尚陶艺服饰配件[M], 武汉: 武汉理工大学出版社, 2015.

[34] 耿琴玉. 纺织纤维与产品-上-基础理论[M]. 苏州: 苏州大学出版社, 2007.

[35] 曾丽. 服饰设计[M]. 上海: 上海交通大学出版社, 2013.

[36] 张海晨. 服饰配件设计[M]. 上海: 上海交通大学出版社, 2004.

[37] 丘美玲. 现代纤维艺术中纤维材料语言的研究[J]. 华南理工大学, 2012.

[38] 邬烈炎, 周庆. 材料表现[M]. 北京: 中国美术学院出版社, 2012.

[39] 王峰. 设计材料美感的视觉体现[J]. 南京艺术学院报, 2006.

[40] 郭鸿旭. 纤维材料—现代首饰设计的新媒介[J]. 上海工艺美术, 2011.

[41] 步洪双. 软雕塑艺术在时装首饰设计中的应用研究[J]. 东华大学, 2012.

[42] 谢琴. 服装材料设计与应用[M]. 北京: 中国纺织出版社, 2015.

[43] 张夫也. 外国工艺美术史[M]. 北京: 高等教育出版社, 2015.